P9-AQJ-874
St. Olaf College
MAY 2 1994
Science Library

The undivided universe

The undivided universe

An ontological interpretation of quantum theory

D. Bohm and B.J. Hiley

London and New York

QC
174.12
.B6326
1993

First published 1993
by Routledge
11 New Fetter Lane, London EC4P 4EE

Simultaneously published in the USA and Canada
by Routledge
29 West 35th Street, New York, NY 10001

© 1993 D. Bohm and B. J. Hiley

Typeset from authors' TeX disks
in 10/12pt Computer Modern Roman

Printed and bound in Great Britain by
Mackays of Chatham PLC, Chatham, Kent

All rights reserved. No part of this book may be reprinted or reproduced or utilized in any form or by any electronic, mechanical, or other means, now known or hereafter invented, including photocopying and recording, or in any information storage or retrieval system, without permission in writing from the publishers.

British Library Cataloguing in Publication Data
A catalogue record for this book is available from the British Library.

Library of Congress Cataloging in Publication Data
Bohm, David.
The Undivided Universe: an ontological interpretation of quantum theory/by David Bohm and Basil Hiley.
p. cm.
Includes bibliographical references and index.
1. Quantum theory. I. Hiley, B. J. (Basil J.) II. Title.
QC174.12.B6326 1991
530.1'2–dc20 91–21387
CIP

ISBN 0–415–06588–7

24009979

Contents

	Preface	**xi**
1	**Introduction**	**1**
1.1	*Why an ontological interpretation is called for*	1
1.2	*Brief summary of contents of the book*	5
1.3	*References*	11
2	**Ontological versus epistemological interpretations of the quantum theory**	**13**
2.1	*Classical ontology*	13
2.2	*Quantum epistemology*	13
2.3	*The quantum state*	17
2.4	*von Neumann's approach to quantum theory*	19
2.5	*Are Bohr's conclusions inevitable?*	24
2.6	*References*	26
3	**Causal interpretation of the one-body system**	**28**
3.1	*The main points of the causal interpretation*	28
3.2	*New concepts implied by the ontological interpretation*	31
3.3	*Comparison with de Broglie's idea of a double solution*	38
3.4	*On the role of probability in the quantum theory*	40
3.5	*Stationary states*	42
3.6	*Non-stationary states*	45
3.7	*Are energy and momentum conserved in non-stationary states?*	47
3.8	*The Aharonov-Bohm effect*	50
3.9	*References*	54
4	**The many-body system**	**56**
4.1	*The ontological interpretation of the many-body system*	56
4.2	*More on the notion of active information*	59
4.3	*Further applications of many-body wave functions*	62
4.3.1	The chemical bond	63
4.3.2	Superfluidity and superconductivity	65
4.4	*References*	71

5 Transition processes considered as independent of observation **73**
5.1 *The example of barrier penetration* 74
5.2 *Active and inactive information* 78
5.3 *Quantum transitions discussed independently of measurement* 83
5.4 *On the possibility of bifurcation points in more complex cases* 89
5.5 *The quantum process of 'capture' of particles (fusion-fission)* 92
5.6 *Summary and conclusions* 95
5.7 *References* 95

6 Measurement as a special case of quantum process **97**
6.1 *Brief treatment of the measurement process* 98
6.2 *Loss of information in the unoccupied wave packets* 104
6.3 *Quantum properties not generally attributable to the observed system alone* 106
6.4 *The meaning of the uncertainty principle* 114
6.5 *On proofs of the impossibility of hidden variables in the quantum theory* 116
6.6 *No measurement is a measurement* 123
6.7 *The Schrödinger cat paradox* 125
6.8 *Delayed choice experiments* 127
6.9 *The watchdog effect and Zeno's paradox* 131
6.10 *References* 133

7 Nonlocality **134**
7.1 *Introduction* 134
7.2 *Nonlocality in the conventional interpretations* 134
7.3 *Bell's inequalities* 140
7.4 *Extension of Bell's theorem to three particles* 146
7.5 *The EPR experiment according to the causal interpretation* 147
7.6 *Loss of nonlocality in the classical limit* 151
7.7 *Symmetry and antisymmetry as an EPR correlation* 153
7.8 *On objections to the concept of nonlocality* 157
7.9 *References* 158

8 The large scale world and the classical limit of the quantum theory **160**
8.1 *Introduction* 160
8.2 *In what way is classical physics contained as a limit of the quantum theory?* 161
8.3 *Large scale objects in the classical limit* 165
8.4 *Illustration of the destruction of interference in terms of streams of particles* 172
8.5 *The extent of non-classical behaviour* 174

8.6 *The quantum world and its classical sub-world* 176
8.7 *References* 180

9 The role of statistics in the ontological interpretation of quantum theory **181**
9.1 *Introduction* 181
9.2 *Chaotic behaviour of particles in many-body systems* 182
9.3 *Statistics of wave functions* 185
9.4 *The density matrix as determining all the physically significant results* 191
9.5 *The stochastic explanation of quantum probabilities* 194
9.6 *Detailed mathematical treatment of the stochastic model* 196
9.7 *Stochastic treatment of the many-body system* 201
9.8 *References* 203

10 The ontological interpretation of the Pauli equation **204**
10.1 *Introduction* 204
10.2 *The Bohm, Schiller and Tiomno model* 205
10.3 *The many-body Pauli equation* 208
10.4 *The Pauli equation as the non-relativistic limit of the Dirac equation* 214
10.5 *Extension to the many-body system* 222
10.6 *The EPR experiment for two particles of spin one-half* 225
10.7 *References* 228

11 The ontological interpretation of boson fields **230**
11.1 *Introduction* 230
11.2 *Reasons for starting with quantum field theories for bosonic fields* 232
11.3 *An ontological interpretation of bosonic fields considered from the non-relativistic standpoint* 238
11.4 *Analysis into normal modes and the ground state of the field* 242
11.5 *The excited state of the field* 247
11.6 *Coherent states and the classical limit* 252
11.7 *The concept of a photon* 255
11.8 *Interference experiments* 260
11.8.1 The treatment of interference 260
11.8.2 The two-slit interference pattern 261
11.8.3 The Pfleegor-Mandel experiment 262
11.9 *The EPR experiment* 264
11.10 *Appendix: the destruction of interference by radiation* 267
11.11 *References* 269

12 On the relativistic invariance of our ontological interpretation **271**
12.1 *Introduction* 271
12.2 *The one-particle Dirac equation* 272
12.3 *The ontological interpretation of the Dirac equation for a many-body system* 274
12.4 *Lorentz invariance of the many-body Dirac equation* 276
12.5 *The multiple time formalism* 278
12.6 *On the question of Lorentz invariance of the ontological interpretation of the Dirac equation* 279
12.7 *On the Lorentz invariance of the ontological interpretation of boson field theories* 286
12.8 *On the meaning of non-Lorentz invariance of processes involving individual beables* 288
12.9 *References* 295

13 On the many-worlds interpretation **296**
13.1 *Introduction* 296
13.2 *Everett's approach to the many-world interpretation* 297
13.3 *Comparison of Everett's approach to that of DeWitt* 302
13.4 *Probabilities in the many-worlds interpretation* 305
13.5 *On the preferred basis and the classical limit* 309
13.6 *Comparison between our interpretation and that of the many-worlds* 314
13.7 *References* 316

14 Extension of ontological theories beyond the domain of quantum mechanics **319**
14.1 *Introduction* 319
14.2 *Our general world view, including our attitude to physical and mathematical aspects of basic concepts* 319
14.3 *The Ghirardi, Rimini and Weber approach* 326
14.4 *Stapp's suggestions for regarding quantum theory as describing an actual process* 329
14.5 *The cosmological approach of Gell-Mann and Hartle* 331
14.5.1 Histories 332
14.5.2 Decohering histories 334
14.5.3 On the meaning of decoherence 335
14.5.4 The quasi-classical world 337
14.5.5 Measurement and observation 339
14.5.6 Comparison with our approach 342
14.6 *Extension of our approach beyond the domain of current quantum theory* 345
14.7 *References* 348

15 Quantum theory and the implicate order **350**
15.1 *Introduction* 350
15.2 *Relativity and quantum theory as indicators of a new order for physics* 351
15.3 *Qualitative introduction to the implicate order* 353
15.4 *Further illustrative example of the implicate order* 357
15.5 *On the distinction between implicate and explicate orders* 361
15.6 *More general notion of order* 362
15.7 *The algebra of the implicate order* 363
15.8 *A Hilbert space model of the trajectories in our interpretation of quantum mechanics* 367
15.9 *An example of trajectories arising from wave structures* 368
15.10 *Pre-space and the implicate order* 374
15.11 *The super implicate order* 378
15.12 *The implicate order of consciousness* 381
15.13 *Further extension to an overall approach* 388
15.14 *References* 390

Index 393

Preface

Over the last twenty years or so David Bohm and I have spent many hours discussing the main theme of this book, namely, whether it is possible to provide an ontological interpretation of quantum mechanics, an issue that has been fiercely debated since its inception. It was quite clear from the outset that it was not going to be possible to return to the concepts of classical physics and we found it necessary to make some radical new proposals concerning the nature of reality in order to provide a coherent ontology. The results of our deliberations form the content of this book.

Just as the final touches were being put to the manuscript, David died suddenly. The week before his death we had a series of meetings in which a few outstanding questions were resolved. Indeed in our last meeting David expressed general satisfaction with the final draft and was anxious to proceed quickly with publication. I have included the results of those discussions, but have resisted the temptation to make any significant alterations while preparing the final manuscript.

I hope this book will be a fitting testimony to this very radical and original thinker who rejected the view of conventional quantum mechanics, not for ideological reasons, but because it did not provide a coherent overall view of nature, a feature that David felt an essential ingredient of any physical theory. It was like Escher's "The Waterfall", a fascinating picture in which region by region appeared to be carefully constructed and consistent, but when one stepped back to perceive the whole, a contradiction was there for all to see. Indeed the most radical view to emerge from our deliberations was the concept of wholeness, a notion in which a system formed a totality whose overall behaviour was richer than could be obtained from the sum of its parts. In the ontological theory that we present here, this wholeness is made manifest through the notion of nonlocality, a notion that is seemingly denied by relativity. Yet there is no observational disagreement with experiment. Nevertheless nonlocality does not fit comfortably within a space-time structure that is taken as a priori given and described by a differential manifold. Thus in the final chapter we present some radically new ideas which take us beyond the present paradigm. It was these ideas that David and I were anxious to develop further. Unfortunately his deep perceptions of these new possibilities will be sadly missed.

Finally I should like to warmly thank all our colleagues for helpful discussions, particularly Chris Dewdney, Henry Folse, Sheldon Goldstein, Dipankar Home, Pan Kalolerou, Abner Shimony and Henry Stapp. I should also like to thank Nora Leonard for her exceptional patience and skilful preparation of the manuscript.

B. J. Hiley
January 1993

Chapter 1
Introduction

1.1 Why an ontological interpretation is called for

The formalism of the quantum theory leads to results that agree with experiment with great accuracy and covers an extremely wide range of phenomena. As yet there are no experimental indications of any domain in which it might break down. Nevertheless, there still remain a number of basic questions concerning its fundamental significance which are obscure and confused. Thus for example one of the leading physicists of our time, M. Gell-Mann [1], has said "Quantum mechanics, that mysterious, confusing discipline, which none of us really understands but which we know how to use".

Just what the points are that are not clear will be specified in detail throughout this book, especially in chapters 6, 7, 8 and 14. We can however outline a few of them here in a preliminary way.

1. Though the quantum theory treats statistical ensembles in a satisfactory way, we are unable to describe individual quantum processes without bringing in unsatisfactory assumptions, such as the collapse of the wave function.

2. There is by now the well-known nonlocality that has been brought out by Bell [2] in connection with the EPR experiment.

3. There is the mysterious 'wave-particle duality' in the properties of matter that is demonstrated in a quantum interference experiment.

4. Above all, there is the inability to give a clear notion of what the reality of a quantum system could be.

All that is clear about the quantum theory is that it contains an algorithm for computing the probabilites of experimental results. But it gives no physical account of individual quantum processes. Indeed, without the measuring instruments in which the predicted results appear, the equations of the quantum theory would be just pure mathematics that would have no physical meaning at all. And thus quantum theory merely gives us (generally statistical) knowledge of how our instruments will function. And from this we can make inferences that contribute to our knowledge, for example, of how to carry out various technical processes. That is to say, it seems, as indeed Bohr [3] and Heisenberg [4] have implied, that quantum theory is concerned only with our *knowledge* of reality and especially of how to predict and control the behaviour of this reality, at least as far as this may be possible. Or to put it in more philosophical terms, it may be said that quantum theory is primarily directed towards *epistemology* which is the study that focuses on the question of how we obtain our knowledge (and possibly on what we can do with it).

It follows from this that quantum mechanics can say little or nothing about reality itself. In philosophical terminology, it does not give what can be called an *ontology* for a quantum system. Ontology is concerned primarily with that which *is* and only secondarily with how we obtain our knowledge about this (in the sense, for example, that the process of observation would be treated as an interaction between the observed system and the observing apparatus regarded as existing together in a way that does not depend significantly on whether these are known or not).

We have chosen as the subtitle of our book "An Ontological Interpretation of Quantum Theory" because it gives the clearest and most accurate description of what the book is about. The original papers in which the ideas were first proposed were entitled "An Interpretation in Terms of Hidden Variables" [5] and later they were referred to as a "Causal Interpretation" [6]. However, we now feel that these terms are too restrictive. First of all, our variables are not actually hidden. For example, we introduce the concept that the electron *is* a particle with well-defined position and momentum that is, however, profoundly affected by a wave that always accompanies it (see chapter 3). Far from being hidden, this particle is generally what is most directly manifested in an observation. The only point is that its properties cannot be observed with complete precision (within the limits set by the uncertainty principle). Nor is this sort of theory necessarily causal. For, as shown in chapter 9, we can also have a stochastic version of our ontological interpretation. The question of determinism is therefore a secondary one, while the primary question is whether we can have an adequate conception of the reality of a quantum system, be this causal or be it stochastic or be it of any other nature.

In chapter 14 section 14.2 we explain our general attitude to determinism in more detail, but the main point that is relevant here is that we regard all theories as approximations with limited domains of validity. Some theories may be more nearly determinate, while others are less so. The way is open for the constant discovery of new theories, but ultimately these must be related coherently. However, there is no reason to suppose that physical theory is steadily approaching some final truth. It is always open (as has indeed generally been the case) that new theories will have a qualitatively different content within which the older theories may be seen to fit together, perhaps in some approximate way. Since there is no final theory, it cannot be said that the universe is either ultimately deterministic or ultimately indeterministic. Therefore we cannot from physical theories alone draw any conclusions, for example, about the ultimate limits of human freedom.

It will be shown throughout this book that our interpretation gives a coherent treatment of the entire domain covered by the quantum theory. This means that it is able to lead to the same statistical results as do other generally accepted interpretations. In particular these include the Bohr interpretation and variations on this which we shall discuss in chapter 2 (e.g. the interpretations of von Neumann and Wigner). For the sake of convenience we shall put these altogether and call them the conventional interpretation.

Although our main objective in this book is to show that we can give an ontological explanation of the same domain that is covered by the conventional interpretation, we do show in the last two chapters how it is possible in our approach to extend the theory in new ways implying new experimental consequences that go beyond the current quantum theory. Such new theories could be tested only if we could find some domain in which the quantum theory actually breaks down. In the last two chapters we sketch some new theories of this kind and indicate some areas in which one may expect the quantum theory to break down in a way that will allow for a test.

Partly because it has not generally been realised that our interpretation has such new possibilities, the objection has been raised that it has no real content of its own and that it merely recasts the content of the conventional interpretation in a different language. Critics therefore ask: "If this is the case, why should we consider this interpretation at all?"

We can answer this objection on several levels. Firstly we make the general point that the above argument could be turned the other way round. Thus de Broglie proposed very early what is, in essence, the germ of our approach. But this met intense opposition from leading physicists of the day. This was especially manifest at the Solvay Congress of 1927 [7]. This opposition was continued later when in 1952 one of us [5] proposed an extension of the theory which answered all the objections and indeed encouraged de Broglie to take up his ideas again. (For a discussion of the history of this development and the sociological factors behind it, see Cushing [8] and also Pinch [9].)

Let us suppose however that the Solvay Congress had gone the other way and that de Broglie's ideas had eventually been adopted and developed. What then would have happened, if 25 years later some physicists had come along and had proposed the current interpretation (which is at present the conventional one)? Clearly by then there would be a large number of physicists trained in the de Broglie interpretation and these would have found it difficult to change. They would naturally have asked: "What do we concretely gain if we do change, if after all the results are the same?" The proponents of the suggested 'new' approach would then probably have argued that there were nevertheless some subtle gains that it is difficult to weigh concretely. This is the kind of answer that we are giving now to this particular criticism of our own interpretation. To fail to consider such an answer seriously is equivalent to the evidently specious argument that the interpretation that "gets in there first" is the one that should always prevail.

Let us then consider what we regard as the main advantages of our interpretation. Firstly, as we shall explain in more detail throughout the book but especially in chapters 13, 14 and 15, it provides an intuitive grasp of the whole process. This makes the theory much more intelligible than one that is restricted to mathematical equations and statistical rules for using these equations to determine the probable outcomes of experiments. Even though many physicists feel that making such calculations is basically what physics is all about, it is our view that the intuitive and imaginative side which makes the whole theory intelligible is as important in the long run as is the side of mathematical calculation.

Secondly, as we shall see in chapter 8, our interpretation can be shown to contain a classical limit within it which follows in a natural way from the theory itself without the need for any special assumptions. On the other hand, in the conventional interpretation, it is necessary to presuppose a classical level before the quantum theory can have any meaning (see Bohm [10]). The correspondence principle then demonstrates the consistency of the quantum theory with this presupposition. But this does not change the fact that without presupposing a classical level there is no way even to talk about the measuring instruments that are essential in this interpretation to give the quantum theory a meaning.

Because of the need to presuppose the classical level (and perhaps eventually an observer), there is no way in the conventional interpretation to give a consistent account of quantum cosmology. For, as this interpretation now stands, it is always necessary to assume an observer (or his proxy in the form of an instrument) which is not contained in the theory itself. If this theory is intended to apply cosmologically, it is evidently necessary that we should not, from the very outset, assume essential elements that are not capable of being included in the theory. Our interpretation does not suffer from this difficulty because the classical level flows out of the theory itself and does not have to be presupposed from outside.

Finally as we have already pointed out our approach has the potentiality for extension to new theories with new experimental consequences that go beyond the quantum theory.

However, because our interpretation and the many others that have been proposed lead, at least for the present, to the same predictions for the experimental results, there is no way experimentally to decide between them. Arguments may be made in favour or against any of them on various bases, which include not only those that we have given here, but also questions of beauty, elegance, simplicity and economy of hypotheses. However, these latter are somewhat subjective and depend not only on the particular tastes of the individual, but also on socially adopted conventions, consensual opinions and many other such factors which are ultimately imponderable and which can be argued many ways (as we shall indeed point out in more detail especially in chapters 14 and 15).

There does not seem to be any valid reason at this point to decide finally what would be the accepted interpretation. But is there a valid reason why we need to make such a decision at all? Would it not be better to keep all options open and to consider the meaning of each of the interpretations on its own merits, as well as in comparison with others? This implies that there should be a kind of dialogue between different interpretations rather than a struggle to establish the primacy of any one of them. (This point is discussed more fully in Bohm and Peat [11].)

1.2 Brief summary of contents of the book

We complete this chapter by giving a brief summary of the contents of this book.

The book may be divided roughly into four parts. The first part is concerned with the basic formulation of our interpretation in terms of particles. We begin in chapter 2 by discussing something of the historical background of the conventional interpretation, going into the problems and paradoxes that it has raised. In chapter 3 we go on to propose our ontological interpretation for the one-body system which however is restricted to a purely causal form at this stage (see Bohm and Hiley [12]). We are led to a number of new concepts, especially that of *active information*, which help to make the whole approach more intelligible, and we illustrate the approach in terms of a number of key examples.

In chapter 4 we extend this interpretation to the many-body system and we find that this leads to further new concepts. The most important of these are *nonlocality* and *objective wholeness.* That is to say, particles may be strongly connected even when they are far apart, and this arises in a way which implies that the whole cannot be reduced to an analysis in terms of its constituent parts.

In chapter 5 we apply these ideas to study the process of transition. Firstly in terms of the penetration of a barrier and secondly in terms of 'jumps' of an atom from one quantum state to another. In both cases we see that these transitions can be treated objectively without reference to observation or measurement. Moreover the process of transition can in principle be followed in detail, at least conceptually, in a way that makes the process intelligible (whereas in the conventional interpretation, as shown in chapter 2, no such account is possible). This sort of insight into the process enables us to understand, for example, how quantum transitions can take place in a time that is very much shorter than the mean life time of the quantum state.

In the next part of the book we discuss some of the more general implications of our approach. Thus in chapter 6 we go into the theory of measurement. We treat this as an objective process in which the measuring instrument and what is observed interact in a well-defined way. We show that after the interaction is over, the system enters into one of a set of 'channels', each of which corresponds to the possible results of the measurement. The other channels are shown to become inoperative. There is never a 'collapse' of the wave function. And yet everything behaves as if the wave function had collapsed to one of the channels.

The probability of a particular result of the interaction between the instrument and the observed object is shown to be exactly the same as that assumed in the conventional interpretation. But the key new feature here is that of the *undivided wholeness* of the measuring instrument and the observed object, which is a special case of the wholeness to which we have alluded in connection with quantum processes in general. Because of this, it is no longer appropriate, in measurements to a quantum level of accuracy, to say that we are simply 'measuring' an intrinsic property of the observed system. Rather what actually happens is that the process of interaction reveals a property involving the whole context in an inseparable way. Indeed it may be said that the measuring apparatus and that which is observed *participate irreducibly* in each other, so that the ordinary classical and common sense idea of measurement is no longer relevant.

The many paradoxes that have arisen out of the attempt to formulate a measurement theory in the conventional interpretation are shown not to arise in our interpretation. These include the treatment of negative measurements (i.e. results following from the non-firing of the detector), the Schrödinger cat paradox [13], the delayed choice experiments [14] and the watchdog effect (Zeno's paradox)[15].

In chapter 7 we work out the implications of nonlocality in the framework of our interpretation. We include a discussion of the Bell inequality [2] and the EPR experiment [16]. We then go on to discuss how nonlocality disappears in the classical limit, except in the special case of the symmetry and antisymmetry of the wave function for which there is a superselection rule, implying that EPR correlations can be maintained indefinitely even at the large scale. This explains how the Pauli exclusion principle can be understood in our interpretation. Finally we discuss and answer objections to the concept of nonlocality.

In chapter 8 we discuss how the classical limit of the quantum theory emerges in the large scale level, without any break in the whole process either mathematically or conceptually. Thus, as we have already explained earlier, we do not need to presuppose the classical level as required in the conventional interpretation.

In the next part of the book we extend our approach in several ways. Firstly in chapter 9, we discuss the role of statistics in our interpretation. We show that in typical situations the particles behave chaotically in a many-body system. From this we can infer that our originally assumed probability density, $P = |\psi|^2$, will arise naturally from an arbitrary initial probability distribution. We then go on to treat quantum statistical mechanics in our framework and show how the density matrix can be derived as a simplified form that expresses what is essential about the statistical distribution of wave functions.

Finally we discuss an alternative approach to this question which has been explored in the literature [17], i.e. a stochastic explanation in which one assumes that the particle has a random component to its velocity over and above that which it has in the causal interpretation. We show that in this theory, an arbitrary probability distribution again approaches $|\psi|^2$, but now this will happen even for single particle systems that would not, in the causal interpretation, give rise to chaotic motion.

In chapter 10 we develop an ontological interpretation of the Pauli equation. We begin with a discussion of the history of this interpretation, showing that the simple model of a spinning extended body will not work if we wish to generalise our theory to a relativistic context. Instead, we are led to begin with an ontological interpretation of the Dirac equation and to consider its non-relativistic limit. We show that in addition to its usual orbital motion, the particle then has an additional circulatory motion which accounts for its magnetic moment and its spin. We extend our treatment to the many-body system and illustrate this in terms of the EPR experiment for two particles of spin one-half.

In chapter 11 we go on to consider the ontological interpretation of boson fields. We first give reasons showing the necessity for starting with quantum field theories rather than particle theories in extending our interpretation to bosonic systems. We then develop our ontological interpretation in detail, but from a non-relativistic point of view. The key new concept here is that the field variables play the role which the particle variables had in the particle theory, while there is a superwave function of these field variables, that replaces the wave function of the particle variables.

We illustrate this approach with several relevant examples. We then go on to explain why the basically continuous field variables nevertheless deliver quantised amounts of energy to material systems such as atoms. We do this without the introduction of the concept of a photon as a 'bullet-like' particle. Finally we show how our interpretation works in interference experiments of various kinds.

In chapter 12 we discuss the question of the relativistic invariance of our approach. We begin by showing that the interpretation is relativistically invariant for the one-particle Dirac equation. However, for the many-particle Dirac equation, only the statistical predictions are relativistically invariant. Because of nonlocality, the treatment of the individual system requires a particular frame of reference (e.g. the one in which nonlocal connections would be propagated instantaneously). The same is shown to hold in our interpretation for bosonic fields [18].

We finally show, however, that it is possible to obtain a consistent approach by assuming a sub-relativistic level of stochastic movement of particles which contains the ordinary statistical results of the quantum theory as well as the behaviour of the world of large scale experience which is Lorentz covariant. Therefore we are able to explain the covariance of all the experimental observations thus far available (at least for all practical purposes). We point out several situations in which this sort of theory could be tested experimentally and give different results from those of the current theory in any domain in which relativity (and possibly quantum theory) were to break down.

We then come to the final part of the book which is concerned with various other ontological interpretations that have been proposed and with modifications of the quantum theory that are possible in terms of these interpretations. The first of these is the many-worlds interpretation which has recently aroused the interest of people working in cosmology. We begin by pointing out that there is as yet no generally agreed version of this interpretation and that there are two different bodies of opinion about it. One of these starts from Everett's approach [19] and the other from DeWitt's [20].

Though these are frequently regarded as the same, we show that there are important differences of principle between them. We discuss these differences in some detail and also the as yet not entirely successful efforts of other workers in the field to deal with the unresolved problems in these two approaches. Finally we make a comparison between our interpretation and the many-worlds point of view.

The many-worlds interpretation was not explicitly aimed at going beyond the limits of the current quantum theory. In chapter 14 we discuss theories that introduce concepts that do go beyond the current quantum theory, at least in principle. The first of these is the theory of Ghirardi, Rimini and Weber [21] who propose nonlinear, nonlocal modifications of Schrödinger's equation that would cause the wave function actually to collapse. The modifications are so arranged that the collapse process is significant only for large scale systems containing many particles, while for systems containing only a few particles, the results are the same, for all practical purposes, as those of the current linear and local form of Schrödinger's equation.

Even more striking changes are proposed by Stapp [22] and by Gell-Mann and Hartle [23], the latter of whom develop their ideas in considerable detail. They deal with the whole question cosmologically from the very outset by introducing mathematical concepts that enable them to describe actual histories of processes taking place in the cosmos, from the beginning of the universe to the end.

We give a careful analysis of these approaches. Both of them aim to do what the many-worlds interpretation has not yet succeeded in doing adequately, i.e. to show that the quantum theory contains a 'classical world' within it. While they have gone some way towards this goal, it becomes clear that there are still unresolved problems standing in the way of its achievement.

We also give a critical comparison between their approach and ours, pointing out that their histories actually involve a mathematical assumption analogous to that involved in the notion of particles in our interpretation. Therefore it is not basically a question of the number of assumptions. Rather, we suggest that the main advantage claimed over ours, at least implicitly, is that it expresses all concepts in terms of Hilbert space, whereas we introduce a notion of particles that goes outside this framework.

Finally we discuss some proposals of our own going beyond the quantum theory. Basically these are an extension of what we suggested in chapter 12. The idea is that there will be a stochastic sub-quantum and sub-relativistic level in which the current laws of physics will fail. This will probably first be encountered near the Planck length of 10^{-33} cm. However, over longer distances our stochastic interpretation of relativistic quantum theory will be recovered as a limiting case, but as we have suggested earlier, experiments involving shorter times could reveal significant differences from the predictions of the current relativistic quantum theory.

Up to this point we have, in a certain sense, been discussing in the traditional Cartesian framework even though many new concepts have been introduced within this framework. In chapter 15, the final chapter of the book, we introduce a radically new overall framework which we call the implicate or enfolded order. In this chapter we shall give a sketch of these ideas which are in any case only in early stages of development.

We begin by showing that the failure of quantum theory and relativity to cohere conceptually already begins to point to the need for such a new order for physics as a whole. We then introduce the implicate order and explain it in terms of a number of examples which illustrate the enfoldment of a whole structure into each region of space, e.g. as happens in a hologram. We show that the notion of order based on such an enfoldment gives an accurate and intuitive grasp of the meaning of the propagator function of quantum mechanics and, more generally, of Hilbert space itself. We indicate how this notion is contained mathematically in an algebra which is essentially the algebra of quantum mechanics itself.

These ideas are connected with our ontological interpretation by means of a model of a particle as a sequence of incoming and outgoing waves, with successive waves very close to each other. For longer times, this approximates our stochastic trajectories, while for shorter times it leads to a very new concept. What is to be emphasised here is that in this way our trajectory model can be incorporated into the framework of Hilbert space. When this is done, we see that it is part of a larger set of possible theories which include those of Stapp and Gell-Mann and Hartle.

One of the main new ideas implied by this approach is that the geometry and the dynamics have to be in the same framework, i.e. that of the implicate order. In this way we come to a deep unity between quantum theory and geometry in which each is seen to be inherently conformable to the other. We therefore do not begin with traditional Cartesian notions of order and then try to impose the dynamics of quantum theory on this order by using the algorithm of 'quantisation'. Rather quantum theory and geometry are united from the very outset and are seen to emerge together from what may be called pre-space.

Finally we discuss certain analogies between the implicate order and consciousness and suggest an approach in which the physical and the mental sides would be two aspects of a greater order in which they are inherently related.

1.3 References

1. M. Gell-Mann, 'Questions for the Future', in *The Nature of Matter, Wolfson College Lectures 1980*, ed. J.H. Mulvey, Clarendon Press, Oxford, 1981.

2. J. S. Bell, *Speakable and Unspeakable in Quantum Mechanics*, chapter 2, Cambridge University Press, Cambridge, 1987.

3. N. Bohr, *Atomic Physics and Human Knowledge*, Science Editions, New York, 1961.

4. W. Heisenberg, *Physics and Philosophy*, Allen and Unwin, London, 1963.

5. D. Bohm, *Phys. Rev.* **85**, 166–193 (1952).

6. D. Bohm, *Phys. Rev.* **89**, 458–466 (1953).

7. L. de Broglie, *Electrons et Photons, Rapport au Ve Conseil Physique Solvay*, Gauthier-Villiars, Paris, 1930.

8. J. T. Cushing, 'Causal Quantum Theory: Why a Nonstarter?', in *The Wave-Particle Duality*, ed. F. Selleri, Kluwer Academic, Amsterdam, to be published.

9. T. J. Pinch, in *The Social Production of Scientific Knowledge. Sociology of the Sciences Vol. 1*, ed. E. Mendelson, P. Weingart and R. Whitley, Reidel, Dordrecht, 1977, 171–215.

10. D. Bohm, *Quantum Theory*, Prentice-Hall, Englewood Cliffs, New Jersey, 1951.

11. D. Bohm and F. D. Peat, *Science, Order and Creativity*, Bantam Books, Toronto, 1987.

12. D. Bohm and B. J. Hiley, *Phys. Reports* **144**, 323–348 (1987).

13. E. Schrödinger, *Proc. Am. Phil. Soc.* **124**, 323-338 (1980).

14. J. A. Wheeler, in *Mathematical Foundation of Quantum Mechanics*, ed. R. Marlow, Academic Press, New York, 1978, 9–48.

15. B. Misra and E. C. G. Sudarshan, *J. Math. Phys.* **18**, 756–783 (1977).

16. A. Aspect, J. Dalibard and G. Roger, *Phys. Rev. Lett.* **11**, 529–546 (1981).

17. D. Bohm and B. J. Hiley, *Phys. Reports* **172**, 93–122 (1989).

18. D. Bohm, B. J. Hiley and P. N. Kaloyerou, *Phys. Reports* **144**, 349–375 (1987).

19. H. Everett, *Rev. Mod. Phys.* **29**, 454–462 (1957).

20. B. S. DeWitt and N. Graham, *The Many-Worlds Interpretation of Quantum Mechanics*, Princeton University Press, Princeton, New Jersey, 1973, 155–165.

21. G. C. Ghirardi, A. Rimini and T. Weber, *Phys. Rev.* **D34**, 470–491 (1986).

22. H. P. Stapp, 'Einstein Time and Process Time', in *Physics and the Ultimate Significance of Time*, ed. D. R. Griffin, State University Press, New York, 1986, 264–270.

23. M. Gell-Mann and J. B. Hartle, 'Quantum Mechanics in the Light of Quantum Cosmology', in *Proc. 3rd Int. Symp. Found. of Quantum Mechanics*, ed. S. Kobyashi, Physical Society of Japan, Tokyo, 1989.

Chapter 2
Ontological versus epistemological interpretations of the quantum theory

2.1 Classical ontology

In classical physics there was never a serious problem either about the ontology, or about the epistemology. With regard to the ontology, one assumed the existence of particles and fields which were taken to be essentially independent of the human observer. The epistemology was then almost self-evident because the observing apparatus was supposed to obey the same objective laws as the observed system, so that the measurement process could be understood as a special case of the general laws applying to the entire universe.

2.2 Quantum epistemology

As we have already brought out in chapter 1, in quantum mechanics this simple approach to ontology and epistemology was found to be no longer applicable. In the present chapter we shall go into this question in more detail especially in connection with the way this subject is treated in the conventional interpretation.

Let us begin with the fact that quantum mechanics was introduced as an essentially statistical theory. Of course statistical theories in general are capable of being given a straightforward ontological interpretation, for example, in terms of an objective stochastic process. The epistemology could then be worked out along the same lines as for a deterministic theory such as classical mechanics. But Bohr and Heisenberg raised further questions about the validity of such an approach in the quantum theory. Their argument was based on two postulates: (a) the indivisibility of the

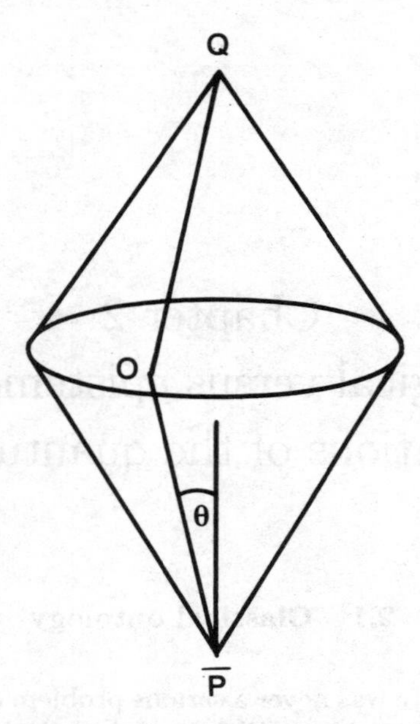

Figure 2.1: Sketch of Heisenberg microscope

quantum of action and (b) the unpredictability and uncontrollability of its consequences in each individual case.

It follows from the above assumptions, as we shall show in more detail presently, that in the measurement of p and x, for example, there is a maximum possible accuracy given by the uncertainty principle $\Delta p \Delta x > h$. This is clearly a limitation on the possible accuracy and relevance of our knowledge of the observed system. However, this has been taken not as a purely epistemological limitation on our knowledge, but also as an ontological limitation on the possibility of defining the state of being of the observed system itself.

To bring out what is meant here let us briefly review the Heisenberg microscope argument. A particle at some point P (see figure 2.1) scatters a quantum of energy $h\nu$ which follows the path POQ to arrive at the focal point Q of the lens. From a knowledge of this point Q there is an ambiguity in our ability to attribute the location of the point P to within the resolving power of the lens $\Delta x = \lambda / \sin \alpha$ where λ is the wave length and α is the aperture angle of the lens. This follows from the wave nature of the quantum that links P to Q. But because the light has a particle nature as well, the quantum has a momentum $h\nu/c$ and it produces a change of momentum in the particle $\Delta p = h\nu \sin \theta / c$ where θ is the angle through which the quantum has been scattered by the particle. The indivisibility of the quantum guarantees that its momentum cannot be reduced below this value, while the assumed unpredictability and uncontrollability of the scattering process within $\theta \leq \alpha$ guarantees that we cannot make an unambiguous attribution of momentum to the particle within the range $\Delta p = h\nu \sin \alpha / c$. And it is well known that from this we obtain Heisenberg's uncertainty relation $\Delta p \Delta x \geq h$.

Relationships of this kind implied, for Heisenberg and Bohr, that the basic properties of the particle, i.e. its position and momentum, are not merely *uncertain to us*, but rather that there is no way to give them a meaning beyond the limit set by Heisenberg's principle. They inferred from this that there is, as we have already pointed out, an inherent ambiguity in the state of being of the particle. And this in turn implied that, at the quantum level of accuracy, there is no way to say what the electron *is* and what it *does*, such concepts being applicable approximately only in the classical (correspondence) limit.

This evidently represented a totally new situation in physics and Bohr felt that what was called for was a correspondingly new way of describing an experiment in which the entire *phenomenon* was regarded as a single and unanalysable whole [1]. In order to bring out the full meaning of Bohr's very subtle thoughts on this point, let us contrast his view of the quantum phenomenon with the ordinary approach to the classical phenomenon. To do this we may take the classical counterpart to the Heisenberg microscope as an example. The relevant phenomena can be described by first of all giving the overall experimental arrangement (the lens, the photographic plate, the scattering block and the incident light). Secondly one has to specify the experimental result (the spot on the photographic plate). But of course this result by itself would be of very little interest. The main point of the phenomenon is to give the *meaning* of the result. (In this case the location of the particle that scattered the light.) Evidently this is possible only if we know the behaviour of the light that links the experimental result to this meaning. Classically this behaviour is well defined since it follows from the wave nature of light. Since the light can be made arbitrarily weak and of arbitrarily short wave length, there is clearly no limit to the possible accuracy of the link between the experimental result and its meaning. That is to say the disturbance of the 'particle' and the ambiguity of its properties can be made negligible. This implies that the particle can then be considered to be essentially independent of the rest of the phenomenon in which its properties were determined. Therefore it is quite coherent to use the customary language which says that we have established a state of being of this independent particle as having been observed, and so that the measurement could then be left entirely out of the account in discussing the behaviour of the particle from this point on.

In the corresponding quantum phenomenon there is an entirely different state of affairs. For, as we have already pointed out, the quantum link connecting the experimental result with its meaning is indivisible, unpredictable and uncontrollable. The meaning of such a result can, therefore, no longer be coherently described as referring unambiguously to the properties of a particle that exists independently of the rest of the phenomenon. Instead this meaning has to be regarded as an inseparable feature of the entire phenomenon itself. Or, to put it more succinctly, the form of the experimental conditions and the content (meaning) of the experimental results *are* a whole, not further analysable. It is this whole that, according to Bohr, constitutes the quantum phenomenon.

It follows from this that given a different experimental arrangement (e.g. one needed to measure a complementary variable more accurately) we would have a different total phenomenon. The two phenomena are mutually exclusive in the sense that the conditions needed to determine one are incompatible with those needed to determine the other (whereas classically the two sets of conditions are, in principle, compatible).

Bohr emphasises that incompatible phenomena of this kind actually complement each other in the sense that together they provide a complete though ambiguous description of the 'atomic object'. These complementary descriptions "cannot be combined into a single picture by means of ordinary concepts, they represent equally essential aspects of any knowledge of the object in question that can be obtained in this domain" [2]. Classically two such concepts can always be combined in a single unambiguous picture. This enables us to form a well-defined concept of an actual process independent of the means of observation (in which, for example, a particle actually moves from one state to another). But at the quantum level where the indivisibility of the quantum of action implies an ambiguity in the distinction between the observed object and observing apparatus, there is no way to talk consistently about such a process. It follows from Bohr's approach that very little can be said about quantum ontology.*

One has at most an unambiguous classical ontology and the quantum theory is reflected in this ontology by requiring basic concepts such as p and x to be ambiguous. One might perhaps suppose that there could be some unambiguous deeper quantum concepts of a new kind. But Bohr would say there is no way to relate these definitely to what we ordinarily regard as objective reality, i.e. the domain in which classical physics is a

*Folse [3] has made it clear that Bohr is not simply a positivist, but that the notion of some kind of independent physical reality underlies all his thinking.

good approximation.

We can summarise Bohr's position as saying that all physical concepts must correspond to phenomenon, i.e. appearances. Each phenomenon is an abstraction. This is also true classically. But because the correspondence between the phenomenon and the independent reality which underlies it may, in principle, be unambiguous, and because all the phenomena are mutually compatible, we may say that the independent reality can be reflected completely in the whole set of phenomena. This means in effect that we can know the independent reality itself. But quantum mechanically we cannot apply all relevant abstractions together in an unambiguous way and therefore whatever we say about independent reality is only implicit in this way of using concepts.

What then is the meaning of the mathematics of the quantum theory (which is very well defined indeed)? Bohr describes this as the *quantum algorithm* which gives the probabilities of the possible results for each kind of experimental arrangement [4]. Clearly this means that the mathematics must not be regarded as reflecting an independent quantum reality that is well defined, but rather that it constitutes in essence only ***knowledge about the statistics of the quantum phenomena.***

All this, as we have already pointed out, is a consequence of the indivisibility of the quantum of action which is very well verified experimentally. Bohr therefore does not regard his notion of complementary as based on philosophical assumptions. Rather it has for him an ontological significance in the sense that it says something about reality, i.e. that it is ambiguously related to the phenomena. He would probably say that attempts to define the ontology in more detail would be contradictory.

2.3 The quantum state

Bohr's view seems to have had a very widespread influence, but his ideas do not appear to have been well understood by the majority of physicists. Rather the latter generally thought in terms of a different approach along lines initiated by Dirac, and von Neumann, in which the concept of a *quantum state* played a key role (whereas with Bohr this concept was hardly even mentioned and was certainly not a fundamental part of his ideas).

To understand what is meant by a quantum state we can begin with Dirac's notion that each physical quantity is represented by an Hermitean operator which is called an observable [5]. When this is measured by a suitable apparatus the system is left with a wave function corresponding to an eigenfunction of this observable. In general such a measurement will, in agreement with Heisenberg's principle, alter this wave function in an uncontrollable and unpredictable way. But the probability of a certain result n is $|C_n|^2$ where C_n is the coefficient of the nth eigenfunction in the expansion of the total wave function.

Once we obtain such an eigenfunction we can measure the same observable again and again, in principle, in a time so short that the wave function does not change significantly (except for a phase factor which is not relevant). Each measurement will then reproduce the same result. In terms of the 'naive' ontology that pervades ordinary experience, this leads one to suppose that, between measurements of the same observable, the system continues to exist with the same wave function ψ_n (again, except for a phase factor). Therefore one could say that during this time the system is in a certain state of being, i.e. it *stands* independently of its being observed. Of course, this state might change in longer times of its own accord and, in addition, it would also change if a different observable were measured.

In contrast, Bohr would never allow the type of language that admitted the independent existence of any kind of quantum object which could be said to be in a certain state. That is to say, he would not regard it as meaningful to talk about, for example, a particle existing between quantum measurements even if the same results were obtained for a given observable in a sequence of such measurements. Rather, as we have seen, he considered the experimental arrangement and the content (meaning) of the result to be a single unanalysable whole. To talk of a state in abstraction from such an experimental arrangement would, for Bohr, make no sense.

This general point can be clarified by considering what is in essence an intermediate approach adopted by Heisenberg.† He suggested that the wave function represented, not an actual reality, but rather a set of *potentialities* that could be realised according to the experimental conditions. A helpful analogy may be obtained by considering a seed, which is evidently not an actual plant, but which determines potentialities for realising various possible forms of the plant according to conditions of soil, rain, sunlight, wind, etc. Thus when the measurement of a given observable was repeated, this would correspond to a plant producing a seed, which growing under the same conditions, produced the same form of plant again (so that there was no continuously existent plant). Measurement of another observable would correspond to changing the experimental conditions, and this could produce a statistical range of possible plants of different forms. Returning to the quantum theory, it is clear that in this approach the apparatus is regarded as actually helping to 'create' the observed results.

It must be emphasised, however, that Bohr specifically rejected this

†This point of view was indeed proposed earlier by Bohm [6].

suggestion which he probably felt gave too much independent reality to whatever is supposed to be represented by the wave function. (As we recall he regarded this as only part of a calculus for predicting the statistics of experimental results.) Thus he states "I warned especially against phrases, often found in the physical literature, such as 'disturbing of phenomena by observation' or 'creating physical attributes to atomic objects by measurement'. Such phrases are ... apt to cause confusion, ..." [7]. Bohr is evidently saying here essentially what we have said before, i.e. that for him it has no meaning to talk of a quantum object with its attributes apart from the unanalysable whole phenomenon in which it is actually observed.

It is thus clear, as we have indeed already pointed out earlier, that Bohr's objection to the potentiality approach, as well as to taking the concept of quantum state too literally, does not represent for him a purely philosophical question. Indeed in his discussion of the Einstein, Podolsky and Rosen experiment [EPR], it was just this point that was crucial in his answer to the challenge presented by EPR. As we shall show in more detail in chapter 7, Bohr would say that the EPR paradox was based on an inadmissible attribution of properties to a second particle solely on the basis of measurements that could be carried out on the first particle.

2.4 von Neumann's approach to quantum theory

It is clear then that there is an important distinction between Bohr's approach and that of Heisenberg with his notion of potentiality, and perhaps an even greater difference from that of most physicists, who give a basic significance to the concept of quantum state. The notion of quantum state has indeed been most systematically and extensively developed by von Neumann, who not only gave it a precise mathematical formulation, but who also attempted, in his own way, to come to grips with the philosophical issues to which this approach gave rise.

It was a key part of this development to give a proof claiming to show that quantum mechanics had an intrinsic logical closure (in the sense, that no further concepts, e.g. involving 'hidden variables', could be introduced that would make possible a more detailed description of the state of the system than is afforded by the wave function). On this basis he concluded that the wave function yielded the most complete possible description of what we have been calling quantum reality, which is thus totally contained in the concept of a quantum state.

In order to clarify the physical meaning of these notions he developed a more detailed theory of measurements. This theory still gave a basic significance to epistemology because the only meaning attributed to the wave function was that it gave probabilities for the results of possible measurements (i.e. it did not begin with the assumption of an independently existing universe that would have meaning apart from the process in which its properties were measured). Nevertheless this theory gave more significance to ontology than Bohr did because it assumed the quantum system existed in a certain quantum state.

This state could only be manifested in phenomena at a large scale (classical) level. Thus he was led to make a distinction between the quantum and the classical levels. Between them, he said there was a 'cut' [8]. This is, of course, purely abstract because von Neumann admitted, along with physicists in general, that the quantum and classical levels had to exist in what was basically one world. However, for the sake of analysis one could talk about these two different levels and treat them as being in interaction. The effect of this interaction was to produce at the classical level a certain observable experimental result. The probability of the nth result was, of course, $|C_n|^2$, where the original wave function was $\psi = \sum C_n \psi_n$ and ψ_n is an eigenfunction of the operator being measured. But reciprocally, this interaction produced an effect on the quantum level; that is, the wave function changed from its original form ψ to ψ_n, where n is the actual result of the measurement obtained at the classical level. This change has been described as a 'collapse' of the wave function. Such a collapse would violate Schrödinger's equation, which must hold for any quantum system. However, this does not seem to have disturbed von Neumann unduly, probably because one could think that in its interaction with the classical level such a system need not satisfy the laws that apply when it is isolated.

One difficulty with this theory is that the location of the cut between quantum and classical level is to a large extent arbitrary. For example, one may include the apparatus and the observed object as part of a single combined system, which is to be treated quantum mechanically. We then observe this combined system with the aid of yet another apparatus which is, however, treated as being in the classical level. The 'cut' has then been moved to some point between the first apparatus and the second.

Von Neumann has given a mathematical treatment of this experiment which we shall sketch here. Let O be the operator that is to be measured. Let O_n be its eigenvalues and $\psi_n(x)$ the corresponding eigenfunctions in the x-representation. The initial wave function is, as we have already stated,

$$\Psi = \sum_n C_n \psi_n(x).$$

The apparatus may have a large number of coordinates, but it will be sufficient to consider one of these, y, representing, for example, a pointer from whose points one can read the result of the measurement. Initially the apparatus is in a fairly well-defined state represented by a wave packet $\phi_0(y)$. The initial wave function of the combined system is then

$$\Psi_0 = \phi_0(y) \sum_n C_n \psi_n(x). \tag{2.1}$$

We then assume an interaction between the observed system and the apparatus which lasts only for a time Δt. For the purpose of explaining the principles involved, it will be sufficient to consider what is called an impulsive measurement, i.e. one in which the interaction is so strong that throughout the period in which it works, the changes in the observed system and the observing apparatus that would occur independently of the interaction may be neglected.

The interaction Hamiltonian may be chosen as

$$H_I = \lambda O i\hbar \frac{\partial}{\partial y} \tag{2.2}$$

where λ is a suitable constant. Since this is the same as the total Hamiltonian during this period, we can easily solve Schrödinger's equation,

$$i\hbar \frac{\partial \Psi}{\partial t} = i\hbar \lambda O \frac{\partial \Psi}{\partial y} \tag{2.3}$$

to obtain for the wave function after an interval Δt,

$$\Psi = \sum_n C_n \psi_n(x) \phi_0(y - \lambda O_n \Delta t). \tag{2.4}$$

If the interaction is chosen so that

$$\lambda \Delta t \Delta O_n \gg 1 \tag{2.5}$$

where ΔO_n is the change of O_n for successive values of n, then it follows that the wave packets multiplying different $\psi_n(x)$ will not overlap. To each $\phi_0(y - \lambda O_n \Delta t)$ there will correspond a wave function $C_n \psi_n(x)$.

If we now observe this system with the aid of a second piece of apparatus, then in accordance with the postulates that have been described earlier, the latter will register the value of y. But it will now have to be in one of the packets, while $\psi_n(x)$ will then represent the corresponding state of the original quantum system. In effect the total wave function has 'collapsed' from the original linear combination of products to a particular product $\psi_n(x)\phi_0(y - \lambda O_n \Delta t)$. The probability that this happens can be shown to be $|C_n|^2$ exactly as it was when we had only one measuring apparatus.

In a 'naive' view of this process, one could readily say that this collapse represented merely an improvement of our knowledge of the state of the system which resulted from its being measured by the second apparatus. Indeed in the application of classical probability in physics such 'collapses' are quite common. Thus before one has observed a specified ensemble, the probability of a certain result n may be P_n. When one observes the result s, the probability suddenly collapses from P_n to δ_{ns}.

But this interpretation is not valid here because in the classical situation we have a linear combination of *probabilities* of each of the results, whereas quantum mechanically we have a linear combination of *wave functions*, while the probability depends quadratically on these wave functions. Before the second measuring apparatus has functioned, we therefore cannot say that the system is definitely in one of the n states with probability $|C_n|^2$. For, a whole range of subtle physical properties exist which depend on the linear combination of wave functions. Thus although the wave packets corresponding to different values of n do not overlap, they could, in principle, be made to do so once again by means of further interactions. For example, one could introduce a suitable term in the Hamiltonian that brought such an overlap about. Moreover one could have subtle observables corresponding to operators that couple the combined states of both systems and the mean values of these would depend on the existence of linear combinations of the kind we have discussed above. (This point is discussed in some detail in Bohm [9].) All of this means that such linear combinations have an ontological significance and do not merely describe our knowledge of the probabilities of possible values of n which could be the result of this measurement.

Actually a similar problem was present even when we had only one piece of apparatus. But this is not generally felt to be disturbing because of the tacit assumption that the quantum of action that connects the observed system and the observing apparatus could readily introduce significant physical changes in a microsystem such as an atom (along the general lines described in connection with the Heisenberg microscope experiment). However, we are now led to the conclusion that observation could also introduce significant changes of this kind in a macrosystem which includes the first piece of apparatus. Or, to put it differently, we may readily accept the notion that in an observation, the quantum state of a microsystem undergoes a real change when the wave function 'collapses' from a linear combination $\Psi = \sum C_n \psi_n(x)$ down to one of the eigenfunctions $\psi_s(x)$. It is not clear however what it means to say that there is a similar real change in a macrosystem when the wave function collapses from $\Psi = \sum C_n \psi_n(x)\phi_0(y - \lambda O_n \Delta t)$ to a single state $\psi_s(x)\phi_0(y - \lambda O_s \Delta t)$.

This difficulty arises in essence because von Neumann introduced the basically ontological notion that the wave function represents a quantum state that somehow 'stands on its own' (although, of course, in interaction with the classical level). Bohr avoids this problem by never speaking of a quantum object that could stand on its own, but rather by speaking only of a *phenomenon* which is an unanalysable whole. The question of interaction between a quantum level and a classical level thus cannot arise. Therefore, in this sense, he is more consistent than von Neumann.

At first sight one might be inclined to regard these questions as not very important. For after all the cut is only an abstraction and one can see that the statistical results do not depend on where it is placed. However, in so far as von Neumann effectively gave the quantum state a certain ontological significance, the net result was to produce a confused and unsatisfactory ontology. This ontology is such as to imply that the collapse of the wave function must also have an ontological significance (whereas for Bohr it merely represents a feature of the quantum algorithm which arises in the treatment of a new experiment). To show the extent of this difficulty, one could, for example, introduce a third apparatus that would measure a system that consisted of the observed object and the first two pieces of apparatus. For this situation the collapse would take place between the second and third piece of apparatus. One could go on with this sort of sequence indefinitely to include, for example, a computer recording of the results on a disc. In this case the collapse would take place when the disc was read, perhaps even a year or so later. (In which case the whole system would be in a certain quantum state represented by a linear combination of wave functions over this whole period of time.) And, as von Neumann himself pointed out, one could even include parts of the human brain within the total quantum system, so that the collapse could be brought about as a function of the brain.

It is evident that this whole situation is unsatisfactory because the ontological process of collapse is itself highly ambiguous. Perhaps Bohr's rather more limited ambiguity may seem preferable to von Neumann's indefinitely proliferating ambiguity.

Wigner has carried this argument further and has suggested that the above ambiguity of the collapse can be removed by assuming that this process is definitely a consequence of the interaction of matter and mind [10]. Thus he is, in effect, placing the cut between these two and implying that mind is not limited by quantum theory. (Pauli has also felt for different reasons that mind plays a key role in this context [11].)

We can see several difficulties in the attempt to bring in the direct action of the mind to give an ontological interpretation of the current physical laws of the quantum theory. Thus in a laboratory, it is hard to believe that the human mind is actually significantly affecting the results of the functioning of the instruments (which may, as we have already pointed out, be recorded on a computer that is not even examined for a long time). Moreover quantum theory is currently applied to cosmology, and it is difficult to believe that the evolution of the universe before the appearance of human beings depended fundamentally on the human mind (e.g. to make its wave function 'collapse' in an appropriate way). Of course one could avoid this difficulty by assuming a universal mind. But if we know little about the human mind, we know a great deal less about the universal mind. Such an assumption replaces one mystery by an even greater one.

One may ask why physicists have felt the need to bring in mind in their attempts to make sense of the quantum theory. Such a need is, indeed, implied in the work of those following along the lines of von Neumann. These want to say that the wave function has an ontological significance, i.e. as representing the quantum state, and at the same time to assume that it is a complete description of reality. However, as we have already pointed out, Bohr has claimed (apparently with greater consistency) that any ontology whatsoever is ruled out by the very nature of reality as revealed throughout the quantum theory. This would suggest that it would be better to adopt Bohr's point of view.

2.5 Are Bohr's conclusions inevitable?

But does the fact that the quantum theory has been applied so successfully lead inevitably to Bohr's conclusion concerning the nature of reality? Clearly it does not. For as we have already explained, it involves certain assumptions about real physical processes. In order to examine these assumptions properly, let us repeat them. Firstly the quantum of action is taken to be indivisible and secondly it is assumed to be unpredictable and uncontrollable. From this, Bohr draws the conclusion that the state of being is inherently ambiguous at the quantum level of accuracy.

It is essential to look more carefully at this conclusion which is based, in part, on a tacit identification of determinism with predictability and controllability. This identification is clearly characteristic of positivist philosophy. In this philosophy, science is not regarded as dealing with *what is*, so that concepts cannot be regarded as reflecting reality. Rather they have to be defined empirically, i.e. in relation to their manifestation in observation and experience. In such a philosophy it follows that determinism can have no meaning beyond predictability and controllability.

Since the quantum theory was first formulated, the relationship of determinism to predictability and controllability has been clarified by the discovery that a very general class of deterministic systems (i.e. those having unstable and chaotic motions) are neither predictable nor controllable, as has been discussed in some detail by Penrose [12]. Thus the identification of determinism with predictability and controllability has been invalidated. It follows that the mere uncontrollability and unpredictability of quantum phenomena does not necessarily imply that there can be no quantum world, which could in itself be determinate.

What about the indivisibility and unanalysability of the quantum of action? It is true that in some sense, at least, the quantum of action is neither divisible nor analysable at the level of the *phenomena*. Consider, for example, an atom emitting a quantum of light. We have two distinct states. (a) The atom in an excited state and no quantum present. And (b) the atom in its ground state and a quantum present.

The process of going from (a) to (b) is said to be a 'quantum jump' in the sense that there are no phenomena which correspond to any state in between. Of course, we may try to find such phenomena by observing the system in its process of transition. But as implied in Bohr's views, this would constitute an entirely different experimental arrangement that would be incompatible with the process of transition that we are considering.

From this however it does not follow that there is no more complete description perhaps at a deeper more complex level in which this process can be treated as continuous and analysable. One can indeed easily conceive of such a process in general terms. For example, the same kind of nonlinear equations that give rise to unstable and chaotic motions can also lead to what are called stable limit cycles in which the system stays near a certain state of motion. But more generally this stability may be limited so that the system can 'jump' from one such limit cycle to another, in a movement so fast and unstable that it could neither be predicted, controlled nor followed. Thus it would not appear in the phenomena. Indeed as we shall see in chapter 5, our interpretation of the quantum theory implies just such 'jumps' between what are in essence stable limit cycles. In this way one may explain processes that in the quantum theory are called "unpredictable, uncontrollable and indivisible quantum transitions between discrete orbits".

But as we recall, Bohr's entire position depended crucially on his assumptions of the nature of the quantum of action. Therefore from what we have said above it follows that there is no inherent necessity to adopt Bohr's position, and that there is nothing in Bohr's analysis that could rule out a quantum ontology. But of course, this latter would require the introduction of new concepts beyond that of the wave function and the quantum state. We would have to begin by simply assuming the new concepts and defining them through their participation in the laws of physics.

In doing this we have to differ from Bohr who at least tacitly required that all basic physical concepts be defined by referring them to specific phenomena in which they are measured. In contrast we *derive* the possible phenomena as forms on the overall structure of concepts and their relationships. An example of this is given by Einstein's derivation of particles obeying the usual laws of motion, either as singularities, or as very strong static pulses in a continuous non-linear field. The test of the theory is then to see whether the derived phenomena, not only explain the general form of the observed phenomena, but also their detailed relationships. In such an approach, the epistemology follows naturally from the ontology (just as it does in classical physics).

At this point, however, we have to return to von Neumann who, as we have already pointed out, believed that the wave function contained the most complete possible description of reality, thus implying that there was no way to do what we have suggested above. Von Neumann based this belief on his theorem to which we have alluded earlier, that claimed to show that a more detailed description would not be compatible with the laws of the quantum theory [13]. This proof was however questioned by Bohm [14] in 1952 and later by Bell [15]. A number of those who followed along von Neumann's lines, refined his arguments in several ways, but these refinements were also shown by Bell to make tacit assumptions about ontological theories that are too limited. (All of this will be discussed in more detail in chapter 7.)

We conclude that there are no sound reasons against seeking an ontological interpretation of the quantum theory. This book presents in essence the first complete ontological interpretation that has been proposed. As indicated in the introduction, there have been several other ontological interpretations since then. In chapter 14 we shall discuss these and compare and contrast them with the interpretation given in this book.

2.6 References

1. N. Bohr, *Atomic Physics and Human Knowledge*, Science Editions, New York, 1961, 50–51.
2. *Ibid.*, p.26.
3. H. J. Folse, *The Philosophy of Niels Bohr, the Framework of Complementarity*, North-Holland, Amsterdam, 1985.
4. N. Bohr, *Atomic Physics and Human Knowledge*, Science Editions, New York, 1961, 71.

5. P. A. M. Dirac, *The Principles of Quantum Mechanics*, Clarendon Press, Oxford, 1947.

6. D. Bohm, *Quantum Theory*, Prentice-Hall, Englewood Cliffs, New Jersey, 1951. Also available in Dover Publications, New York, 1989.

7. N. Bohr, *Atomic Physics and Human Knowledge*, Science Editions, New York, 1961, 63–64.

8. J. von Neumann, *Mathematical Foundations of Quantum Mechanics*, Princeton University Press, Princeton, 1955.

9. D. Bohm, *Quantum Theory*, Prentice-Hall, Englewood Cliffs, New Jersey, 1951, chapter 22.

10. E. Wigner, in *The Scientist Speculates*, ed. I. J. Good, Heinemann, London, 1961.

11. K. V. Laurikainen, *Beyond the Atom. The Philosophical Thought of Wolfgang Pauli*, Springer-Verlag, Heidelberg, 1988.

12. R. Penrose, *The Emperor's New Mind*, Oxford University Press, Oxford, 1989.

13. J. von Neumann, *Mathematical Foundations of Quantum Mechanics*, Princeton University Press, Princeton, 1955, 324.

14. D. Bohm, *Phys. Rev.* **85**, 166–193 (1952).

15. J. S. Bell, *Rev. Mod. Phys.* **38**, 447–452 (1966), and also in *Speakable and Unspeakable in Quantum Mechanics*, Cambridge University Press, Cambridge, 1987.

Chapter 3
Causal interpretation of the one-body system

In this chapter we develop the basic principles of our ontological interpretation of the quantum theory in the context of a one-body system (while the many-body system will be treated in the next chapter).

3.1 The main points of the causal interpretation

Let us begin by considering the standard WKB approximation for the classical limit in quantum mechanics. To do this we write the wave function in polar form $\psi = R\exp(iS/\hbar)$. We insert this form into Schrödinger's equation

$$i\hbar\frac{\partial\psi}{\partial t} = -\frac{\hbar^2}{2m}\nabla^2\psi + V\psi \tag{3.1}$$

where V is the classical potential. This gives rise to two equations

$$\frac{\partial S}{\partial t} + \frac{(\nabla S)^2}{2m} + V - \frac{\hbar^2}{2m}\frac{\nabla^2 R}{R} = 0 \tag{3.2}$$

and

$$\frac{\partial R^2}{\partial t} + \nabla\cdot\left(R^2\frac{\nabla S}{m}\right) = 0. \tag{3.3}$$

In order to obtain the WKB approximation, we note that in the classical limit in which there is a wave packet of width much greater than the wave length, λ, the term $-\hbar^2\nabla^2 R/2mR$ will be very small compared with the term $(\nabla S)^2/2m$. We therefore neglect it and obtain

$$\frac{\partial S_c}{\partial t} + \frac{(\nabla S_c)^2}{2m} + V = 0. \tag{3.4}$$

In the above we have written $S = S_c$ to indicate that we are dealing with the classical Hamilton-Jacobi equation representing a particle with momentum

$$\boldsymbol{p} = \nabla S_c \tag{3.5}$$

which moves normal to the wave front S_c = const. It follows then that equation (3.2) can be regarded as a conservation equation for the probability in an ensemble of such particles, all moving normal to the same wave front with a probability density $P = R^2$.

All of this may seem familiar but nevertheless there is something very noteworthy here. For though we have started with the quantum theory with all its ambiguities about the nature of a quantum system, we have somehow 'slipped over' into what is in essence the ordinary classical ontology. It seems natural at this point to ask whether this kind of ontology could not be extended to the quantum domain. Thus we note the quantum equation (3.2) differs from the classical equation (3.4) only by the term $-\hbar^2/2m(\nabla^2 R/R)$ which evidently can be regarded as playing the role of an additional potential in what we may call the quantum Hamilton-Jacobi equation. To bring this out we shall define what we call the quantum potential:

$$Q = -\frac{\hbar^2}{2m}\frac{\nabla^2 R}{R}. \tag{3.6}$$

The quantum Hamilton-Jacobi equation then becomes

$$\frac{\partial S}{\partial t} + \frac{(\nabla S)^2}{2m} + V + Q = 0. \tag{3.7}$$

The equation (3.3) still expresses the conservation of probability, but for an ensemble of particles which satisfies (3.7) rather than (3.4).

Let us now discuss this ontology in a more systematic way. Its key points are:

1. The electron actually *is* a particle with a well-defined position $\boldsymbol{x}(t)$ which varies continuously and is causally determined.

2. This particle is never separate from a new type of quantum field that fundamentally affects it. This field is given by R and S or alternatively by $\psi = R\exp(iS/\hbar)$. ψ then satisfies Schrödinger's equation (rather than, for example, Maxwell's equation), so that it too changes continuously and is causally determined.

3. The particle has an equation of motion

$$m\frac{\mathrm{d}\boldsymbol{v}}{\mathrm{d}t} = -\nabla(V) - \nabla(Q). \tag{3.8}$$

This means that the forces acting on it are not only the classical force, $-\nabla V$, but also the quantum force, $-\nabla Q$.

4. The particle momentum is restricted to $\boldsymbol{p} = \nabla S$. Since the quantum field ψ is single valued it follows (as can easily be shown) that

$$\oint p \, \mathrm{d}x = nh. \tag{3.9}$$

This resembles the old Bohr-Sommerfeld condition, but differs from the latter in which $\boldsymbol{p} = \nabla S_c$ where S_c is the solution of the classical Hamilton-Jacobi equation (3.4) rather than the quantum Hamilton-Jacobi equation (3.7).

5. In a statistical ensemble of particles, selected so that all have the same quantum field ψ, the probability density is $P = R^2$. We shall discuss the significance of this in more detail in a later section, but we can now note that if $P = R^2$ holds initially, then the conservation equation (3.3) guarantees that it will hold for all time.

Given that the particle is always accompanied by its quantum field ψ, we may say that the combined system of particle plus field is causally determined. (The statistics merely apply to an ensemble of causally determined trajectories.) For this reason the above proposals have been called the 'causal interpretation' in some of the earlier papers (though it must be emphasised that the basic ontological point of view given here can be extended, as we shall indeed do in chapter 9, to a more general stochastic context).

Finally it should be pointed out that unlike what happens with Maxwell's equations for example, the Schrödinger equation for the quantum field does not have sources, nor does it have any other way by which the field could be directly affected by the conditions of the particles. This of course constitutes an important difference between quantum fields and other fields that have thus far been used. As we shall see, however, the quantum theory can be understood completely in terms of the assumption that the quantum field has no sources or other forms of dependence on the particles. We shall in chapter 14, section 14.6, go into what it would mean to have such dependence and we shall see that this would imply that the quantum theory is an approximation with a limited domain of validity. In this way, as well as in other ways, we will see that our ontological interpretation permits a generalisation of the laws of physics going beyond the quantum theory, yet approaching the quantum theory as a suitable limit within which physics has thus far been contained.

3.2 New concepts implied by the ontological interpretation

At first sight it may seem that to consider the electron, for example, as some kind of particle that is affected by the quantum field ψ is just a return to older classical ideas. Such a notion is however generally felt to have long since been proved to be inadequate in understanding quantum theory which possesses so many features that are very different from those of classical mechanics. However, closer inspection shows that we are not actually reducing quantum mechanics in this way to an explanation in terms of classical ideas. For the quantum potential has a number of strikingly new features which do not cohere with what is generally accepted as the essential structure of classical physics.

In this connection it is important to note that the form of Newton's laws alone is not enough to determine that the general structure of classical physics shall hold. For example, a great deal of work has been done showing that to obtain determinism (which is surely an essential feature of classical physics) we require further assumptions on the nature of the forces [1]. For example, if infinite particle velocities or signal velocities are allowed, it has been demonstrated that determinism may fail. But no one has claimed to have given an exhaustive treatment of all the requirements even for determinism alone. Moreover it seems reasonable to suppose that other features of classical physics may also depend on further (largely tacit) assumptions about the nature of the forces. Indeed as we shall see, the new qualitative features of the quantum potential that we have mentioned above are just such as to imply the new properties of matter that are revealed by the quantum theory.

The first of these new properties can be seen by noting that the quantum potential is not changed when we multiply the field ψ by an arbitrary constant. (This is because ψ appears both in the numerator and the denominator of Q.) This means that the effect of the quantum potential is independent of the strength (i.e. the intensity) of the quantum field but depends only on its *form*. By contrast, classical waves, which act mechanically (i.e. to transfer energy and momentum, for example, to push a floating object), always produce effects that are more or less proportional to the strength of the wave. For example one may consider a water wave which causes a cork to bob. The further the cork is from the centre of the wave the less it will move. But with the quantum field, it is as if the cork could bob with full strength even far from the source of the wave.

Such behaviour would seem strange from the point of view of classical physics. Yet it is fairly common at the level of ordinary experience. For example we may consider a ship on automatic pilot being guided by radio waves. Here, too, the effect of the radio waves is independent of their intensity and depends only on their form. The essential point is that the ship is moving with its own energy, and that the *form* of the radio waves is taken up to direct the much greater energy of the ship. We may therefore propose that an electron too moves under its own energy, and that the *form* of the quantum wave directs the energy of the electron.

This introduces several new features into the movement. First of all, it means that particles moving in empty space under the action of no classical forces need not travel uniformly in straight lines. This is a radical departure from classical Newtonian theory. Moreover, since the effect of the wave does not necessarily fall off with the distance, even remote features of the environment can profoundly affect the movement.

As an example, let us consider the interference experiment. This involves a system of two slits. A particle is incident on this system, along with its quantum wave. While the particle can only go through one slit or the other, the wave goes through both. On the outgoing side of the slit system, the waves interfere to produce a complex quantum potential which does not in general fall off with the distance from the slits [2]. This potential is shown in figure 3.1. Note the deep 'valleys' and broad 'plateaux'. In the regions where the quantum potential changes rapidly there is a strong force on the particle. The particle is thus deflected, even though no classical force is acting.

We now consider a statistical ensemble of particles which may be obtained, for example, by having electrons 'boiled out' of a hot filament in a random way. Each electron has its own quantum field, but with the aid of suitable collimators and velocity selectors, we choose only those electrons with quantum fields corresponding approximately to waves with given direction and wave number that are incident on the slits in the manner described above. While all the electrons now have essentially the same form of the quantum field and therefore of the quantum potential, they will all approach the slit system from different starting points. As will be shown in section 3.4 as well as in chapter 9, for this case we may expect an essentially random distribution of such incident electrons. The resulting trajectories which are shown in figure 3.2 are then bunched into a series of dense and rare regions. These evidently constitute what are commonly called interference fringes.

If, however, just one slit had been open, the quantum field, ψ, would only pass through this slit, so that beyond the slit system there would have been a different quantum field and therefore a different potential. This would produce a more nearly uniform distribution of particles arriving at the screen, rather than a set of fringe-like regions. In this way we explain why the opening of a second slit can prevent particles from arriving at

Figure 3.1: Trajectories for two Gaussian slit systems

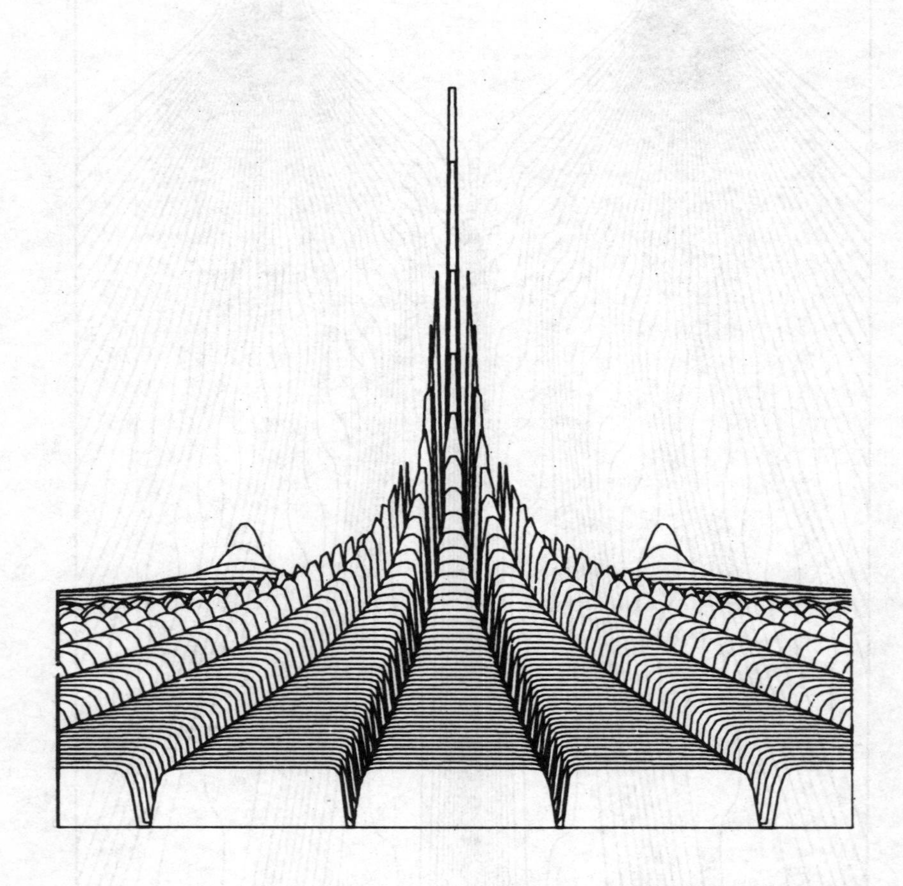

Figure 3.2: Quantum potential for two Gaussian slits

points to which they would not have come if only one slit had been open.

In this explanation of the quantum properties of the electron, the fact that the quantum potential depends only on the form and not on the amplitude of the quantum field is evidently of crucial significance. As we have already suggested, although at first sight such behaviour seems to be totally outside of our common experience, a little reflection shows that this is not so. Effects of this kind are indeed frequently encountered in ordinary experience wherever we are dealing with *information*. Thus in the example of the ship guided by radio waves, one may say that these waves carry information about what is in the environment of the ship and that this information enters into the movements of the ship through its being taken up in the mechanism of the automatic pilot. Similarly we explain the interference properties by saying that the quantum field contains information, for example about the slits, and that this information is taken up in the movements of the particle. In effect we have in this way introduced a concept that is new in the context of physics—a concept that we shall call *active information*. The basic idea of active information is that a form having very little energy enters into and directs a much greater energy. The activity of the latter is in this way given a form similar to that of the smaller energy.

It is important to distinguish our concept of active information from the more technical definition of information commonly adopted in physics in terms of, for example, Shannon's ideas [3] implying that there is a quantitative measure of information that represents the way in which the state of a system is *uncertain to us* (e.g. that we can only specify probabilities of various states). It is true that such concepts have been used to calculate objective properties of systems in thermodynamics and even black holes etc. [4], but we wish to propose here a quite different notion of information that is not essentially related to our own knowledge or lack of it. Rather in the case that we are discussing, for example, it will be information that is relevant to determining the movement of the electron itself. We emphasise again that it is our thesis that this sort of usage of the word information is actually encountered in a wide range of areas of experience. What is crucial here is that we are calling attention to the literal meaning of the word, i.e. to in-form, which is actively to put form into something or to imbue something with form.

As a simple example of what we mean, consider a radio wave whose form carries a signal. The sound energy we hear in the radio does not come directly from the radio wave itself which is too weak to be detected by our senses. It comes from the power plug or batteries which provide an essentially *un*formed energy that can be given form (i.e. in-formed) by the pattern carried by the radio wave. This process is evidently entirely objective and has nothing to do with our knowing the details of how this happens. The information in the radio wave is *potentially* active everywhere, but it is actually active, only where and when it can give form to the electrical energy which, in this case, is in the radio.

A more developed example of such a situation is given by considering the computer. The information content in a silicon chip can determine a whole range of potential activities which may be actualised by giving form to the electrical energy coming from a power source. Which of these possibilities will be actualised in a given case depends on a wider context and the responses of a computer operator.

Although the above examples do indicate what we mean by the objective significance of active information, nevertheless they still depend on structures (like the radio set and the computer) which were originally designed and put together by human beings and so may be felt to retain a trace of subjectivity. An example that does not involve structures set up by human beings is the function of the DNA molecule. The DNA is said to constitute a code, that is to say, a language. The form of the DNA molecule is considered as information content for this code, while the 'meaning' is expressed in terms of various processes; e.g. those involving RNA molecules, which 'read' the DNA code, and carry out the protein-making activities that are implied by particular sections of the DNA molecule. The comparison to our notion of objective and active information is very close. Thus, in the process of cell growth it is only the form of the DNA molecule that counts, while the energy is supplied by the rest of the cell (and indeed ultimately by the environment as a whole). Moreover, at any moment, only a part of the DNA molecule is being 'read' and giving rise to activity. The rest is potentially active and may become actually active according to the total situation in which the cell finds itself.

While we are bringing out above the objective aspects of information, we do not intend to deny its importance in subjective human experience. However, we wish to point out that even in this domain, the notion of active information still applies. A simple example is to be found in reading a map. In this activity we apprehend the information content of this map through our own mental energy. And by a whole set of virtual or potential activities in the imagination, we can see the possible significance of this map. Thus the information is immediately active in arousing the imagination, but this activity is still evidently inward within the brain and nervous system. If we are actually travelling in the territory itself then, at any moment, some particular aspect may be further actualised through our physical energies, acting in that territory (according to a broader context, including what the human being knows and what he is perceiving at that moment).

We therefore emphasise once again that even the information held by human beings is, in general, active rather than passive, not merely reflecting something outside itself but actually, or at least potentially, capable of participating in the thing to which it refers. Passive information may in fact be regarded as a limiting case in which we abstract from the activity of information. This is essentially the kind of information that is currently used in information theory, e.g. as used by Shannon. The puzzle in this approach is that of how information that is merely passive within us is able to determine actual objective processes outside of us. We suggest that passive information is rather like a map reflecting something of these processes which can guide us to organise them conveniently for our use, e.g. by means of algorithms that enable us to calculate entropy and other such properties.

If the notion of active information applies both objectively and subjectively, it may well be that all information is at least potentially active and that complete passivity is never more than an abstraction valid in certain limited circumstances. In this context our proposals to use the concept of active information at the quantum level does not seem to be unnatural.

To show how these ideas work out in more detail, we can go once again into the example that we gave earlier of the electron in an interference experiment. We could say that this particle has the ability to do work. This ability is released by active information in the quantum field, which is measured by the quantum potential. As the particle reaches certain points in front of the slits, it is 'in-formed' to accelerate or decelerate accordingly, sometimes quite violently.

Although equation (3.8) may look like a classical law implying pushing or pulling by the quantum potential, this would not be understandable because a very weak field can produce the full effect which depends only on the form of the wave. We therefore emphasise that the quantum field is not pushing or pulling the particle mechanically, any more than the radio wave is pushing or pulling the ship that it guides. So the ability to do work does not originate in the quantum field, but must have some other origin (a suggestion which we shall discuss presently).

The fact that the particle is moving under its own energy, but being guided by the information in the quantum field, suggests that an electron or any other elementary particle has a complex and subtle inner structure (e.g. perhaps even comparable to that of a radio). This notion goes against the whole tradition of modern physics which assumes that as we analyse matter into smaller and smaller parts its behaviour always grows more and more elementary. But our interpretation of the quantum theory indicates that nature is far more subtle and strange than previously thought. However, this sort of inner complexity is perhaps not as implausible as may appear at first sight. For example, a large crowd of people can be treated by simple statistical laws, whereas individually, their behaviour is immensely more subtle and complex. Similarly, large masses of matter reduce approximately to a simple Newtonian behaviour, whereas the molecules and atoms out of which matter is built have a more complex inner structure.

To make this suggestion yet more plausible, we note that between the shortest distances now measurable in physics (of the order of 10^{-16} cm) and the shortest distances in which current notions of space-time probably have meaning which is of the order of 10^{-33} cm, there is a vast range of scale in which an immense amount of yet undiscovered structure could be contained. Indeed, this range of scale is comparable to that which exists between our own size and that of the elementary particle. Moreover, since the vacuum is generally regarded as full (see Wheeler [5] and Hiley [6] for example) with an immense energy of fluctuation, revealed for example in the Casimir effect [7], it may be further suggested that ultimately the energy of this particle comes from this source. (Some of it may also come from the rest energy of the particle.)

It should be added here that (as happens with the radio wave) the quantum information field may also have some energy. However, as has been made clear in the many analogies given here, this must be negligible in comparison to the energy of the particle which it guides. (Though perhaps these might be detectable in new kinds of experiments of the sort suggested in chapter 13, section 13.6 and chapter 14, section 14.6, which would go beyond the domain of current quantum theory.)

A very important further implication of the notion of active information is that in a certain sense an entire experiment has to be regarded as a single undivided whole. This arises because the motion of the particles can be strongly affected by distant features of the environment such as the slits. On the other hand, in the corresponding classical experiment the slit system can be ignored once the particle has passed through it. And as has been pointed out, a different slit system would produce a different quantum potential which would affect the motion of the particles in a different way. Therefore the motion of the particles cannot properly be discussed in abstraction from the total experimental arrangement. This is reminiscent of Bohr's notion of wholeness, but it differs in that the entire process is open to our 'conceptual gaze' and can therefore be analysed in thought, even if it cannot be divided in actuality without radically changing its nature.

3.3 Comparison with de Broglie's idea of a double solution

The idea of a 'pilot wave' that guides the movement of the electron was first suggested by de Broglie [8] in 1927, but only in connection with the one-body system. De Broglie presented this idea at the 1927 Solvay Congress where it was strongly criticised by Pauli [9]. His most important criticism was that, in a two-body scattering process, the model could not be applied coherently. In consequence de Broglie abandoned his suggestion. The idea of a pilot wave was proposed again in 1952 by Bohm [10] in which an interpretation for the many-body system was given. This latter made it possible to answer Pauli's criticism and indeed opened the way to a coherent interpretation including a theory of measurement which was applicable over a wide range of quantum phenomena. As a result de Broglie [11] took up his original ideas again and continued to develop them in various ways.

An important part of de Broglie's early approach was to try to explain the assumptions underlying the pilot wave interpretation in terms of what he called the theory of the 'double solution'. This was based on the assumption of a non-linear field equation which, in the linear approximation, approached the ordinary Schrödinger equation. However, for large amplitude, the non-linearity became important. He suggested that there would exist solutions which would correspond to a stable singularity or pulse when the amplitude was high and would gradually shade off into solutions of the linear Schrödinger equation at larger distances. The pulse would evidently correspond to a particle. He then gave arguments aimed at showing that in order to obtain continuity of the field, the movement of this pulse would have to be determined by the relation $\boldsymbol{p} = \nabla S$, where S is the phase of the linear part of wave at the location of the pulse.

However, a closer analysis shows that this is actually only a necessary condition and not a sufficient one. Indeed we can see that it cannot be sufficient by considering the fact that energy and momentum conservation are necessary consequences of the kinds of equations discussed by de Broglie. The momentum in the singularity will be very large in comparison with that available in the extremely weak pilot wave. Therefore it will not be possible to obtain solutions of the field equations which would lead to the very great accelerations that are in general implied by the guidance relation (e.g. as seen in our discussion of the two slit interference experiment). Rather, we have seen that to obtain a powerful effect from a very weak field we need something like our concept of active information. For the phase, S, clearly depends only on the *form* of the field and not on the amplitude. In our approach, it is this form which 'in-forms' the energy of the self-movement of the particle. Therefore the key difference of our idea from that of de Broglie is that we do not attempt to explain the guidance relation in a simple mechanical way as an effect of non-linear propagation of fields. Instead we are appealing to the notion that a particle has a rich and complex inner structure which can respond to information and direct its self-motion accordingly.

In addition it should be noted that the concept of the double solution has never yet been extended to the many-body system, nor is there any idea available as to how this is to be done. On the other hand, as will be seen in later chapters, the notion of active information can quite naturally be extended to the many-body system. For these reasons as well as for those given earlier it seems that the general idea of something like active information is strongly indicated as needed for an ontological explanation of quantum theory.

Returning to a consideration of the guidance relationship, we emphasise that this by itself is not enough to determine the behaviour of the particle. It is necessary also to have some equation that determines the changes of S. In our case this is, of course, the Schrödinger equation. And as we have seen, it follows that the equation of motion is given by the modified form of Newton's law (3.8) which contains the quantum potential.

If we start from the guidance relations, it is then implied that the quantum potential, Q, will contribute to the acceleration in the way described above. However, we emphasise once again that the significance of this potential is not that it represents the mechanical effects of attraction and repulsion of the particle by various features of its environment, e.g. the slit system. Rather its significance is that it represents a contribution to the acceleration of the particle in its self-movement. But it must also be stressed that if the field satisfies an equation other than that of Schrödinger, then there will be a different formula for this contribution. (For example as we shall see in chapters 10 and 12, this actually happens in the case of the Pauli equation and the Dirac equation.)

3.4 On the role of probability in the quantum theory

Thus far we have mainly been considering the individual particle. We shall now discuss in more detail how the concept of probability is used in the theory, in a suitable statistical ensemble.

Let us first note that the function $P = R^2$ has two interpretations, one through the quantum potential and the other through the probability density. It is our proposal that the more fundamental meaning of R (and therefore indirectly of P) is that it determines the quantum potential. In contradistinction to the usual interpretation its meaning as a probability is only secondary. Its significance in this respect is that it gives the probability for the particle to *be* at a certain position. Here we differ again from the usual epistemological interpretations which suppose it is the probability of *finding* a particle there in a suitable measurement. Indeed as will be brought out in more detail in chapter 6, the measurement process itself has to be interpreted as a particular application of the theory. This theory is formulated basically in terms of what Bell [12] has called 'beables' rather than of 'observables'. These beables are assumed to have a reality that is independent of being observed or known in any other way. The observables therefore do not have a fundamental significance in our theory but rather are treated as statistical functions of the beables that are involved in what is currently called a measurement.

The above implies that $|\psi|^2$ has no necessary relationship to probability. That is to say, the two concepts are basically independent. It will be shown in chapter 9, however, that under typical chaotic conditions that prevail in most situations an arbitrary probability distribution, P, will approach and remain equal to $|\psi|^2$, the latter being an equilibrium distribution. The relationship between P and $|\psi|^2$ is in this way seen to be contingent. We will indeed discuss, for example, a possibility that $P \neq |\psi|^2$ under certain new kinds of conditions which could in principle ultimately be investigated, even though P would be equal to $|\psi|^2$ in all those situations that we have been thus far able to investigate. Moreover a stochastic model of the particle trajectories will also be given in chapter 9, which gives an additional possible explanation of why P approaches $|\psi|^2$.

To illustrate these ideas let us consider a statistical ensemble of particles in which the experimental situation is so arranged that all the particles have the same wave function. In the usual interpretation, this situation is called a pure state. (We discuss the mixed state in chapter 9.) As an example, we may consider our previous discussion of the two slit experiment in which electrons boil out of a hot filament and are then suitably selected to have the same quantum field ψ corresponding approximately to a plane wave with a definite wave vector $\boldsymbol{k}$ incident on the slits. This is the process that Dirac [13] calls the 'preparation' of the measurement (and for Bohr [14] it is, of course, part of the experimental conditions).

Suppose then that electrons enter the system one-by-one. Each one will have its own quantum field, but the forms of all these fields will be essentially the same. However, the particles themselves will have a chaotically varying distribution of initial positions with a probability $P = |\psi|^2$. It should be added that, as we shall show in chapter 9, there is no way to control or predict these conditions even though the motion is determinate in each individual case.

In the experiment under discussion, our ensemble of particles with the same quantum field, ψ, will pass through the slit system to be detected at the screen. Each individual result is determined by the initial conditions of the corresponding particle. As we have already pointed out, such a particle must pass through one slit or the other, but its motion is determined by active information coming from quantum fields that have passed through both slits. Therefore particles can be kept away from certain points at which they could have arrived if only one of the slits had been open. As the results accumulate, one will then obtain the bunching of particles that produces the interference pattern in the way that has been described.

To sum up then, the idea of probability that we have introduced here is clearly not essentially different from that used in classical statistical ensembles. Thus in no sense is probability being regarded as a fundamental concept. Rather the properties of the individual system are taken as primary, and probabilities are interpreted in terms of these. The main new quantum properties of matter follow not from the use of the probability theory, but rather from the qualitatively new features of the quantum potential which, for example, imply a novel quantum wholeness such that the behaviour of a particle may depend crucially on distant features of the environment.

3.5 Stationary states

We shall now go on to show how our interpretation can be applied to a number of specific examples which help to bring out the meaning of this interpretation. We shall begin by considering stationary states. In a stationary state the wave function oscillates harmonically with the time. It can be written as

$$\psi(\boldsymbol{x},t) = \psi_0(\boldsymbol{x}) \exp[-iEt/\hbar]. \tag{3.10}$$

The quantum Hamilton-Jacobi equation (3.8) becomes

$$E = (\nabla S)^2/2m + V + Q. \tag{3.11}$$

Let us first consider an atom in an s-state. In this case the wave function $\psi_0(\boldsymbol{x})$ is real so that one can write $S = 0$. This implies that

$$\boldsymbol{p} = 0 \quad \text{and} \quad E = V + Q. \tag{3.12}$$

The above result means that in an s-state the electron is at rest. This may seem surprising, since classical physics leads us to expect a condition of dynamical equilibrium for a state of definite energy as an explanation of why the potential does not cause the particle to fall into the nucleus. However, equation (3.12) shows that in terms of our interpretation, the stability of this state arises from the fact that the quantum potential cancels out the space variation of the classical potential leaving a constant energy E that is independent of position.

It must be remembered, however, that we are interpreting the quantum forces in terms of the new notion of active information that we have introduced in the previous chapter. This information 'informs' the particle in its self-motion to contribute an acceleration $-\nabla Q/m$ which, in a stationary state, balances the classical acceleration $-\nabla V/m$. The particle is therefore able to remain in a fixed position and the reason it does not fall into the nucleus is that this is prevented by the outward acceleration due to the quantum potential.

There is no state more closely bound to the nucleus than the lowest state. Indeed if we try to find one, we would discover that the repulsive effect of the quantum potential would be much stronger than the attractive effect of the classical potential. This explanation must be contrasted with the one usually given which attributes the stability of the lowest state to a kind of 'pressure' due to the random motion that is said to be implied by the uncertainty principle [15].

More generally however, in a non-stationary state, the balance between $-\nabla Q$ and $-\nabla V$ will not hold. Moreover if this system is coupled to one with which it can exchange energy, then a transition is possible from one stationary state to another. We shall discuss this process in chapter 5 where it will be seen that the probability that the particle will arrive at the point $\boldsymbol{x}$ is $|\psi(\boldsymbol{x})|^2$. In this way we understand how such a distribution of static particles can arise.

Nevertheless one may ask how the stationary values of $\boldsymbol{x}$ are related to what is implied by the usual interpretation for this case, i.e. the probability of obtaining a certain momentum $\boldsymbol{p}$ in a measurement is $|C_p|^2$ where C_p is the Fourier coefficient of the wave function. It will be shown, in fact, in chapter 6 dealing with the theory of measurement according to our interpretation, that if we measure the momentum of the particle in a stationary state, the resulting disturbance will change the quantum field and cause the particle to be accelerated by the quantum potential in a way that depends on its initial location. The probability of obtaining a net momentum $\boldsymbol{p}$ in this process will be shown to be just $|C_p|^2$. This means that for this case we have reproduced the results of the usual interpretation, but that we have given a different account of the meaning of these results. In chapter 6 we shall go into this question in more detail and show that our account of this process is consistent.

Moreover if one is not aesthetically satisfied with this picture of a static electron in a stationary state, one can go to the stochastic model given in chapter 9. In this model the particle will have a random motion round an average $\boldsymbol{p} = \nabla S$, and the net probability density in this random motion comes out as $P = |\psi(\boldsymbol{x})|^2$.

In principle all stationary state wave functions can be taken as real, so that the above considerations apply. However, if the energy levels are degenerate, it is possible to form stationary states that are complex linear combinations of these real wave functions. As an example, let us consider the three p-states. In their real forms, the basic wave functions are

$$\begin{aligned} \psi_1 &= x\, f(r)/r \\ \psi_2 &= y\, f(r)/r \\ \psi_3 &= z\, f(r)/r, \end{aligned} \tag{3.13}$$

where r is the radial distance from the nucleus. From these we can form linear combinations having a definite angular momentum in any direction which may with sufficient generality be taken as that of z. The three corresponding wave functions are then

$$\begin{aligned} \psi_+ &= (x+iy)\, f(r)/r \\ \psi_0 &= z\, f(r)/r \\ \psi_- &= (x-iy)\, f(r)/r, \end{aligned} \tag{3.14}$$

which represent, respectively, angular momentum projections in the z direction of $+\hbar$, 0 and $-\hbar$. The phases of these wave functions are respectively

$$\begin{aligned} S_+ &= \hbar\phi \\ S_0 &= 0 \\ S_- &= -\hbar\phi \end{aligned} \tag{3.15}$$

where ϕ is the polar angle round the z-axis. The particle momenta for these phases are

$$\begin{aligned} p_+ &= \hbar\hat{\phi}/\rho \\ p_0 &= 0 \\ p_- &= -\hbar\hat{\phi}/\rho \end{aligned} \tag{3.16}$$

where ρ is the radius in cylindrical polar coordinates around the z-axis.

Clearly the particle with z-component of angular momentum zero is at rest. But p_+ and p_- correspond to motion in a circle around the z-axis at an arbitrary value of z, with angular momentum $\pm\hbar$. (See figure 3.3.) The velocity becomes infinite at $\rho = 0$, but as is well known the wave function vanishes on this line so that there is no probability that a particle will be there.

The energy is

$$\begin{aligned} E &= (\nabla S)^2/2m + V(r) + Q \\ E &= \left(\frac{\partial S}{\partial z}\right)^2 + \left(\frac{\partial S}{\partial \rho}\right)^2 + \frac{1}{\rho}\left(\frac{\partial S}{\partial \phi}\right)^2 + V + Q \end{aligned}$$

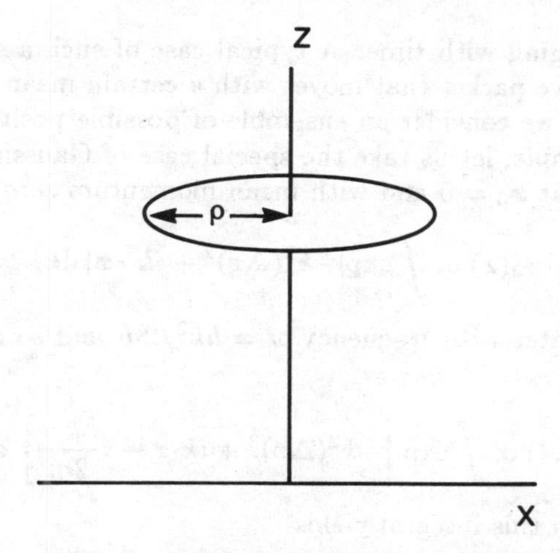

Figure 3.3: Possible particle orbit for the p_+ state

with

$$\frac{\partial S}{\partial z} = \frac{\partial S}{\partial \rho} = 0$$

so that

$$E = \frac{\hbar^2}{\rho^2} + V(r) + Q. \tag{3.17}$$

In this case part of the energy $\hbar^2/\rho^2$ is kinetic, while the quantum potential balances $V(r) + \hbar^2/\rho^2$ only in the directions of z and ρ.

We see then that by forming complex combinations of real stationary state wave functions, we obtain situations in which some of the energy is kinetic. In the case of a free particle, the two complex wave functions corresponding to an energy $E = h^2k^2/2m$ are $\exp(i\boldsymbol{k}\cdot\boldsymbol{x})$ and $\exp(-i\boldsymbol{k}\cdot\boldsymbol{x})$. These correspond respectively to particle momenta $\boldsymbol{p} = h\boldsymbol{k}$ and $\boldsymbol{p} = -h\boldsymbol{k}$. This is an extreme case in which, as one can readily verify, the quantum potential is zero and this is why all the energy is kinetic. To understand our interpretation properly it is necessary therefore to keep in mind that a part of what is thought of in classical intuition as kinetic energy is now treated as an energy associated with the quantum potential.

3.6 Non-stationary states

The general wave function is built out of a linear combination of stationary state wave functions and represents a situation in which the quantum potential is changing with time. A typical case of such a situation is represented by a wave packet that moves with a certain mean velocity. Inside this wave packet we consider an ensemble of possible positions of the particle. As an example, let us take the special case of Gaussian wave packet, initially centred at $\boldsymbol{x}_0 = 0$ and with mean momentum zero.

$$\psi_0(\boldsymbol{x}) \propto \int \exp[-k^2(\Delta x)^2 + i\boldsymbol{k}\cdot\boldsymbol{x}]\,\mathrm{d}\boldsymbol{k}. \tag{3.18}$$

Each wave oscillates with frequency $\omega = \hbar k^2/2m$ and so as a function of time, we have

$$\psi(\boldsymbol{x}, t) \propto \int \exp\left[-k^2(\Delta x)^2 + i\boldsymbol{k}\cdot\boldsymbol{x} - i\frac{\hbar k^2 t}{2m}\right]\mathrm{d}\boldsymbol{k}. \tag{3.19}$$

The evaluation of this integral yields

$$\begin{aligned}\psi(\boldsymbol{x}, t) \quad &\propto \quad \left[\Delta x + \frac{i\hbar t}{2m\Delta x}\right]^{-\frac{1}{2}} \exp\left\{\frac{-x^2\Delta x^2}{4\left[\Delta x^4 + \hbar^2 t^2/4m^2\right]}\right\} \\ &\qquad \times \exp\left\{\frac{(i\hbar t/2m)x^2}{4\Delta x^4 + \hbar^2 t^2/m^2}\right\}.\end{aligned} \tag{3.20}$$

The above represents a wave packet which remains centred at the origin but which spreads so that for large values of t, its width is

$$\sim \frac{\hbar t}{m\Delta x} \sim \frac{\Delta p t}{m} \sim \Delta v t.$$

This means that eventually the width of the packet corresponds to the spread of distances covered by the particles which is in turn determined by the spread of velocities which is equal to $\Delta \boldsymbol{v}$. What the usual picture gives for this process is that there actually is an initial spread of velocities $\Delta \boldsymbol{v} = \Delta \boldsymbol{p}/m$ which is implied by the uncertainty principle $\Delta \boldsymbol{p} \sim \hbar/\Delta \boldsymbol{x}$. But in the causal interpretation, the velocity is always well defined at each point. Its value is

$$v = \frac{\hbar}{m}\frac{\partial \phi}{\partial x}$$

where ϕ is the phase of the wave function. Since

$$\phi = \frac{\hbar x^2 t/2m}{4\Delta x^4 + \hbar^2 t^2/m^2} \tag{3.21}$$

we obtain

$$v = \frac{\hbar^2}{m^2}\left[\frac{xt}{4\Delta x^4 + \hbar^2 t^2/m^2}\right].$$

It is clear then that the particles are accelerated from zero velocity, and for large values of t the velocity is $v = x/t$. This acceleration is evidently a result of the quantum potential which is

$$Q = -\frac{\hbar^2}{2m}\frac{\nabla^2 R}{R} = \frac{\hbar^2}{m}\left[\frac{\Delta x^2}{4\Delta x^4 + \hbar^2 t^2/m^2} - \frac{2\Delta x^2 x^2}{(4\Delta x^4 + \hbar^2 t^2/m^2)^2}\right]. \quad (3.22)$$

Evidently the quantum potential decreases as the wave packet spreads, falling eventually to zero.

The picture is then that as the wave packet spreads, the particle gains kinetic energy, the amount depending upon where it was initially in the packet. This clearly denies the common idea to which we have already alluded that the spread of velocities was there from the start and given by the uncertainty principle. A similar situation was seen to arise in the case of stationary states with real wave functions, where likewise we had a zero velocity. The measurement process disturbed the system so that the energy represented by the quantum potential was turned into kinetic energy. However, the difference is that, in the case of the wave packet, we do not need an external disturbance such as that produced by a measurement process to initiate the acceleration of the particles.

3.7 Are energy and momentum conserved in non-stationary states?

It is clear that in a non-stationary state, the amplitude of the wave function R, and therefore the quantum potential associated with it, will be a function of time. This evidently means that the energy of the particle, $\partial S/\partial t$, will not be conserved in detail. It will however be conserved on the average as can be seen by considering

$$E = \int P(x)\frac{\partial S}{\partial t}\,\mathrm{d}x = \int R^2\left[\frac{(\nabla S)^2}{2m} + V - \frac{\hbar^2}{2m}\frac{\nabla^2 R}{R}\right]\mathrm{d}x.$$

Putting $\psi = R\exp(iS/\hbar)$ in the above, we readily obtain

$$\bar{E} = \int\left[\frac{\hbar}{2m}\psi^*\nabla^2\psi + \psi^* V\psi\right]\mathrm{d}x.$$

This is just the ordinary expression for the mean value of the Hamiltonian as given in the usual interpretation. And, as is well known, this is conserved as a consequence of Schrödinger's equation. Moreover this is enough to ensure that there will be conservation of energy in the classical limit where we neglect Heisenberg's uncertainty principle.

Even in the usual interpretation, energy need not be conserved in detail when there is a wave packet. Rather the correct statement would be that energy is conserved only within the range defined by Heisenberg's principle, i.e. only within the mean width, ΔE, of the wave packet. A similar situation can be found in what are called virtual states in the usual interpretation. These do not conserve the original energy, but in compensation they can last only for a time $h/\Delta E$ where ΔE is the extent of non-conservation [16].

In our interpretation, however, the particle energy is definite but constantly changing. Thus in the simple case of the Gaussian packet, we saw that the energy of the particle changes monotonically. More generally, however, the motion will be more complex and the particle may be expected to 'jiggle' in an irregular way. But whatever the motion, the range of variation in the energy will generally be of the order of the mean width, ΔE, of the packet. To see that this is so, let us write for the wave packet

$$\Psi = \exp\left[\frac{-iE_0t}{\hbar}\right]\sum_{\Delta E} C_{\Delta E}\psi_{E+\Delta E}(x)\exp\left[\frac{-i\Delta Et}{\hbar}\right]$$

where E_0 is the mean energy of the packet. Then

$$\begin{aligned} -\frac{\partial S}{\partial t} &= \hbar i\left[\frac{\partial\psi/\partial t}{\psi}\right] \\ &= E_0 + \left\{\sum_{\Delta E}\frac{\Delta E C_E\psi_{E+\Delta E}(x)\exp[-i\Delta Et/\hbar]}{\sum C_E\psi_{E+\Delta E}(x)\exp[-i\Delta Et/\hbar]}\right\}. \end{aligned}$$

It is clear that on the average the time dependent part of $\partial S/\partial t$ will be of the order of ΔE.

Our picture is then that each individual particle moves with an energy close to the average E_0, and with changes of energy of the order of ΔE (which may in general give rise to a complicated pattern of motion).

A similar treatment can be given for momentum, and one will find that a wave packet, even for a free particle, undergoes corresponding changes of momentum as well. However, on the whole these remain more or less within the limits set by Heisenberg's principle.

Where does this fluctuating energy and momentum come from? Evidently it can be attributed to the quantum potential which is now a function of time. But as we have seen earlier the quantum potential is implied by the guidance condition, $\boldsymbol{p} = \nabla S$, which, we recall, is to be interpreted as being brought about by the activity of the information in the quantum field. The energy and momentum then come from the self-movement of the particle and, as we have suggested earlier, may ultimately originate in the vacuum fluctuations. In such movement there is no intrinsic reason why energy and momentum should always be conserved. To be sure, this conservation holds for an isolated system in the classical limit, but it also holds there in our approach. Its failure to hold quantum mechanically is, as we have seen, of exactly the order as it is in the usual interpretation. In our approach one may however say that the electron viewed as a particle is never completely isolated because it is always affected by the quantum field and possibly by the vacuum fluctuations (so that the degree of isolation of any given system is, in general, relative and limited). The energy of the particle is, in fact, strictly conserved only in special cases, i.e. stationary states.

We must emphasise again that the one-body case is at best a simplification and an abstraction. For example, in the case of the slits, the particle may be strongly affected by the system through which it has gone, even when it is far away. If we treat the slits and the electron as a single combined system (as we shall show in chapter 4, section 4.2) it would then be possible to regard the resulting many-body wave function as expressing a direct nonlocal interaction between the electron and the slit (along lines to be discussed in more detail in the next chapter). In this case the total momentum of the combined system would be conserved because the slit system would take up the momentum lost by the particle. (This would evidently also happen in the conventional interpretation (see Bohr [14]).) Since the slit system is so heavy, its response would be negligible in comparison with the statistical fluctuations of the momentum so that this transfer would not show up in experiment.

Indeed whenever we abstract a system from a larger one, the wave function will reflect the overall system from which the abstraction has been taken. Therefore the goal of obtaining complete conservation for what we have wrongly been calling an isolated system will prove to be unattainable. For example, even if we take the slits and the electron as a combined system, the incident quantum field would reflect the collimating slits and velocity selectors which determined the beam incident on the slit system in question. And so, because the quantum potential can be large even when the quantum field is small, it follows that we will inevitably have non-conservation of both energy and momentum resulting from the fact that complete isolation of any quantum system is actually impossible.

This behaviour is reminiscent of Bohr's argument mentioned in chapter 2, namely that we cannot discuss the properties of a particular system apart from the context of the entire experimental arrangement with the aid of which these properties are observed. But of course, the difference is that we have given a conceptual analysis that explains why all this happens, whereas for Bohr nothing more can be said that can go beyond the description of the experimental phenomenon itself.

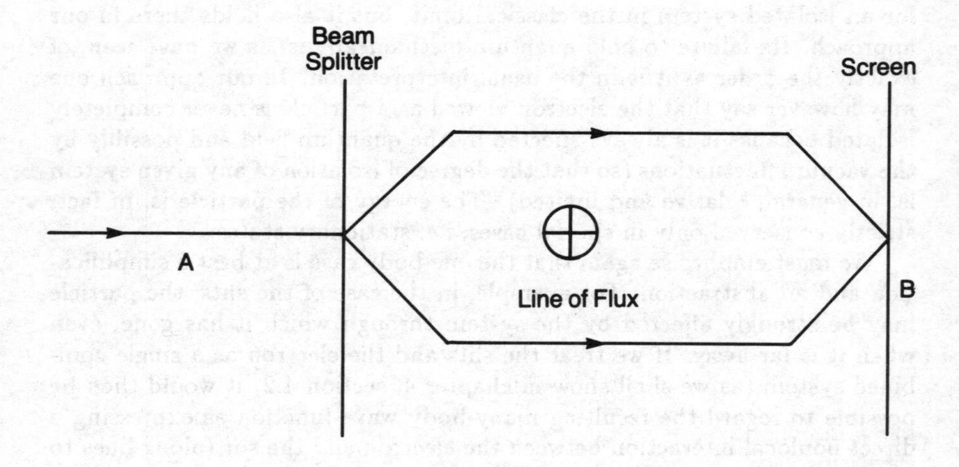

Figure 3.4: Schematic AB effect

3.8 The Aharonov-Bohm effect

A particularly interesting application of our interpretation is to the Aharonov-Bohm (AB) effect [17]. This effect demonstrates that a magnetic line of flux can produce observable modifications of an electron interference pattern, even though the electrons themselves never encounter a magnetic field. This is often felt to be mysterious because it is generally assumed that only the field $\boldsymbol{B}(x)$ is physically active and that the vector potential, $\boldsymbol{A}$, from which the field is derived via $\boldsymbol{B} = (\nabla \times \boldsymbol{A})$, has itself no further physical significance beyond that of being a mathematical auxiliary in the description of the field. If no magnetic field acts on the electrons how then can the interference pattern be changed?

We now give a brief description of the experiment. An electron beam from the source and collimating system, A, is incident on a beam splitter. The two resulting beams go round a line of flux and are recombined to meet again at a screen B, as shown in figure 3.4. When the flux is zero, a certain interference pattern is obtained. But for a finite value of the flux, this pattern is shifted. This shift was predicted from the quantum theory and it has been verified experimentally [18].

The ordinary quantum mechanical treatment begins with the Schrödinger equation for the case where there is an electromagnetic vector potential,

$\boldsymbol{A}$. This is

$$i\hbar\frac{\partial\psi}{\partial t}=\frac{1}{2m}\left(-i\hbar\nabla-\frac{e}{c}\boldsymbol{A}\right)^2\psi.$$

The above equation has a property called gauge invariance. That is, if we replace ψ by $\psi'\exp[i\chi(x)]$ and $\boldsymbol{A}$ by $\boldsymbol{A}'+\nabla\chi(x)$ then we find that ψ' satisfies the same Schrödinger equation with $\boldsymbol{A}$ replaced by $\boldsymbol{A}'$. It is evident that the magnetic field $\boldsymbol{B}=\nabla\times\boldsymbol{A}'$ is not changed in this transformation. Thus the vector potential is not fully defined by the fields, but may be changed arbitrarily by a gauge transformation. It is for this reason that the vector potential is regarded as a mathematical auxiliary rather than a genuine physical concept. However, although $\boldsymbol{A}$ itself has no unique physical definition, we note that

$$\oint_C \boldsymbol{A}\cdot \mathrm{d}\boldsymbol{l}=\int_S \nabla\times\boldsymbol{A}\cdot\mathrm{d}\boldsymbol{s}$$

has a unique physical significance for it is not affected by the gauge transformation $\boldsymbol{A}=\boldsymbol{A}'+\nabla\chi$ because

$$\oint\nabla\chi\cdot\mathrm{d}\boldsymbol{l}$$

vanishes. With a line of flux of strength Φ, the vector potential is

$$\boldsymbol{A}=\frac{\phi}{2\pi r}\hat{\phi}$$

where $\hat{\phi}$ is the unit vector in the ϕ-direction.

Since the magnetic field is zero outside the line of flux, it follows that $\boldsymbol{A}$ should be derivable from a potential in this region. However, this potential cannot be single valued or else the flux, Φ, inside the circuit would have to be zero. Let λ represent this potential. We have

$$\oint\nabla\lambda\cdot\mathrm{d}\boldsymbol{r}=\Phi.$$

This implies that there must be a cut somewhere in this space which we are not allowed to cross so that in this restricted space, the potential λ is single valued. As long as we restrict ourselves in this way, we may write Schrödinger's equation as

$$i\hbar\frac{\partial\psi}{\partial t}=\frac{1}{2m}\left(-i\hbar\nabla-\frac{e}{c}\nabla\lambda\right)^2\psi.$$

If we make the gauge transformation $\psi=\psi'\exp[ie\lambda/c]$ we find that ψ' satisfies the simple Schrödinger equation

$$i\hbar\frac{\partial\psi'}{\partial t}=\frac{1}{2m}\left[-i\hbar\nabla\right]^2\psi'.$$

We can therefore write for the actual solution

$$\psi = \psi' \exp\left[\frac{ie\lambda(x)}{c}\right].$$

This means that the flux line will affect the phase of the solution by a factor $\exp[ie\lambda(x)/c]$. In general $\lambda(x)$ will be different in the two beams and when the two beams come together, the relative phases will therefore have been altered by

$$\frac{e}{c}\oint \nabla\lambda \cdot \mathrm{d}\boldsymbol{l} = \frac{e}{c}\Phi.$$

In this way we predict a fringe shift, which we cannot explain as long as we believe that only the magnetic field can affect the electron. (The above treatment is an approximation but an exact treatment gives the same result [17].)

However, in our interpretation there can be additional quantum forces. Indeed if we define the velocity as

$$\boldsymbol{v} = \frac{\boldsymbol{p}}{m} - \frac{e}{mc}\boldsymbol{A} = \frac{\nabla S}{m} - \frac{e}{mc}\boldsymbol{A}$$

we can show from Schrödinger's equation that

$$m\frac{\mathrm{d}\boldsymbol{v}}{\mathrm{d}t} = m\frac{\mathrm{d}\boldsymbol{v}}{\mathrm{d}t} + m(\boldsymbol{v}\cdot\nabla)\boldsymbol{v} = \frac{e}{c}\boldsymbol{v}\times(\nabla\times\boldsymbol{A}) - \nabla Q$$

where Q is the quantum potential defined in the usual way as $-(\hbar^2/2m)$-$\nabla^2 R/R$. Evidently R is not altered by the gauge transformation which affects only the phase and so Q is gauge invariant.

We see then that in general, in addition to the magnetic force, there is a quantum force which is present even when the magnetic field is zero. This is basically how the AB effect is to be explained. It is clear that Q will be large only where R is changing rapidly. There will, of course, be a small quantum potential at the edge of the beam, but the particle deflections due to this correspond only to what is, in the usual interpretation, called diffraction of the beam at its edges. The main variation of R will occur where the two beams overlap to produce interference. Here the quantum potential will be large; indeed it is just what is responsible for the interference (as we have seen, for example, in the two slit experiment). The phase shift will then alter the quantum potential in a significant way and this will explain the origin in the shift in the interference pattern.

This pattern has been calculated for a case in which the beam is split by a pair of slits with the line of magnetic flux in the geometric shadow between them. The trajectories are shown in figure 3.5. One can see how

Figure 3.5: Trajectories for the AB effect

Figure 3.6: Quantum potential for the AB effect

the orbits are shifted so as to displace the net interference pattern. The corresponding quantum potential is shown in figure 3.6 and can be seen to be similarly shifted (compared with, for example, figure 3.1).

The above completes the explanation of the AB effect. However, it should be noted that this explanation has an interesting further implication; i.e. that the vector potential actually has a physical effect at the quantum level. The vector potential enters into Schrödinger's equation in such a way as to alter the wave function and this leads to the corresponding force in the acceleration of the particle. However, in the classical limit this effect becomes negligible and here it may be said that the vector potential is physically not effective. The mistake in the common way of talking about this is the tacit assumption that the classical idea about the vector potential also holds quantum mechanically.

3.9 References

1. J. Earman, *A Primer on Determinism*, Reidel, Dordrecht, 1986.

2. C. Philippidis, C. Dewdney and B. J. Hiley, *Nuovo Cimento* **52B**, 15–28 (1979).

3. C. E. Shannon and W. Weaver, *The Mathematical Theory of Communication*, The University of Illinois Press, Chicago, Illinois, 1959.

4. J. D. Bekenstein, *Phys. Rev.* **D7**, 2333–2346 (1973).

5. J. A. Wheeler, 'Superspace and Quantum Geometrodynamics', in *Battelle Rencontres*, ed. C. M. DeWitt and J. A. Wheeler, Benjamin, New York, 1968, 242–307.

6. B. J. Hiley, 'Vacuum or Holomovement', in *The Philosophy of the Vacuum*, ed. S. Saunders and H. Brown, Oxford University Press, Oxford, 1991.

7. H. B. S. Casimir, *Proc. Kon. Ned. Akad. Wet.* **51**, 793 (1948).

8. L. de Broglie, *J. Physique, 6e série* **8**, 225 (1927).

9. W. Pauli, in *Reports on the 1927 Solvay Congress*, Gauthiers-Villiars et Cie, Paris (1928), 280.

10. D. Bohm, *Phys. Rev.* **85**, 166–193 (1952).

11. L. de Broglie, *Une interprétation causale et non linéaire de la mécanique ondulatoire: la théorie de la double solution*, Gauthiers-Villiars, Paris (1956). English translation: Elsevier, Amsterdam (1960).

12. J. S. Bell, 'Speakable and Unspeakable' in *Quantum Mechanics*, chapter 19, Cambridge University Press, Cambridge, 1987.

13. P. A. M. Dirac, *Quantum Mechanics*, Oxford University Press, Oxford, 1947.

14. N. Bohr, *Atomic Physics and Human Knowledge*, Science Editions, New York, 1961, 50.

15. D. Bohm, *Quantum Theory*, chapter 5, section 5, Prentice-Hall, Englewood Cliffs, New Jersey, 1951.

16. *Ibid.*, chapter 5, section 11.

17. Y. Aharonov and D. Bohm, *Phys. Rev.* **115**, 485–491 (1959).

18. M. Peshkin and A. Tonomura, *The Aharonov-Bohm Effect*, Springer-Verlag, Berlin, 1989.

Chapter 4
The many-body system

We shall now go on to consider the many-body system in which we shall see, in several striking ways, a further development of the difference between the classical and quantum ontologies, along with new concepts that are required for the latter.

4.1 The ontological interpretation of the many-body system

Let us begin by considering the two-body system. The wave function $\psi(\boldsymbol{r}_1, \boldsymbol{r}_2, t)$ satisfies the Schrödinger equation

$$i\hbar\frac{\partial\psi}{\partial t} = \left[-\frac{\hbar^2}{2m}\left(\nabla_1^2 + \nabla_2^2\right) + V\right]\Psi \tag{4.1}$$

where ∇_1 and ∇_2 refer to particles 1 and 2 respectively. Writing $\psi = R\exp(iS/\hbar)$ and defining $P = R^2 = \psi^*\psi$ we obtain

$$\frac{\partial S}{\partial t} + \frac{(\nabla_1 S)^2}{2m} + \frac{(\nabla_2 S)^2}{2m} + V + Q = 0 \tag{4.2}$$

where

$$Q = -\frac{\hbar^2}{2m}\frac{\left(\nabla_1^2 + \nabla_2^2\right)R}{R} \tag{4.3}$$

and

$$\frac{\partial P}{\partial t} + \nabla_1 \cdot (P\nabla_1 S/m) + \nabla_2 \cdot (P\nabla_2 S/m) = 0. \tag{4.4}$$

As in the case of the one-body system, equation (4.2) can be interpreted as the quantum Hamilton-Jacobi equation with the momenta of the two particles being respectively

$$\boldsymbol{p}_1 = \nabla_1 S \qquad \text{and} \qquad \boldsymbol{p}_2 = \nabla_2 S. \tag{4.5}$$

The above is evidently an extension of the guidance relationship for the two-body system. It implies that the particles are guided in a correlated way. To see the significance of this correlation, one can consider the quantum potential, Q, which follows from the assumption that Ψ satisfies Schrödinger's equation.

As can be seen from equation (4.3), this potential contains R both in the denominator and the numerator, so that it does not necessarily fall off with the distance. We have already seen that for the one-body system this means that the particle can depend strongly on distant features of the environment. In the two-body system we can have a similar dependence on the environment, but in addition, the two particles can also be strongly coupled at long distances. Their interaction can therefore be described as *nonlocal.* This is the first of the new concepts implied by the causal interpretation that comes out in the many-body case.

Going on to the N-body system, we would obtain in a similar way

$$Q = Q(\boldsymbol{r}_1, \boldsymbol{r}_2, \ldots, \boldsymbol{r}_N, t)$$

so that the behaviour of each particle may depend nonlocally on the configuration of all the others, no matter how far away they may be.

As in the one-body case, we may take $P = R^2$ as the probability density, but this now is in the configuration space of all the particles. The Schrödinger equation will imply the conservation of this probability (see equation (4.4)). Therefore if $P = R^2$ initially, this will hold for all time. And as explained for the one-body case, if $P \neq R^2$ initially, typical chaotic movements will bring about $P = R^2$ in the long run as an equilibrium distribution (as we will show in detail in chapter 9).

For several centuries, there has been a strong feeling that nonlocal theories are not acceptable in physics. It is well known, for example, that Newton felt very uneasy about action-at-a-distance [1] and that Einstein regarded it as 'spooky' [2]. One can understand this feeling, but if one reflects deeply and seriously on this subject one can see nothing basically irrational about such an idea. Rather it seems to be most reasonable to keep an open mind on the subject and therefore allow oneself to explore this possibility. If the price of avoiding nonlocality is to make an intuitive explanation impossible, one has to ask whether the cost is not too great.

The only serious objection we can see to nonlocality is that, at first sight, it does not seem to be compatible with relativity, because nonlocal connections in general would allow a transmission of signals faster than the speed of light. In later chapters we extend the causal interpretation to a relativistic context and show that although nonlocality is still present, it does not introduce any inconsistencies into the theory, e.g. it does *not* imply that we can use the quantum potential to transmit a signal faster than light.

While nonlocality as described above is an important new feature of the quantum theory, there is yet another new feature that implies an even more radical departure from the classical ontology, to which little attention has generally been paid thus far. This is that the quantum potential, Q, depends on the 'quantum state' of the whole system in a way that cannot be defined simply as a pre-assigned interaction between all the particles.

To illustrate what this means, we may consider the example of the hydrogen atom, whose wave function is a product of $f(\boldsymbol{x})$, where $\boldsymbol{x}$ is the centre-of-mass coordinate, and $g(\boldsymbol{r})$ where $\boldsymbol{r}$ is the relative coordinate

$$\Psi = f(\boldsymbol{x})g(\boldsymbol{r}).$$

The quantum potential will contain a term representing the interaction of electron and proton

$$Q = -\frac{\hbar^2}{2\mu}\frac{\nabla^2 g(\boldsymbol{r})}{g(\boldsymbol{r})}$$

where μ is the reduced mass of the electron.

In the s-state, Q is a function only of $\boldsymbol{r}$ itself, while with a linear combination of s- and p-wave functions it is easily shown to depend on the relative angles, θ and ϕ, as well. Evidently, it is impossible to find a single pre-assigned function of r, which would simultaneously represent the interaction of electron and proton in both s- and p-states. And, of course, the problem would be still more sharply expressed if we brought in all the other states (d, f, etc.).

The relationship between parts of a system described above implies a new quality of *wholeness* of the entire system going beyond anything that can be specified solely in terms of the actual spatial relationships of all the particles. This is indeed the feature which makes the quantum theory go beyond mechanism of any kind. For it is the essence of mechanism to say that basic reality consists of the parts of a system which are in a preassigned interaction. The concept of the whole, then, has only a secondary significance, in the sense that it is only a way of looking at certain overall aspects of what is in reality the behaviour of the parts. In our interpretation of the quantum theory, we see that the interaction of parts is determined by something that cannot be described solely in terms of these parts and their preassigned interrelationships. Rather it depends on the many-body wave function (which, in the usual interpretation, is said to determine the quantum state of the system). This many-body wave function evolves according to Schrödinger's equation. Something with this kind of dynamical significance that refers directly to the whole system is thus playing a key role in the theory. We emphasise that *this is the most fundamentally new aspect* of the quantum theory.

The above-described feature should, in principle, apply to the entire universe. At first sight this might suggest that we could never disentangle one part of the universe from the rest, so that there would be no way to do science as we know it or even to obtain knowledge by the traditional methods of finding systems that can be regarded as at least approximately isolated from their surroundings. However, it is actually possible to obtain such approximate separation in spite of the quantum wholeness that we have described above. To see how this can come about, let us consider the special case in which the wave function $\Psi(\boldsymbol{r}_1, \boldsymbol{r}_2, t)$ can be written as the product

$$\psi(\boldsymbol{r}_1, \boldsymbol{r}_2, t) = \phi_A(\boldsymbol{r}_1, t)\phi_B(\boldsymbol{r}_2, t).$$

The quantum potential then becomes the sum of two terms

$$Q(\boldsymbol{r}_1, \boldsymbol{r}_2, t) = Q_A(\boldsymbol{r}_1, t) + Q_B(\boldsymbol{r}_2, t),$$

where

$$Q_A(\boldsymbol{r}_1, t) = -\frac{\hbar^2}{2m}\frac{\nabla_1^2 R_A(\boldsymbol{r}_1, t)}{R_A(\boldsymbol{r}_1, t)}$$

and

$$Q_B(\boldsymbol{r}_2, t) = -\frac{\hbar^2}{2m}\frac{\nabla_2^2 R_B(\boldsymbol{r}_2, t)}{R_B(\boldsymbol{r}_2, t)}.$$

In this case the two systems evidently behave independently. In chapter 8 we will show that there is a widespread tendency for such factorised wave functions to arise in typical situations that prevail in the present stage of the development of the universe. Moreover we shall also show that in the classical limit where the quantum potential is negligible, nonlocal interactions will for this reason not be significant.

We may say that while the basic law refers inseparably to the whole universe, this law is such as to imply that the universe tends to fall into a large number of relatively independent parts, each of which may, however, be constituted of further sub-units that are nonlocally connected. Therefore we can deal with these relatively independent parts in the traditional way as we do our experiments.

4.2 More on the notion of active information

We have seen that quantum theory implies a radically new behaviour of matter in several respects. It has been shown that all of this can be understood in terms of our notion of active information applied, not only to the one-body system, but also to the many-body system. Of course in a purely logical sense, the theory could be said to be defined without this notion and could, of course, be thus expressed in a logically consistent way. But it is a key part of our intention in this book to help make the theory more intelligible in an intuitive sense, and not merely regard it as a system of equations from which could be derived algorithms permitting a calculation of interesting results. We feel that at least something like the notion of active information would be needed in any attempt to do this; e.g. to account for quantum interference and the peculiar nonlocal properties of the many-body system in an intuitively understandable way. Here we emphasise again what we have said in chapter 3, that the notion of active information corresponds to a tremendous range of common experience. We have generally devalued this sort of experience as far as physics is concerned and have assumed that physical laws should contain only mechanical concepts such as position, momentum, force, etc. When it is found that these do not apply coherently in a quantum context, then as explained in chapter 2, it is effectively assumed that all we can do is to use the quantum formalism as an algorithm to calculate the probabilities of experimental outcomes. But what we are pointing out here is that if we suitably extend the kind of concepts that we are willing to admit into physics (e.g. to include active information) then we can obtain a much better intuitive apprehension of the theory. This, in turn, may help guide our thinking in physics into new directions (some possibilities of which we shall discuss in chapters 14 and 15).

Let us now consider the many-body system from this point of view. The wave function is defined in the configuration space of all the particles. Its effect on the particles is determined by the phase S and the quantum potential Q, both of which depend only on the form of the wave function and not on its amplitude. Along the lines discussed in chapter 3, we may therefore regard the wave function as containing active information. But this information is ordered in the configuration space rather than in the ordinary space of three dimensions.

The fact that the wave function is in configuration space implies that we have to look more carefully into the meaning of active information in such a context. First of all we may consider its implications for all the motions of the particles. These now respond in a correlated way to what is, in effect, a common pool of information. This information guides the particles according to the condition $p_i = \nabla_i S(\boldsymbol{r}_1, \ldots, \boldsymbol{r}_N, t)/m$ which, of course, leads to a generally nonlocal quantum potential that we have already discussed. Different linear combinations of the wave functions will give rise to different pools of information, which in turn will give rise to corresponding differences in the behaviour of the system.

When the wave function is factorised into independent products, this will correspond to having independent pools of information. There is therefore an objective difference between systems which are wholes guided by a pool of common information and systems constituted of independent parts guided by separate pools of information. We may contrast this possibility of objective wholeness with what happens in classical mechanics, in which the notion of a whole is a subjective convenience for describing the behaviour of what are in reality a set of independently existent parts in interaction.

The fact that the wave function is in configuration space clearly prevents us from regarding the quantum field as one that carries energy and momentum that was simply transferred to the particles with which it interacted (thus effectively pushing or pulling mechanically on the latter). This is a further factor in addition to the form dependence of the activity of the field which leads us to consider the interpretation of this field as active information. The multidimensional nature of this field need not then be so mysterious, since information can be organised into as many sets of dimension as may be needed.

We may illustrate these notions by looking at the interference experiment as a two-body system. To simplify the discussions, we assume that the slit system is rigid and can be described in terms of a single centre of mass coordinate, $\boldsymbol{y}$. Clearly the mass, M, of the slit system is much greater than the mass, m, of the electron. In this case it is well known that the wave function can be approximated as

$$\Psi = \psi(\boldsymbol{x}-\boldsymbol{y})F(\boldsymbol{y}).$$

The quantum potential is

$$Q = -\frac{\hbar^2}{2}\left[\frac{\nabla_x^2 R(\boldsymbol{x}-\boldsymbol{y})}{mR(\boldsymbol{x}-\boldsymbol{y})} + \frac{\nabla_y^2|F(\boldsymbol{y})|}{M|F(\boldsymbol{y})|} + \frac{2\nabla_y R(\boldsymbol{x}-\boldsymbol{y})\,\nabla_y|F(\boldsymbol{y})|}{MR(\boldsymbol{x}-\boldsymbol{y})\quad|F(\boldsymbol{y})|}\right].$$

In the limit that $M \gg m$, the part of Q that acts on the electron reduces to

$$Q_1 = -\frac{\hbar^2}{2m}\frac{\nabla_x^2 R(\boldsymbol{x}-\boldsymbol{y})}{R(\boldsymbol{x}-\boldsymbol{y})}.$$

The above is essentially the same as that obtained in the one-body system, but the important difference is that it is a function of $(\boldsymbol{x}-\boldsymbol{y})$. Therefore, as has indeed already been explained in chapter 3, section 3.7, the energy and momentum are not coming from the quantum field. Rather the electron and the slit system are 'in-formed' to respond in a correlated way by the pool of information that is common to both. The slit system is, of course, so heavy that its response can be neglected. But still from the conceptual point of view, we have to look at this process as a kind of interaction between the electron and the slit system rather than as an effect of the quantum field in space-time. The one-body treatment is therefore an abstraction which leaves out this essential feature.

Finally we wish to point out that even the classical potential can be looked at as representing the effects of active information. To do this we recall that, for example, in a two-body system each particle is guided by the total phase $S(x_1, x_2)$ of the wave function. As we have seen earlier, this guidance condition combined with Schrödinger's equation implies the extended quantum Hamilton-Jacobi equation

$$\frac{\partial S}{\partial t} + \frac{\hbar^2}{2m}\left[(\nabla_1 S)^2 + (\nabla_2 S)^2\right] + V + Q = 0.$$

In the above equation, V and Q appear in essentially the same way. In fact both describe certain effects of the guidance condition and therefore of active information in the wave function. What is then the difference? The difference is that Q describes a part of the change with time of the active information that depends in a nonlocal way and in detail on the whole wave function (i.e. what is commonly called the quantum state). On the other hand, V describes a part of the change of this information that is independent of all this and that is generally local in the way that it acts. This part is what manifests in the classical limit.

Physics has already developed in such a way that the classical potential, V, can be regarded as an actual existent field distributed throughout three dimensional space. This field can be regarded as the source of the change of information with time brought about by V in the Hamilton-Jacobi equation. It obeys a wave equation implying the detailed conservation of energy and momentum. Therefore we can, if we wish, attribute the energy and momentum to the field and thus imagine it is flowing throughout space and into and out of the particle. But as we have seen, this picture could not be coherently applied to the quantum potential. Moreover as we shall see in chapter 11 when we consider our interpretation of the quantum field theory, the picture of a continuously conserved flow of energy and momentum will not also apply except as a limiting case holding only in the classical approximation.

4.3 Further applications of many-body wave functions

We shall now give two further applications of the notions that have been described in this chapter which will make the physical meaning of our interpretation clearer.

4.3.1 The chemical bond

We begin by considering the chemical bond. Classically it is incomprehensible why chemical bonds can form, and furthermore, why there are so many specific kinds of such bonds. It is well known that the quantum theory has explained this in a generally satisfactory way. But this has been achieved at the expense of a loss of intuitive comprehension of what is involved physically in such a bond. In our interpretation we obtain a more intelligible way of understanding this.

We shall illustrate these ideas in terms of the hydrogen molecule with its covalent bond. The molecule consists of two protons whose coordinates we label by $\boldsymbol{x}_A$ and $\boldsymbol{x}_B$, and two electrons whose coordinates are represented by $\boldsymbol{r}_1$ and $\boldsymbol{r}_2$. The interaction between the particles are of the Coulomb type. This can be expressed as

$$\begin{aligned} V_I &= e^2\left[\frac{1}{|\boldsymbol{x}_A-\boldsymbol{x}_B|}+\frac{1}{|\boldsymbol{r}_1-\boldsymbol{r}_2|}-\frac{1}{|\boldsymbol{r}_1-\boldsymbol{r}_A|}\right. \\ &\left.\quad -\frac{1}{|\boldsymbol{r}_1-\boldsymbol{r}_B|}-\frac{1}{|\boldsymbol{r}_2-\boldsymbol{r}_A|}-\frac{1}{|\boldsymbol{r}_2-\boldsymbol{r}_B|}\right]. \end{aligned}$$

One of the usual procedures (due to Heitler and London) is to solve the Schrödinger equation for this case approximately, beginning with functions that are solutions when the nuclei are very far apart. We then have two separate hydrogen atoms, each in an s-state. Because the protons are so heavy we can, in the first approximation, regard them as fixed in specified positions. One approximate solution of the combined system is then

$$\Psi_I = \phi(\boldsymbol{r}_1-\boldsymbol{x}_A)\phi(\boldsymbol{r}_2-\boldsymbol{x}_B). \tag{4.6}$$

This represents two separated hydrogen atoms A and B in their lowest s-states in which the first electron is in atom A, while the second is in atom B. Another solution is

$$\Psi_{II} = \phi(\boldsymbol{r}_2-\boldsymbol{x}_A)\phi(\boldsymbol{r}_1-\boldsymbol{x}_B) \tag{4.7}$$

in which the first electron is in atom B, while the second is in atom A. These two states evidently have the same energy. And as is well known from standard perturbation theory, one has to start from a suitable linear combination of Ψ_I and Ψ_{II} in order to obtain a series of approximations to the wave function when the particles come closer together.

Because of the symmetry of V_I between $\boldsymbol{r}_1$ and $\boldsymbol{r}_2$ it follows, as can easily be shown, that the appropriate linear combinations are

$$\Psi_\pm = 1/\sqrt{2}(\Psi_I \pm \Psi_{II}). \tag{4.8}$$

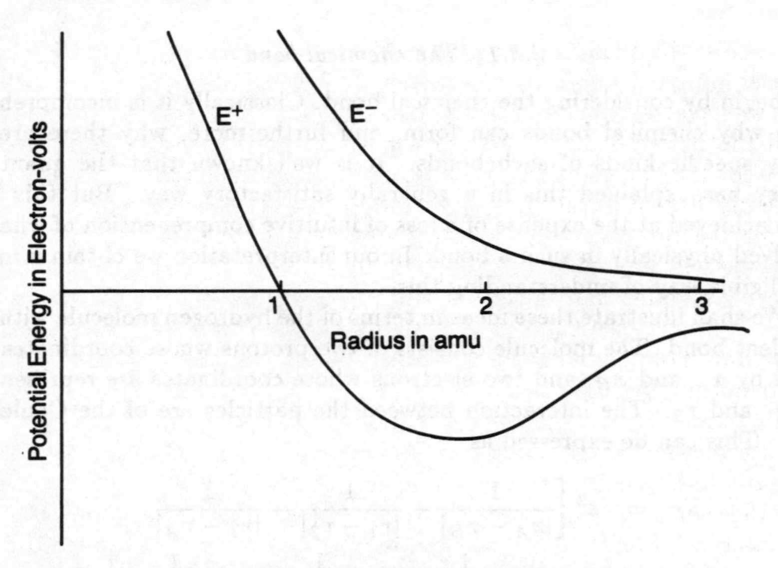

Figure 4.1: Possible bonding energies of two hydrogen atoms

From this we obtain:

$$R^2 = 1/2(\Psi_I^2 + \Psi_{II}^2 \pm 2\Psi_I\Psi_{II}) \tag{4.9}$$

(noting that ψ_I and ψ_{II} are real because ϕ represents the lowest s-state). The energy of these states comes out as a function of the separations $\eta = |\boldsymbol{x}_A - \boldsymbol{x}_B|$. They are shown in figure 4.1.

Evidently the state corresponding to E^+ gives rise to an attraction between the protons at medium distances, which turns into a repulsion at short distances. On the other hand, the state corresponding to E^- leads everywhere to a repulsion. There is no way to understand this energy difference solely in terms of a pre-assigned interaction of the particles constituting the molecule. For, since the wave function is real, the momenta of all the particles are zero and the energy is just the sum of the classical interaction energy V_I and the quantum potential

$$E^{\pm} = V_I + Q_{\pm}. \tag{4.10}$$

To obtain the quantum potential we must first obtain an expression for R. From (4.9), (4.6) and (4.7) we see that

$$\begin{aligned} R^2 \quad = \quad & \frac{1}{2}[\phi^2(\boldsymbol{r}_1 - \boldsymbol{x}_A)\phi^2(\boldsymbol{r}_2 - \boldsymbol{x}_B) + \phi^2(\boldsymbol{r}_1 - \boldsymbol{x}_B)\phi^2(\boldsymbol{r}_2 - \boldsymbol{x}_A) \\ & + 2\phi(\boldsymbol{r}_1 - \boldsymbol{x}_A)\phi(\boldsymbol{r}_2 - \boldsymbol{x}_B)\phi(\boldsymbol{r}_2 - \boldsymbol{x}_A)\phi(\boldsymbol{r}_1 - \boldsymbol{x}_B)]. \end{aligned}$$

This shows the quantum potential will be a function of all the particle coordinates implying a nonlocal interaction with the kind of quantum wholeness that we have described in the previous section. This latter means that even if all the particles are in the same positions and have the same velocity $v = 0$ for the two solutions, their energies will be different. This difference evidently has to be attributed to the quantum potential which is different for the two cases. As we have already seen, for the case of Ψ_+, the molecule will be stable while for Ψ_- it is not. The force that holds the molecule together and that determines the specificity of the chemical properties is thus a particular case of the nonlocality and wholeness which is, as we have pointed out earlier, the most essentially new feature implied by the quantum mechanics.

In terms of the usual interpretation of the quantum theory, we have no intuitive notion of the difference between the states Ψ_+ and Ψ_-, nor indeed what it means to form a linear combination of states. Yet it is just this which is responsible for a vast range of chemical properties. In our interpretation these properties are understood through the notion that the wave function of a molecule constitutes a common pool of information that guides the activity of all its particles in related ways along the lines discussed at the end of the previous section. Different wave functions, constituting different pools of information, will then give rise to well-defined differences in the behaviour of the system.

4.3.2 Superfluidity and superconductivity

Superfluidity and superconductivity provide a very good illustration of how a common pool of information can give rise to strikingly new properties of a many-body system.

It is well known that at low temperature many substances can go into a new state in which currents (either of atoms or electrons) can flow indefinitely without viscosity or resistance. This state is stable only up to a certain temperature at which the property disappears. These superflow properties have been explained as a consequence of the many-body Schrödinger equation, but this explanation is very abstract and gives little intuitive understanding of how the phenomena in question come about.

Our discussion is aimed at illustrating how the quantum potential makes superflow possible in its more common forms such as ordinary low temperature superconductivity and superfluidity in He^4. It is not however intended to be a detailed treatment of all the complexities involved in the above nor will we consider the less common forms of superflow such as high temperature and gapless superconductivity and superflow in He^3.

We begin by discussing the case of superfluidity because this is easier both conceptually and mathematically, but, as we shall see later, basically the same considerations apply to superconductivity as we have restricted this term above. What is essential here is that helium nuclei are bosons and that the electron pairs that are responsible for superconductivity are also effectively bosons.

At first sight the problem of superflow in He^4 seems to be almost impossibly difficult because the helium nuclei repel each other with such strong forces that it would appear that no reasonable approximation could be made. However, it has been shown by D. Pines [3] that, as a consequence of the short range particle correlations, the system could be described as a collection of 'quasi-particles' with only weak soft-core repulsions between them. In the first approximation we can neglect these latter, so that a good starting point for the condensate wave function is one in which all the particles are moving at the same velocity $\boldsymbol{v} = \boldsymbol{K}/m$. This wave function, which is evidently symmetric, is

$$\Psi_0 \propto \exp\left[i\boldsymbol{K} \cdot \sum_j \boldsymbol{r}_j\right] \tag{4.11}$$

where $\boldsymbol{r}_j$ represents the coordinate of the jth particle and where we are assuming units in which $\hbar = 1$.

However, the fact that all the particles are moving at the same speed is not sufficient by itself to account for the properties of superfluidity. What is required, in addition, is that this state be *stable* under external perturbation. That is to say, under an external disturbance, a particle should not be able to drop out of the common velocity $\boldsymbol{v} = \boldsymbol{K}/m$, by giving up energy and going to a lower velocity $\boldsymbol{v}' = (\boldsymbol{K} - \boldsymbol{q}_0)/m$. To account for this we follow along the lines of the explanation given by Landau [4], Bogolubov [5] and Feynman [6].

A key point for such stability is that it should not be possible to change the momentum of a single particle in this system from $\boldsymbol{K}$ to $\boldsymbol{K} - \boldsymbol{q}_0$ without affecting the others in such a way as to bring about a net increase in energy of the whole system. Basically this stability arises because of the symmetry of the wave function. Indeed it is easily seen that without this symmetry there can be no such stability. Thus consider, for example, what would happen if the wave function had no particular symmetry. We could then obtain an excited wave function by giving momentum to a single particle, say the ith, which would give rise to a wave function

$$\Psi_i = \exp\left[-i\boldsymbol{q}_0 \cdot \boldsymbol{r}_i\right] \Psi_0.$$

Or

$$\Psi_i = \exp\left[iK \cdot \sum_{j \neq i} r_j\right] \exp\left[i(K - q_0)r_i\right]. \tag{4.12}$$

Clearly this will be a possible solution of Schrödinger's equation and so there is nothing to prevent this loss of momentum by slowing down one of the particles independently of others.

However, for bosons, the wave function has to be symmetric. The lowest state of excitation corresponding to a loss of momentum, q_0, will be

$$\Psi = \sum_j \exp\left[-iq_0 \cdot r_j\right] \Psi_0. \tag{4.13}$$

If we neglect interactions, this too would permit the same kind of loss of momentum we described above. However, as shown by Bogolubov and by Feynman, when interactions are taken into account, the system is capable of supporting sound waves with frequency $\omega = \omega(q)$. The term $\rho(q_0) = \sum_j \exp\left[-q_0 \cdot r_j\right]$ is well known to be the Fourier coefficient of the particle density. In such a case it is

$$\rho(x) = \sum_j \delta(x - r_j). \tag{4.14}$$

Therefore the wave function (4.13) represents the excitation of a sound wave (while the equation (4.12), which is a wave function of no particular symmetry, represents no change of particle density and does not represent such a wave). $\rho(q_0)$ thus describes a collective deviation from equilibrium whose energy is $E = \omega(q_0)$. At low frequencies, $\omega(q_0)$ will satisfy the linear relationship

$$\omega(q_0) = qv_s,$$

where v_s is the velocity of sound, while at higher frequencies it deviates from this relation. (As explained by Feynman [6], these deviations are associated with rotons.) Stability will be achieved if the energy required to create a wave of excitation is greater than the kinetic energy given off by the particle for all q_0. This requires that

$$\begin{aligned} E - E_0 &= \frac{(K - q_0)^2}{2m} - \frac{K^2}{2m} + \omega(q_0) \\ &= -\frac{K \cdot q_0}{m} + \frac{q_0^2}{2m} + \omega(q_0) > 0. \end{aligned} \tag{4.15}$$

The above will follow if $\omega(q_0) > K \cdot q_0/m$ or if $v = K/m < \omega(q_0)/q_0$, where v is the velocity of flow. This is, of course, just the well-known condition obtained by Landau [4].

It has to be pointed out that the Landau condition does not determine the actual maximum velocity in typical experimental situations such as flow in tubes. Indeed this is in general considerably less than that given by the Landau condition and is determined by further factors such as the existence of rotons and vortex structures of various kinds (Feynman [6]). However, the Landau condition is evidently still satisfied in such cases. Moreover it is relevant to understanding the existence of superflow because if it is not satisfied, the particles can be scattered out of the common velocity thus making superflow impossible. Clearly no matter what the detailed structure of the vortices may be, the understanding of superflow depends basically on this sort of explanation of why the particles move stably together with a common velocity.

To explain this stability intuitively, we shall have to show how the quantum potential implied by the wave function (4.13) actually constitutes active information that keeps all the particles moving together in spite of perturbations that would otherwise scatter them. To do this we shall have to calculate how the wave function (4.13) changes when a perturbation is introduced.

To simplify the problem we assume a small perturbing potential, V, which, of course, has to be a symmetric function of all the particles. We therefore write

$$V = \sum_j V(\boldsymbol{r}_j). \tag{4.16}$$

With the aid of perturbation theory, we find that

$$\Psi = \Psi_0 + \Delta\Psi$$

where

$$\Delta\Psi = \sum_j \sum_{q \neq 0} \frac{V(\boldsymbol{q}) \exp[-i\boldsymbol{q} \cdot \boldsymbol{r}_j]\Psi_0}{\Delta E_q}. \tag{4.17}$$

In computing ΔE_q it will be sufficient for the illustrative purposes of this discussion to restrict ourselves to values of $\boldsymbol{q}$ small enough so that the collective excitations will have the linear frequency relation $\omega(\boldsymbol{q}) = \boldsymbol{q}v_s$, where v_s is the velocity of sound. This gives us

$$\Delta E_q = E_0 - E = -\left[qv_s + \frac{q^2}{2m} - \frac{\boldsymbol{K} \cdot \boldsymbol{q}}{m}\right]. \tag{4.18}$$

Evidently as long as $v_s > K/m$ this denominator will never vanish and so there will be no real transitions, i.e. the perturbation involves only what are, in the usual interpretation, called virtual states. (When rotons are taken into account the limiting velocity of superflow will in general be less than v_s but the above condition will still be satisfied.) Evidently we may reasonably assume that $V(q)$ becomes negligible for $q/m > v_s$ (this means that the potential is 'soft' which is implied by the fact that its high Fourier coefficients are small). We then obtain

$$\Delta\Psi = -\sum_i \sum_q \frac{V(q)\exp[-i\boldsymbol{q}\cdot\boldsymbol{r}_i]}{qv_s[1-(\boldsymbol{K}\cdot\hat{\boldsymbol{q}})/mv_s]}\Psi_0. \tag{4.19}$$

Once again for the purposes of illustration we may assume that $K/m \ll v_s$. This gives

$$\Delta\Psi = -\sum_i \sum_q \frac{V(q)}{qv_s}\exp[-i\boldsymbol{q}\cdot\boldsymbol{r}_i]\left[1+\frac{\boldsymbol{K}\cdot\hat{\boldsymbol{q}}}{mv_s}\right]\Psi_0. \tag{4.20}$$

To further simplify the calculation we assume a $V(r)$ that gives a Fourier coefficient $V(\boldsymbol{q}) = \boldsymbol{q}\exp[-q\lambda]$. This implies that

$$\begin{aligned} V(r) &\propto \frac{\partial^2}{\partial\lambda^2}\left[\frac{1}{\lambda^2+r^2}\right] \\ &\propto \frac{1}{(\lambda^2+r^2)^2} - \frac{4\lambda^2}{(\lambda^2+r^2)^3}. \end{aligned} \tag{4.21}$$

Writing

$$M_i = \frac{A}{(\lambda^2+r_i^2)^2 v_s} \quad \text{and} \quad N_i = \frac{B}{(\lambda^2+r_i^2)mv_s^2}$$

where A and B are suitable constants, we obtain

$$\Psi = \Psi_0\left[1-\sum_i(M_i + i\boldsymbol{K}\cdot\nabla_i N_i)\right]. \tag{4.22}$$

The probability density is, in the first approximation,

$$P = |\Psi_0|^2\left[1-2\sum_i M_i\right] \tag{4.23}$$

and the current density of the ith particle in the direction $\boldsymbol{q}$ will be

$$\hat{\boldsymbol{q}}\cdot\boldsymbol{j}_i = |\Psi_0|^2\left\{\frac{\boldsymbol{K}\cdot\hat{\boldsymbol{q}}}{m} - \frac{2B}{(mv_s)^2}\hat{\boldsymbol{q}}\cdot\nabla_i\left[\frac{\boldsymbol{K}\cdot\boldsymbol{r}_i}{(\lambda^2+r_i^2)^2}\right]\right\}. \tag{4.24}$$

The velocity in the direction $\boldsymbol{q}$ will be given by

$$v_c = \frac{\hat{\boldsymbol{q}}\cdot\boldsymbol{j}_i}{P} = \frac{\hat{\boldsymbol{q}}\cdot\boldsymbol{j}_i}{|\Psi_0|^2[1-2\sum_i M_i]}. \tag{4.25}$$

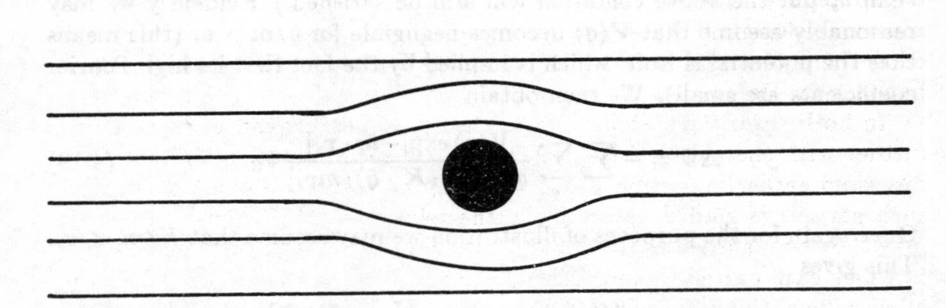

Figure 4.2: Stream lines around obstacle in liquid helium

According to equation (4.23), the distribution is altered in the neighbourhood of the perturbation. If the perturbation is repulsive, the particles will on the whole be pushed away and vice versa if it is attractive. According to equation (4.24) the current will change in a corresponding way. This implies that for a repulsive potential, the particle will first move away from the perturbing obstacle. But because the perturbed wave function vanishes at infinity, the particle will eventually move back and will be left with its original velocity, thus implying there is no scattering (as shown in figure 4.2).

In ordinary scattering processes, the energy conservation condition between initial and final states can be satisfied and we therefore get a real transition into a scattered wave. In the case of a superfluid however energy conservation cannot be satisfied so we get only virtual contributions to the wave function. If we evaluated the quantum potential, it would be seen that these latter contributions imply an additional force on the particle which accounts for why it is first accelerated away from the obstacle and then brought back to its original velocity. Since every particle will behave in essentially the same way, it follows that the liquid as a whole will flow around the obstacle and then reconstitute its flow in its original direction.

This is basically our intuitive explanation of why the state of superflow is stable. This is evidently a purely quantum mechanical response of the whole system. It involves not only nonlocal interactions due to the quantum potential, but also the irreducible quantum wholeness implied by the fact that this interaction cannot be expressed in terms of a pre-assigned function of the particle coordinates.

All that we have said here also applies to superconductivity in which the charge carriers are electron pairs that act like bosons. The details can be found elsewhere [7], but the essential point is that, as with He^4, there is a 'gap' representing the energy needed to remove an electron pair from the common velocity. It will follow then that if there is an obstacle, the charge carrier will go round it and reform without scattering in the same way as happens for He^4.

In both cases the stability of the state of superflow can be understood further with the aid of the notion of the active information that we have discussed earlier in section 4.2 of this chapter, where we pointed out that each particle is guided according to the relation $\boldsymbol{v}_i = \nabla_i S(\boldsymbol{r}_1, \ldots \boldsymbol{r}_N)$. It may then be regarded as part of the common pool of information according to which each particle determines its velocity. One can say that in the state of superflow, this information brings about a coordinated movement of all the particles that can be thought of as resembling a 'ballet dance' (in which all the dancers separate in a systematic way to go round an obstacle and then reform their original pattern).

As the temperature goes up the property of superfluidity disappears. In terms of our approach, this is because the wave function breaks up into a set of independent factors which can, as pointed out earlier, be thought of as representing independent pools of information. The electrons will then no longer be guided by a common pool of information and will, therefore, scatter off obstacles rather than go round them and return to their original movement. In this way they begin to behave like an unorganised crowd of people who are acting independently and get in each other's way so that the property of superflow disappears.

This means that the guidance condition and the quantum potential implied by it can, under certain conditions, have the novel quality of being able to organise the activity of an entire set of particles, in a way that depends on a pool of information common to the whole set. This behaviour is reminiscent of what is found in living beings which are similarly organised by a common pool of information that is now thought to reside in the genes (and also society which is evidently organised by common pools of information of yet another kind). Thus a considerable similarity is seen between the microlevel and other levels which we shall discuss in more detail in chapter 15.

4.4 References

1. I. Newton, *Principia Mathematica*, ed. Mothe-Cajori, London, 1713.

2. A. Einstein, in *The Born-Einstein Letters*, ed. I. Born, Macmillian, London, 1971, 158.

3. D. Pines, in *Quantum Implications: Essays in Honour of David Bohm*,

ed. B. J. Hiley and F. D. Peat, Routledge and Kegan Paul, London, 1987, 66–84.

4. L. D. Landau, *J. Phys. (USSR)* **5**, 71 (1941).
5. N. Bogolubov, *J. Phys. (USSR)* **11**, 23 (1947).
6. R. Feynman, *Prog. Low Temp. Phys.* **1**, 17 (1955).
7. J. Bardeen, L. N. Cooper and J. R. Schrieffer, *Phys. Rev.* **106**, 162 (1957), and **108**, 1175 (1957).

Chapter 5
Transition processes considered as independent of observation

In the usual interpretation it is an essential feature that transition processes can only be discussed in a context in which they are observed in a measurement process. In an ontological approach such as ours, it is, of course, necessary to discuss transition processes taking place on their own, independently of their being observed. In this chapter we shall show how this can be done.

We shall begin with the simple example of barrier penetration and then go on to consider transitions of electrons in atoms from one level to another. In developing our ideas we shall also show that in our interpretation, the quantum theory implies two general tendencies that oppose each other:

1. **A tendency for wave functions to spread and become less well defined in space when the system is isolated.**
2. **A tendency for these functions to 'narrow down' the range in which they have an appreciable effect when interaction with a background takes place.**

We shall call this latter tendency 'self-definition'. It is essential to take this into account if we are to understand how the universe can be in a fairly well-defined state at any given moment in spite of the tendency to become less well defined because of the general 'spreading' of wave functions. On this basis we are able to discuss the states of things without any need to refer to a measuring apparatus which, in the usual interpretation, would cause the wave function to 'collapse' to a more definite state.

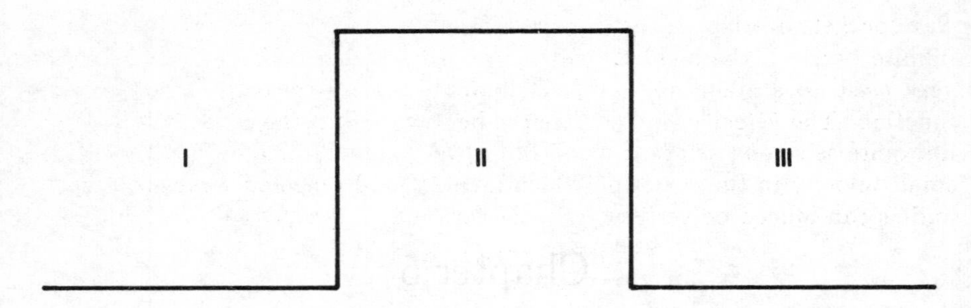

Figure 5.1: Square potential barrier of height V

5.1 The example of barrier penetration

Let us consider a one-dimensional square barrier of height V and thickness a located between $x = 0$ and $x = a$, as shown in figure 5.1.

We assume a particle of momentum p in region I incident from the left with energy $E < V$. In accordance with standard procedure, we also assume a reflected wave in region I with the same momentum. For region III we assume only an outgoing wave, which evidently will also have the momentum p. If we write the wave function in region III as

$$\Psi_{III} = A\exp[ipx], \tag{5.1}$$

again setting $\hbar = 1$, then the solution in region II will be the linear combination

$$\Psi_{II} = B\exp[qx] + C\exp[-qx], \tag{5.2}$$

where $q = \sqrt{2m(V - E)}$.

In region I the solution will be

$$\Psi_I = D\exp[ipx] + E\exp[-ipx]. \tag{5.3}$$

It is necessary then to provide for the continuity of the wave function and its derivative at $x = 0$ and $x = a$. This will determine the ratios of the coefficients and therefore the transmissivity and the reflectivity [1]. If the barrier is high and thick, the transmitted wave will be small and the reflected wave very nearly equal in intensity to the incident wave.

We then form a wave packet by integrating ψ over a small range of p. This packet will be incident from $x = -\infty$ and will come up to the barrier where some of it will be transmitted, but (in general) most will be reflected. There will be a period during which the incident and the reflected wave packets overlap. If the transmissivity were zero (as it would be for an infinite barrier), the incident and reflected packets would form something very close to a standing wave which would have essentially a real wave function. The velocity would therefore be zero. But if the reflected wave is not quite as strong as the transmitted wave there will be a net (in general) small velocity in the region in which incident and reflected waves overlap, and as can indeed be verified from the formula

$$v = \frac{1}{2mi}\left(\frac{\psi^{*}\nabla\psi - \psi\nabla\psi^{*}}{\psi^{*}\psi}\right). \tag{5.4}$$

When the incident wave packet disappears and only the reflected wave packet remains, the speed will once again be p/m, but of course in the negative direction.

There is a complicated variation of velocity inside the barrier, but for the wave packet that has passed through, the speed will once again be p/m, but this time in the positive direction.

The trajectories for the case of a Gaussian wave packet with $E = V/2$ have been calculated for a range of initial positions of the particle within the wave packet [2]. Clearly their behaviour will depend on the quantum potential. This is shown as a function of x and t in figure 5.2. As the packet comes up against the barrier we can see the quantum potential developing as a function of time in the region of overlap of the incident and reflected wave packets. We can also see how the quantum potential modifies the potential inside the barrier in a rather complicated way. In general the quantum potential will lower the barrier height, thus permitting particles to enter it. This of course removes the mystery as to how barrier penetration is possible. For such penetration depends on the total potential $V+Q$ and not just the classical potential V alone. But it also depends on whether the net potential $V+Q$ is greater than E for the whole period in which the particle is inside the barrier. It is clear from the figure that this will in turn depend crucially on the initial location of the particle, because the quantum potential remains low only for certain periods of time and indeed may have spikes in certain places that would tend to reflect back a particle positioned there.

All of this can be brought out clearly by considering the trajectories shown in figure 5.3. We can see that only the particles that happen to be at the front of the wave packet can even enter the barrier. But even most of these will be turned around before they are able to go through. There is however a critical trajectory, which divides the trajectories that go through from those that do not. This evidently resembles a bifurcation point of a kind which is typical of the non-linear equations describing unstable systems. Indeed it is clear that even though the equation for the wave function

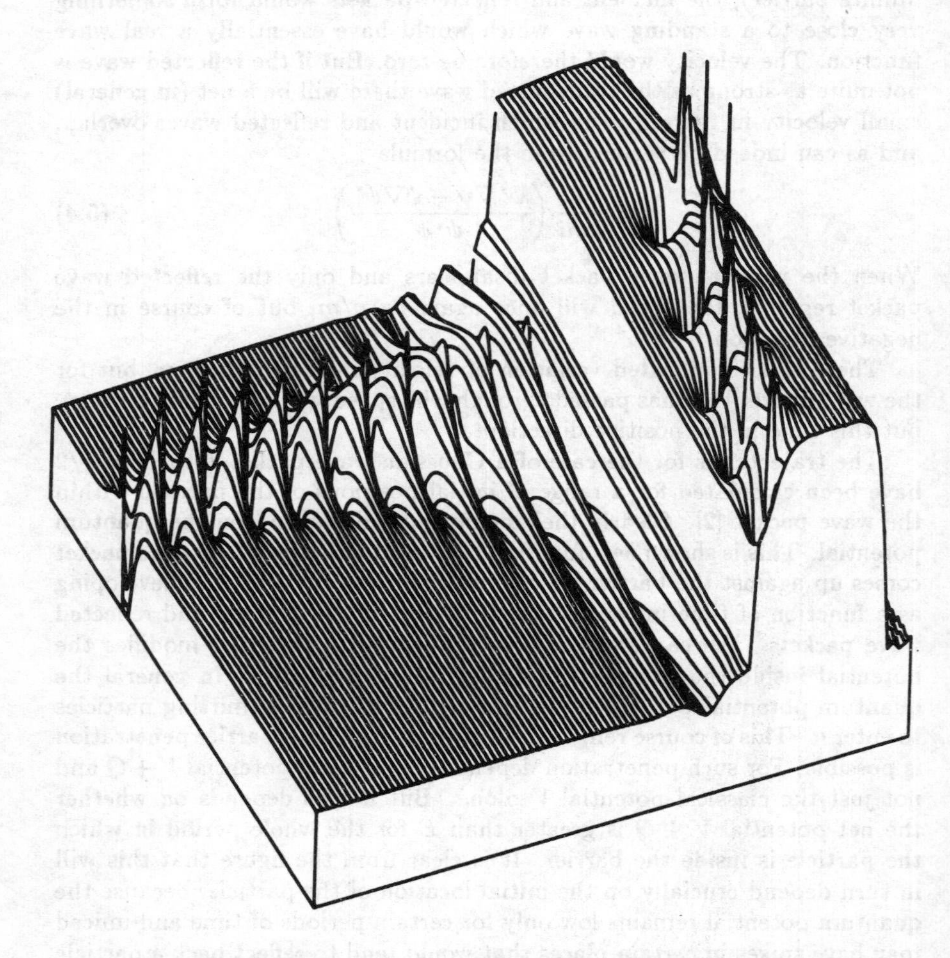

Figure 5.2: Quantum potential for barrier ($E = V/2$)

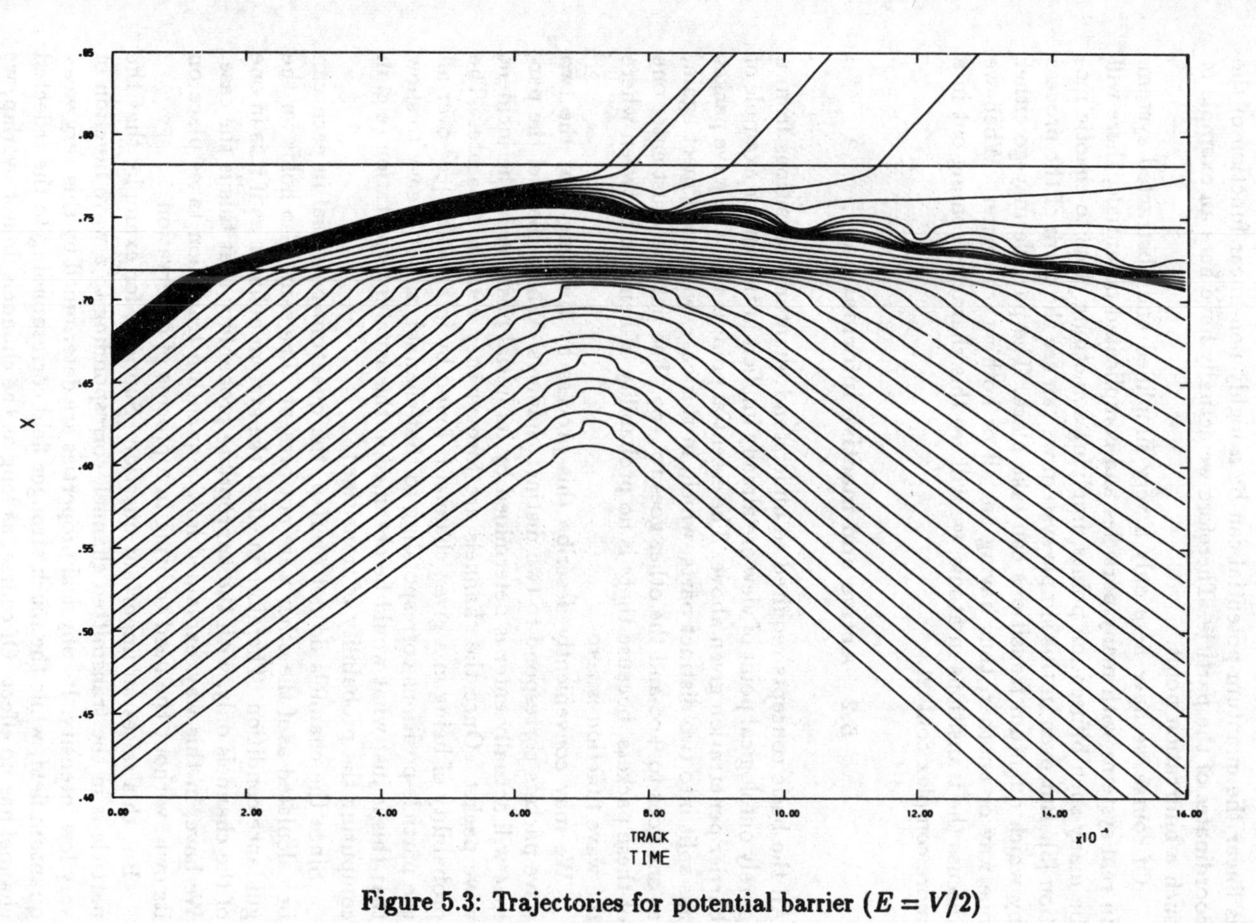

Figure 5.3: Trajectories for potential barrier ($E = V/2$)

is linear the quantum potential can be a highly non-linear function of the coordinates of the particle. Therefore we actually have here an example of such a bifurcation point.

Of course we have here only a very simplified and abstracted system. In real systems with many particles and complicated potentials, there will be many such bifurcation points, implying something close to chaotic motion [3]. But even in this simple system we can see the germ of the process by which quantum transitions can take place. The particle may go either one way or another, thus having an 'all or nothing' character. When we discuss the transitions in atoms, we will see this character coming out in a more complex context.

5.2 Active and inactive information

All the basic concepts required to understand quantum transitions from a purely ontological point of view are already implicit in the simple example of barrier penetration given above. The essential point is that the wave packet has split into two distinct parts, which, in this case, will never meet again, as one goes to $+\infty$ and the other goes to $-\infty$. The particle must enter one of these packets, because there is no probability of being in between where the wave function is zero.

We may conveniently describe this process by thinking that the two wave packets correspond to two distinct channels. Which channel the particle will actually enter is determined by its initial position in the incident wave packet. Once the channels are formed they remain separate. The probability of being in a given channel is found by integrating $|\psi|^2$ over all the particle positions corresponding to that channel, and it is easy to show that this is just what would be obtained in the usual interpretation (e.g. in computing the probability of transition).

Since the channels do not overlap, the quantum potential in each can be calculated as if the other were not present. The same also holds for the guidance condition. Therefore in effect the system behaves as if it is in one of the channels only, with the appropriate probability that this is the case. We have, in this way, explained how a quantum transition takes place on its own, without the need to bring in a process of observation.

But this raises an important question. Suppose, for example, that the particle is in the transmitted channel, corresponding to a wave function of very low intensity. Its physical properties are determined by this very weak quantum field, while the much stronger field corresponding to the reflected channel has no effect. Of course, as long as the channels do not overlap, we can understand this, because the quantum field does not act mechanically, but rather, as we have already discussed, it acts as information that guides the particle in its self-movement and in such action the intensity of the field does not matter.

Nevertheless one may raise the question of what would happen if the channels were made to overlap again. For example, in two or three dimensions the experiment could be arranged to reflect both of the quantum fields into the same region of space. When this happened, the larger field would overwhelm the smaller one and the behaviour of the particle would be radically altered. In other words while the information in an 'empty' channel may not *actually* be active on the particle at a given moment, it may have the *potentiality* to become active later. Therefore one cannot say that the transmitted particle is in a state that is entirely independent of the reflected channel.

This would, at first sight, seem to put in doubt our claim of having explained the transition process without the aid of an observation which would cause the wave function to 'collapse' irreversibly and finally, for example, to the transmitted state. In this connection, however, it must at first be pointed out, that the possibility of interference of the two channels that we have described is an inevitable consequence of the quantum theory whether in our interpretation or in the usual one. As long as we restrict ourselves to the abstraction of a one-body system, this implication neither can nor should be avoided.

It is our thesis that only when we take into account the interaction of this system with a background containing many particles that the transition becomes, in effect, irrevocable. The reason for this is that the information in the unoccupied channel will lose all its potentiality for becoming active so that it will remain permanently inactive after interaction with a sufficiently complex background.

The simplest case in which this happens arises when the particle in question collides with another one and scattering takes place through a short range interaction.

To discuss this process let us consider a three-dimensional system but assume a barrier of finite width that is uniform in the x- and y-directions with a plane wave $\psi = \exp(ik_0 z)$ incident in the z-direction. The calculation of transmission coefficient will go through as in the one-dimensional case.

But let us assume now that an atom with coordinates $\boldsymbol{y}$ is placed in the transmitted beam. To simplify the problem, we suppose that its wave packet has average velocity zero. This packet is given by

$$\phi_0(\boldsymbol{y}) = \sum f(\boldsymbol{k}) \exp[i\boldsymbol{k} \cdot \boldsymbol{y}]. \tag{5.5}$$

Let us further assume that the incident particle is represented by a wave packet bounded in all three directions but travelling only in the z-direction. Then the transmitted wave can be represented by the packet

$$\psi_0(\boldsymbol{x}) = \sum g(\boldsymbol{k}' - \boldsymbol{k}'_0) \exp[i\boldsymbol{k}' \cdot \boldsymbol{x}], \tag{5.6}$$

where $\boldsymbol{k}'_0$ represents a vector in the direction of z and the packet is centred at $\boldsymbol{k}' = \boldsymbol{k}'_0$. Before the particles interact, the combined wave function will be

$$\Psi_0 = \phi_0(\boldsymbol{y})\psi_0(\boldsymbol{x}).$$

The particles will begin to scatter each other when their wave packets overlap. We postulate conservation of momentum and energy in this process. Because of the linearity of Schrödinger's equation, we can deal separately with each of the terms that make up the wave packet. Thus a term that was initially $exp[i\boldsymbol{k}' \cdot \boldsymbol{x}] \exp[i\boldsymbol{k} \cdot \boldsymbol{y}]$ will, because of conservation of momentum, go over into $\exp[i(\boldsymbol{k}' + \boldsymbol{K}) \cdot \boldsymbol{x}] \exp[i(\boldsymbol{k} - \boldsymbol{K}) \cdot \boldsymbol{y}]$. The equation for conservation of energy then takes the form

$$\frac{\boldsymbol{k}'^2}{2m'} + \frac{\boldsymbol{k}^2}{2m} = \frac{(\boldsymbol{k}' + \boldsymbol{K})^2}{2m'} + \frac{(\boldsymbol{k} - \boldsymbol{K})^2}{2m} \tag{5.7}$$

and this introduces a relationship between $\boldsymbol{K}$ and the initial momenta. Because the wave packets in $\boldsymbol{k}$ and $\boldsymbol{k}'$ are fairly narrow, the above condition can be approximated by satisfying it for the values of $\boldsymbol{k}$ and $\boldsymbol{k}'$ corresponding to maximum intensity, which are respectively $\boldsymbol{k} = 0$ and $\boldsymbol{k}' = \boldsymbol{k}'_0$. Thus, to a good approximation, $\boldsymbol{K}$ can be obtained from the equation

$$\frac{\boldsymbol{k}'^2_0}{2m'} = \frac{\boldsymbol{K}^2}{2m} + \frac{(\boldsymbol{k}'_0 + \boldsymbol{K})^2}{2m'}. \tag{5.8}$$

Therefore $\boldsymbol{K}$ is related only to $\boldsymbol{k}'_0$ and not to the precise values of $\boldsymbol{k}$ and $\boldsymbol{k}'$. This will simplify the subsequent discussion in an important way.

We now return to the calculation of the wave function after interaction. This will be (as a function of time)

$$\Psi(\boldsymbol{x}, \boldsymbol{y}, t) = \sum_{\boldsymbol{K}} h(\boldsymbol{K}) A(\boldsymbol{K}, t) B(\boldsymbol{K}, t) \tag{5.9}$$

where

$$A(\boldsymbol{K}, t) = \sum_{\boldsymbol{k}'} g(\boldsymbol{k}' - \boldsymbol{k}'_0) \exp[i(\boldsymbol{k}' + \boldsymbol{K}) \cdot \boldsymbol{x}] \exp\left[\frac{-i}{2m'}(\boldsymbol{k}' + \boldsymbol{K})^2 t\right]$$

$$B(\boldsymbol{K}, t) = \sum_{\boldsymbol{k}} f(\boldsymbol{k}) \exp[i(\boldsymbol{k} - \boldsymbol{K}) \cdot \boldsymbol{y}] \exp\left[\frac{-i}{2m}(\boldsymbol{k} - \boldsymbol{K})^2 t\right]$$

and $h(\boldsymbol{K})$ represents the amplitude that finally results from the process of interaction (which will depend on the potential $V(\boldsymbol{x}, \boldsymbol{y})$, of interaction between two particles). The sum over $\boldsymbol{k}$ must, of course, be restricted by the condition of conservation of energy.

To show what the above means let us consider the factor $B(\boldsymbol{K}, t)$ which we rewrite as

$$\exp\left[\frac{-i}{2m}(\boldsymbol{K}\cdot\boldsymbol{y}+K^2t)\right]\sum_{\boldsymbol{k}} f(\boldsymbol{k})\exp\left[i\left(\boldsymbol{k}\cdot\boldsymbol{y}+\frac{\boldsymbol{k}\cdot\boldsymbol{K}t}{m}\right)\right]\exp\left[\frac{-ik^2t}{2m}\right].$$

For t not too large, this can be written as

$$\exp\left[\frac{i}{2m}(\boldsymbol{K}\cdot\boldsymbol{y}+K^2t)\right]\phi\left(\boldsymbol{y}+\frac{\boldsymbol{K}t}{m}\right).$$

This corresponds to a wave packet moving with a velocity $\boldsymbol{K}/m$. The term $\exp[-ik^2t/2m]$, which we have neglected, describes the spread of the wave packet which may be taken to be small during the period of interest. A similar treatment could be given for $A(\boldsymbol{K}, t)$. This will also correspond to a wave packet, but moving with velocity $-\boldsymbol{K}/m'$.

What this all implies is that after interaction, the two particles are separating with a relative speed proportional to K. If the interaction is such that K is considerably greater than the initial spread of momenta in the wave packets ($\Delta\boldsymbol{k}$ and $\Delta\boldsymbol{k}'$), then after a suitable time, the distances moved by the particles

$$\Delta\boldsymbol{x} = -\boldsymbol{K}t/m' \quad \text{and} \quad \Delta\boldsymbol{y} = \boldsymbol{K}t/m \tag{5.10}$$

will be greater than the widths $\delta\boldsymbol{x}$ and $\delta\boldsymbol{y}$ of the wave packets. When this happens the final wave function of the combined system $\Psi(\boldsymbol{x}, \boldsymbol{y}, t)$ will cease to overlap with the initial wave function $\Psi_0(\boldsymbol{x}, \boldsymbol{y}, t)$. Moreover for different values of $\boldsymbol{K}$, the particles will go in different directions so that the contributions of these different values of $\boldsymbol{K}$ to the total wave function will also cease to overlap.

This means that the original transmitted channel will now break into many parts. First there will be an unmodified transmitted channel in the regions far from the wave packet $\phi_0(\boldsymbol{y})$ and a weaker one close to this wave packet. Secondly there will be modified channels in which both the scattering particle $\boldsymbol{y}$ and the transmitted particle $\boldsymbol{x}$ will be going in various directions. Since all these channels do not overlap, the quantum potential (and guidance condition) for the combined system will be determined only by the channel into which the particle actually enters. All the other channels will then correspond to inactive information.

Because there is such a process of the multiplication of non-overlapping channels corresponding to transmission of the particle, the possibility of bringing about overlap of transmitted and reflected channels in the way described earlier has effectively been eliminated. For in order to have such overlap, it is necessary that this overlap takes place in the *configuration space of all systems which have interacted.* Therefore if we merely have an arrangement that brings reflected and transmitted waves into the same region of space, there will still be no effect on the quantum potential. To have such an effect we would also have to bring together *all* the scattered channels. If this is not done, the information in the reflected channel remains inactive, even though it corresponds to a much more intense quantum field than does the transmitted channel.

Of course one could consider trying to bring back into the same region of space all the channels that resulted from the various scattering processes. Evidently this would be immensely difficult, almost certainly beyond our present technical capacity. Nevertheless it is still a logical possibility. But we can see that this possibility is not relevant here because all the particles concerned will keep on encountering further systems with which they interact, and these will bring about a further multiplication of channels.

Ultimately this process will lead to chaotic motions which are characteristic of thermodynamic equilibrium. As long as we accept the current explanations of the irreversibility of thermodynamic processes in terms of the notion of an approach to statistical equilibrium, for example through scattering, then we must also accept that the interactions that we have described here will similarly be irreversible. Therefore the channels to which we have alluded earlier will be irrevocably distinct. The quantum potential and the guidance condition will be determined only by the channel actually occupied by the system. The remaining channels will then correspond to what we have been calling inactive information, i.e. information not only actually inactive, but also having no potential for becoming active.

All this implies that the channels corresponding to actually or potentially active information are constantly narrowing down as the system in question interacts with its surroundings. But it is implied by Schrödinger's equation that the wave function of a system is constantly spreading (e.g. as shown in detail in chapter 3 for Gaussian wave packet). These two processes evidently oppose each other as we have already indicated earlier, and the net result comes out of the balance between the tendency to spread and the tendency to narrow down to a more definite form.

The narrowing down process produces a net effect that is essentially the same as that produced by the 'collapse' of the wave function in the usual interpretation. Yet there is no collapse here. At no stage do we need to violate our basic postulates in order to obtain the self-definition of the system. We shall discuss the full implications of this in more detail later, but for the present we shall emphasise only that this means that measurement does not have to play a fundamental role in our interpretation. Rather the universe is constantly defining itself and measurement is only a special case of this.

5.3 Quantum transitions discussed independently of measurement

We now go on to discuss the process of transition of an electron in an atom. If such an atom is to jump from one stationary state to another, it is necessary that there be an additional system available that can take up the energy given off. This is usually the electromagnetic field within which, for example, a photon can be created. However, to avoid the complexities arising in our interpretation of the electromagnetic field (which we discuss in chapter 11), we shall suppose that the energy is taken up in an Auger-like effect by an additional particle that was originally also in a bound state in the atom in question.

To make possible a comparatively short mathematical treatment, we shall use here a simplified model of this process in which both particles are initially in bound states within a localised potential well. We shall suppose that they are in *s*-states and that all the transitions can be restricted to *s*-states (i.e. there is negligible probability of going to states of higher angular momentum). To further simplify the model we will suppose that initially the two particles do not interact and that at $t = 0$ a small perturbing potential $V(\boldsymbol{x} - \boldsymbol{y})$ is suddenly turned on (where $\boldsymbol{x}$ is the coordinate of the electron and $\boldsymbol{y}$ that of the particle which carries off the energy). We shall then apply the standard perturbation theory to calculate the changes of the wave function with time. The initial conditions will be determined by the requirement that the $\boldsymbol{x}$-particle is in a stationary state with energy E_0^A and with wave function $\psi_0(\boldsymbol{x})\exp[-iE_0^At]$ (where we continue our notation in which $\hbar = 1$). The $\boldsymbol{y}$-particle will be in a stationary state with energy E_0^B and with wave function $\phi_0(\boldsymbol{y})exp[-iE_0^Bt]$. The net initial wave function will then be

$$\Psi = \psi_0(\boldsymbol{x})\phi_0(\boldsymbol{y})\exp[-iE_0t], \tag{5.11}$$

where $E_0 = E_0^A + E_0^B$.

After $t = 0$ the Schrödinger equation for the combined system will be

$$i\frac{\partial \Psi}{\partial t} = \left[-\frac{1}{2m_A}\nabla_{\boldsymbol{x}}^2 - \frac{1}{2m_B}\nabla_{\boldsymbol{y}}^2 + V(\boldsymbol{x} - \boldsymbol{y})\right]\Psi. \tag{5.12}$$

Since we can restrict ourselves to s-states, it will be convenient to introduce the function

$$\chi = r_A r_B \Psi,$$

where r_A and r_B are the radius vectors of the respective particles. In terms of χ, Schrödinger's equation becomes

$$i\frac{\partial \chi}{\partial t} = \left[-\frac{1}{2m_A}\frac{\partial^2}{\partial r_A^2} - \frac{1}{2m_B}\frac{\partial^2}{\partial r_B^2} + V(r_A - r_B)\right]\chi, \qquad (5.13)$$

with the boundary condition that $\chi = 0$ at $r_A = 0$ and $r_B = 0$. The initial wave function is, of course,

$$\chi = \chi_0(r_A)\lambda_0(r_B)\exp[-iE_0 t],$$

where $\chi_0 = r_A\psi_0(r_A)$ and $\lambda_0 = r_B\phi_0(r_B)$.

For $t > 0$ we then write

$$\chi = \chi_0 + \Delta\chi,$$

$\Delta\chi$ will contain the wave function components in which the x-particle has made a transition to $\chi_f \exp[-iE_f t]$, while the y-particle has become free with absolute value of the momentum p. Since both particles were initially in s-states and since the x-particle is finally in an s-state, it follows that the final angular momentum of the y-particle is zero, so that it too is in an s-state. Recalling that $\chi = 0$ at $r_B = 0$, the wave function of the latter will then be proportional to

$$\sin(pr_B)\exp\left[-\frac{ip^2}{2m_B}t\right].$$

The typical contribution to the net wave function after transition will therefore be proportional to

$$\chi_f(r_A)\sin(pr_B)\exp\left[-i\left(E_f + \frac{p^2}{2m_B}t\right)\right].$$

We then obtain

$$\Delta\chi = \chi_f(r_A)\sum_p c_p \sin(pr_B)\exp\left[-i\left(E_f + \frac{p^2}{2m_B}t\right)\right]. \qquad (5.14)$$

Perturbation theory then yields

$$c_p = \frac{V_{f0,p0}}{\Delta E + (p^2/2m_B)}\left(1 - \exp\left[-i\left(E_f + \frac{p^2}{2m_B}t\right)\right]\right), \qquad (5.15)$$

where $\Delta E = E_f - E_0$ which is the net increase of binding energy resulting from transition, and where the corresponding matrix element is

$$V_{f0,p0} = \int_{-\infty}^{\infty}\int_{-\infty}^{\infty} \psi_f^*(\boldsymbol{x}) \sin(p\boldsymbol{r}_B) V(\boldsymbol{x}-\boldsymbol{y})\psi_0(\boldsymbol{x})\phi_0(\boldsymbol{y})\, \mathrm{d}\boldsymbol{x}\mathrm{d}\boldsymbol{y}. \tag{5.16}$$

We will be interested in finding the form of $\Delta\chi$ for moderately long periods of time; i.e. for times long enough so that the factor

$$\frac{1-\exp\left[-i\left(\Delta E + (p^2/2m_B)\right)t\right]}{\Delta E + (p^2/2m_B)}$$

becomes a sharply peaked function with a maximum at that value of p for which the denominator vanishes. This function evidently has a width inversely proportional to the time t. Let us denote the value for which this function is a maximum by $p = p_0$. In evaluating $\Delta\chi$ we can then replace p by p_0 in all the terms in $\Delta\chi$ that multiply this factor. Also, in evaluating this factor we may write $p = p_0 + \Delta p$ and neglect terms of the order of $(\Delta p)^2$ (because we are considering t great enough so that the factor is large only for very small values of Δp). We then obtain $\Delta E + p^2/2m_B \cong p_0\Delta p/m_B = v_0\Delta p$ where v_0 is the speed of the outgoing particle. With these approximations we find that

$$\Delta\chi = \chi_f(\boldsymbol{r}_A)V_{f0,p0}\sum_p \frac{(1-\exp[iv_0\Delta pt])}{v_0\Delta p}\sin(p\boldsymbol{r}_B). \tag{5.17}$$

For an unbounded system, this sum becomes an integral, which can be evaluated by means of contour integration. The wave intensity will be proportional to

$$\int_{-\infty}^{\infty}\left[\exp[i(p_0+\Delta p)\boldsymbol{r}_B] - \exp[-i(p_0+\Delta p)\boldsymbol{r}_B]\right]\left[\frac{(1-\exp[iv_0\Delta pt])}{v_0\Delta p}\right]\mathrm{d}\Delta p.$$

The integral is finite everywhere including $\Delta p = 0$, even though the separate terms are singular there. We may therefore evaluate it by choosing, in the well-known way, a contour in the complex plane that goes around $\Delta p = 0$ in an infinitesimal semi-circle which is chosen so that only outgoing waves will appear for $t > 0$. A typical term in the integral is proportional to

$$\int_{-\infty}^{\infty}\frac{\exp[i\Delta p\boldsymbol{r}_B]}{v_0\Delta p}\,\mathrm{d}\Delta p.$$

If we choose the contour in the upper half of the complex plane as shown in figure 5.4, it can be seen that the integral vanishes.

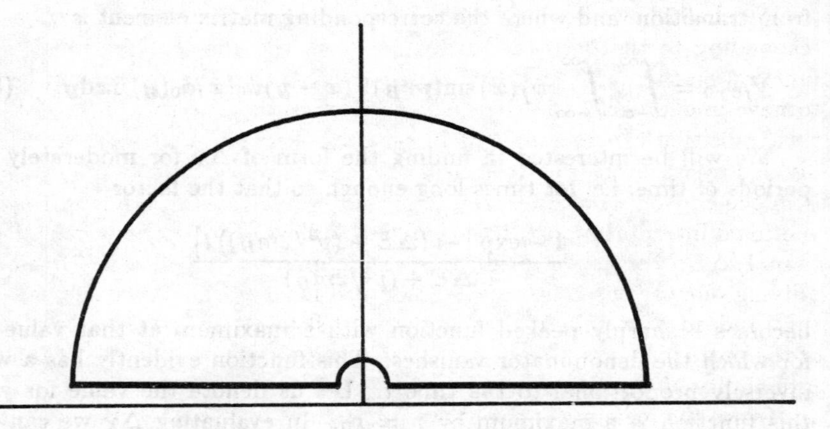

Figure 5.4: Contour to evaluate integral

On the other hand, if we take a term like

$$\int_{-\infty}^{\infty} \frac{\exp[-i\Delta p \boldsymbol{r}_B]}{v_0 \Delta p} \mathrm{d}\Delta p,$$

then by completing the contour in the lower half of the complex plane, we see that it does not vanish, but is equal to the residue at the singularity. By applying this method, we can see that all the terms containing the factor $\exp[i(p_0+\Delta p)\boldsymbol{r}_B]$ will vanish. The remaining two terms will be proportional to

$$G(\boldsymbol{r}_B, t) = S(\boldsymbol{r}_B) - S(\boldsymbol{r}_B - v_0 t),$$

where S is the step function which goes to zero for negative values of its argument and is unity for positive values. This will imply a step-shaped wave that is spreading out at the speed v_0. The total wave function (for large t) is then

$$\chi = \boldsymbol{r}_A \boldsymbol{r}_B \exp[-E_0 t] \left[\chi_0(\boldsymbol{r}_A)\lambda_0(\boldsymbol{r}_B) + \frac{V_{f0,p_0 0}}{v_0} \chi_f(\boldsymbol{r}_A) \exp[ip_0 \boldsymbol{r}_B] G(\boldsymbol{r}_B, t) \right]. \tag{5.18}$$

In the above equation the term $\chi_0(\boldsymbol{r}_A)\lambda_0(\boldsymbol{r}_B)$ represents the initial wave function of the combined system, while the term

$$\frac{V_{f0,p_0 0}}{v_0} \chi_f(\boldsymbol{r}_A) \exp[ip_0 \boldsymbol{r}_B] G(\boldsymbol{r}_B, t)$$

represents the contribution of the perturbation to this wave function.

We shall show presently that if the $\boldsymbol{r}_B$-particle enters the region corresponding to the perturbed wave function, the effect of the initial wave function on the particles becomes negligible so that the system can be said to have undergone a transition to the final state. The total probability of this transition can be found by integrating the square of the second term, divided by $\boldsymbol{r}_B^2$ over $d\boldsymbol{r}_B$. It is clear that this will be proportional to t (as it is in the conventional interpretations). However (as also holds in the conventional interpretations), this result will follow only for times much larger than $1/\Delta E$. For as we recall, this approximation, which arises from simplifying our various contour integrals, will not hold for times shorter than this. Indeed it is evident that at the very shortest times, the contribution of the perturbed wave function must be proportional to t itself, so that the probability of transition is proportional to t^2.

To see the meaning of the above equation in more detail, it is instructive to take the ratio of the perturbed contribution to the wave function to the unperturbed contribution. This is

$$R = \frac{V_{f0,p_0 0}\chi_f(\boldsymbol{r}_A)\exp[ip_0\boldsymbol{r}_B]G(\boldsymbol{r}_B,t)}{v_0\chi_0(\boldsymbol{r}_A)\lambda_0(\boldsymbol{r}_B)}. \tag{5.19}$$

Because the matrix element $V_{f0,p_0 0}$ is, by hypothesis, small, this ratio will also generally be small. Where this is the case, the initial wave function will overwhelm the perturbation, which latter will then at most introduce a very small change in the overall wave function. In such a situation, the state of the particle would undergo negligible change. However, where the denominator in the above ratio is small, the situation can be quite different. Since $\lambda_0(\boldsymbol{r}_B)$ represents a bound state, the ratio will indeed become as small as we please for values of $\boldsymbol{r}_B$ appreciably greater than the mean spread of this wave function. For these values of $\boldsymbol{r}_B$, the initial wave function will be negligible compared with the perturbed wave function. This latter will now dominate and the motion of the particle will no longer be affected significantly by the initial wave function. The motion implied by this wave function will carry the particle further and further from the atom at the speed v_0. The effective wave function of the system will then be proportional to $\chi_f(\boldsymbol{r}_A)\exp[ip_0\boldsymbol{r}_B]G(\boldsymbol{r}_B,t)$. This means that the outgoing particle is free, and that the bound electron now has a wave function $\chi_f(\boldsymbol{r}_A)$. In other words, a transition has taken place from a state in which the electron was bound with the wave function $\chi_0(\boldsymbol{r}_A)$ and the $\boldsymbol{y}$-particle with $\lambda_0(\boldsymbol{r}_B)$, to one in which the electron is bound with the wave function $\chi_f(\boldsymbol{r}_A)$, while the $\boldsymbol{y}$-particle effectively has the free particle wave function $\exp[i\boldsymbol{p}\cdot\boldsymbol{r}_B]$.

It is evident then that this kind of transition can take place when the $\boldsymbol{y}$-particle is initially located at values of $\boldsymbol{r}_B$ large enough so that $\lambda_0(\boldsymbol{r}_B)$ is negligible. Clearly such locations are highly improbable and this explains why only a few members of the ensemble will undergo transition.

To see what happens to particles starting at small values of $\boldsymbol{r}_B$, let us compute the velocity from the guidance condition (which, as we have seen in chapter 3, implies the quantum potential). We find

$$v = \frac{1}{2im_B}\left[\chi^* \frac{\partial \chi}{\partial \boldsymbol{r}_B} - \chi \frac{\partial \chi^*}{\partial \boldsymbol{r}_B}\right] \Big/ \chi^* \chi. \tag{5.20}$$

For values of $\boldsymbol{r}_B$ small enough so that the perturbed wave function has a much lower intensity than the unperturbed one this will be approximately proportional to $p_0/[m_B \lambda_0^*(\boldsymbol{r}_B)\lambda_0(\boldsymbol{r}_B)]$. It is clear that such particles start to move outward at a very low velocity, especially those initially at small radii. They will speed up as they go outward and will eventually enter the region in which the perturbed wave function is dominant. The effective physical state will then be one in which a transition has taken place along the lines that we have described. Ultimately all the particles in the ensemble will be able to undergo transitions, in this way. However, in such an ensemble, the times of transition will be statistically distributed, and it is easy to show that this distribution is exactly what is predicted by the usual interpretation of the quantum theory.

The above holds only in the approximation in which the perturbation theory is valid. In a more exact treatment, e.g. the Wigner-Weisskopf method [4], the initial wave function is found to decay exponentially with some mean life-time τ. It is clear that, depending on the initial conditions, some particles will be emitted in times much shorter than τ. This means that the wave function does not give a complete description of the reality of the transition, because it has no terms in it to discuss the actual and often short time in which a particle approaching the edge of the atom is suddenly 'swept out' to become free in an interval of time that has no essential relationship to τ. Clearly the additional concept of a particle is needed, alongside that of the wave function, to allow this process to be considered ontologically, i.e. as essentially independent of being measured or observed.

A similar situation arises with radioactive decay brought about by barrier penetration. For example, the mean-life of the uranium atom is 2×10^9 years, but once a particle enters the barrier, it may go through in a time of the order of 10^{-22} secs. This speed of the actual process is also something that cannot be described solely in terms of the wave function and its associated concept of mean-life time.

Finally it should be noted that as in the example of barrier penetration discussed in section 5.1 there is still, in principle, the possibility of interference between the initial wave function and that representing particles that have undergone transition. (This is, of course, as valid in the usual interpretation as it is in our interpretation.)

If the outgoing particle undergoes collisions, however, then it becomes steadily more difficult to obtain interference because more and more particles would have to be made to participate in just the right way. As in the case of barrier penetration, this would ultimately become impossible in practice. Simultaneously the domain of active information would narrow down so that the outgoing particle would be affected only by a relatively narrow wave packet, while the information in the remaining parts of the wave function would become inactive.

5.4 On the possibility of bifurcation points in more complex cases

In the simple example of the Auger-like transition discussed in the previous section there are no bifurcation points such as were found in the example discussed in section 5.1. However, if we introduce a number of final states, bifurcation points can be obtained.

To illustrate this, let us consider two possible final wave functions of the electron $\chi_1(\boldsymbol{r}_A)$ and $\chi_2(\boldsymbol{r}_A)$. These will be assumed to have different energies so that the energies transferred to the $\boldsymbol{y}$-particle will be correspondingly different. We will assume that for the wave function $\chi_1(\boldsymbol{r}_A)$, the $\boldsymbol{y}$-particle will have momentum $\boldsymbol{p}_1$ and velocity $\boldsymbol{v}_1$ and for $\chi_2(\boldsymbol{r}_A)$ these will be respectively $\boldsymbol{p}_2$ and $\boldsymbol{v}_2$. The perturbed wave function will then be

$$\begin{aligned}\Delta\chi \;=\; & \boldsymbol{r}_A\boldsymbol{r}_B\exp[-iE_0t]\left[\frac{V_{f_10,p_10}}{v_1}\chi_1(\boldsymbol{r}_A)\exp[ip_1\boldsymbol{r}_B]G_1(\boldsymbol{r}_B,t)\right.\\ & \left.+\frac{V_{f_20,p_20}}{v_2}\chi_2(\boldsymbol{r}_A)\exp[ip_2\boldsymbol{r}_B]G_2(\boldsymbol{r}_B,t)\right]\end{aligned} \tag{5.21}$$

where

$$G_1(\boldsymbol{r}_B,t)=S(\boldsymbol{r}_B)-S(\boldsymbol{r}_B-\boldsymbol{v}_1t)$$

and

$$G_2(\boldsymbol{r}_B,t)=S(\boldsymbol{r}_B)-S(\boldsymbol{r}_B-\boldsymbol{v}_2t).$$

If the particle is in a region beyond that in which the initial wave function is appreciable then, as we have seen earlier, its motion can be treated solely in terms of the perturbed wave function. This latter describes what are effectively two wave packets in $\boldsymbol{r}_B$ space which are multiplied respectively by $\chi_1(\boldsymbol{r}_A)$ and $\chi_2(\boldsymbol{r}_A)$. As shown in figure 5.5, the first packet will have

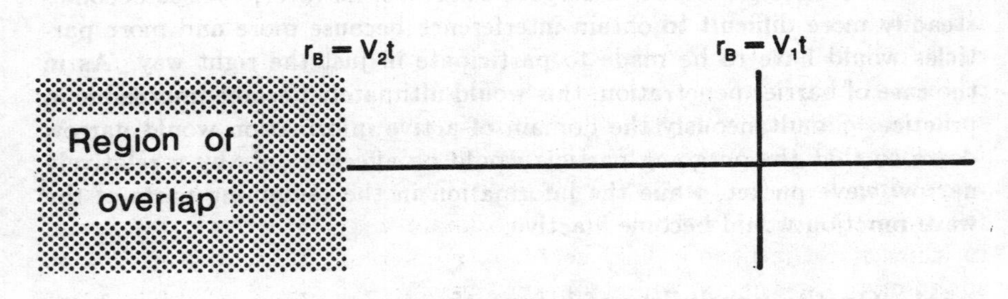

Figure 5.5: Overlap of wave packets

reached $\boldsymbol{r}_B = \boldsymbol{v}_1 t$, while the second will have reached $\boldsymbol{r}_B = \boldsymbol{v}_2 t$. Assuming $\boldsymbol{v}_1 > \boldsymbol{v}_2$, it follows that the second packet will be contained in the first, which latter has an advancing region of width $\Delta \boldsymbol{r}_B = (\boldsymbol{v}_1 - \boldsymbol{v}_2)t$, that does not overlap the first packet. If the particle is in this region, then the second wave packet (containing $\chi_2(\boldsymbol{r}_B)$ as a factor) will not affect its motion. Therefore it can be treated as if the wave function had 'collapsed' to $\chi_1(\boldsymbol{r}_A)\exp[i\boldsymbol{p}_1 \cdot \boldsymbol{r}_B]$. However, if it is in the region of the second packet, there will still be interference between the two terms and we will not yet have a collapse to one term or the other.

How is it possible to obtain a definite result? To answer this, we first note that this state of affairs is a consequence of our approximation in terms of perturbation theory in which the decay of the original wave function is not taken into account. Therefore an indefinite stream of outgoing waves is implied.

However, when this decay is taken into account, e.g. by the Wigner-Weisskopf approximation, it is found that the wave packets will have durations of the order of the mean life-time, τ, of the original wave function. It is clear that when t is great enough, the separation of the packets, $(\boldsymbol{v}_1 - \boldsymbol{v}_2)t$, will be greater than their respective widths $\boldsymbol{v}_1\tau$ and $\boldsymbol{v}_2\tau$, and therefore the packets will separate. Where they have separated, the particle will be in one of these packets, in which case its motion can be treated without reference to the other, as we have done in the examples already given here. The effective wave function will then be either $\chi_1(\boldsymbol{r}_A)\exp[i\boldsymbol{p}_1\boldsymbol{r}_B]$ or $\chi_2(\boldsymbol{r}_A)\exp[i\boldsymbol{p}_2\boldsymbol{r}_B]$. In other words we have shown that there are two mutually exclusive channels into one of which the particle must enter. And as a simple calculation shows, the probabilities of either channel will come out the same as those given by the usual interpretation.

Although the consideration of the finite lifetime shows that the outgoing wave function will ultimately give rise to two separate channels (when t considerably exceeds the mean lifetime), it is necessary to point out here that this can happen in much shorter times if the outgoing particle can undergo collisions with other atoms. As an example, suppose that the energy of the outgoing particle, E_1, corresponding to velocity $\boldsymbol{v}_1$, is great enough to ionise an atom, while that of a particle of energy E_2 and corresponding to velocity $\boldsymbol{v}_2$ is not. After the collision the part of the wave function corresponding to $\chi_1(\boldsymbol{r}_A)$ will then be multiplied by a factor $g(\boldsymbol{z})$ describing the free state of the particle with coordinate $\boldsymbol{z}$ liberated from the atom in which it was originally held, while $\chi_2(\boldsymbol{r}_A)$ is multiplied by $g_0(\boldsymbol{z})$ corresponding to the original bound state of this particle. Clearly, in a relatively short time these wave functions cease to overlap, and subsequent collisions with them bring about a multiplication of channels in the way described in section 5.2. In principle this can happen in times of the order of that required for collisions (perhaps something like 10^{-12} secs). Therefore even if the period of decay is long, we will still generally obtain a separation of channels. However, if no collisions take place, the channels will not separate and we will have interference between the two components of the wave function $\chi_1(\boldsymbol{r}_A)\exp[ip_1\boldsymbol{r}_B]$ and $\chi_2(\boldsymbol{r}_A)\exp[ip_2\boldsymbol{r}_B]$. This interference is indeed an example of the EPR correlation (which we shall discuss in detail in chapter 7). It is an unavoidable prediction of the quantum theory, both in the usual interpretation and in ours.

Whether the channels are formed by separation of wave packets over times of the order of τ or by collisions, it seems evident that there must be some limiting points or curves which divide the initial conditions of the particles that will enter one channel from those that will enter the other (as happened in the case of barrier penetration). Generally speaking however these will be in the overall configuration space and may be quite complex in form. The possible complexity will increase rapidly as we consider more and more stationary states, especially if we go to wave functions of higher angular momentum. In the latter case the stationary states will be characterised, as we have seen in chapter 3, by a set of circular motions, in contrast to the s-state which is characterised by a set of points at which the particle is at rest.

The general behaviour described above is similar to that obtained in the study of non-linear equations whose solution contain what are called stable limit cycles [3]. As we have pointed out earlier the equations for the particle motion in our interpretation are likewise clearly non-linear. It is implied by the discussion above that the solutions also have something like stable limit cycles except that a certain generalisation of this concept is needed here. Thus, as we have already indicated, for p-states,the stable limit cycles would be a set of circles. The difference from the usual kind of non-linear equation is that for each stable motion we have a whole set of possible limit cycles rather than just a single cycle. Each quantum state thus corresponds to a different set of limit cycles and a transition corresponds to an orbit going from one of these to another.

In this discussion the *s*-states should be regarded as degenerate cases of limit cycles which reduce to single points in $\boldsymbol{r}_A$ space and simultaneously to outgoing orbits in $\boldsymbol{r}_B$ space.

Finally it must be emphasised that our equations for the particle motions differ from the non-linear equations that have commonly been considered thus far, in that the forces, as determined by the quantum potential, are not preassigned functions either of external variables or of the particle coordinates. Rather, as was brought out in chapter 4, they depend on a quantum field of active information ordered in the configuration space which involves the whole system in a way that cannot be analysed solely in terms of relationships between the particles. It is this feature which makes it possible for such dramatically different limit cycles to emerge as a result of the evolution of the wave function even for simple systems. The fact that the activity of the quantum field does not necessarily fall off with the intensity of the wave is also crucial for this behaviour, since without it we could not have had some transitions taking place almost immediately even when the perturbed wave function was very weak. It is clear then that the new features of the quantum potential described above are principally what are behind the ability of our extension of Newton's equations of motion to comprehend the new quantum properties of matter. This brings out in further detail the point that we made in chapter 3, section 3.1, i.e. that classical physics does not follow from the form of Newton's equations alone, but also requires certain restrictions on the forces that appear in these equations. When, as is the case of the quantum potential, we go beyond such restrictions, an entirely new kind of physical law is implied.

5.5 The quantum process of 'capture' of particles (fusion-fission)

We now go on to extend the results of the previous section to the case of capture of an electron by an atom. This, combined with the discussion of ionisation that we have just given, will enable us to bring out how quantum mechanics implies the possibility of the formation of objective wholes as well as their dissolution into relatively independent parts and sub-wholes. As we saw in chapter 4, this is one of the principal new properties of matter implied by quantum theory.

We shall discuss this process as a kind of inverse of the one that was treated in the previous section. We consider an atom on which is incident a particle with a plane wave function $\exp(ikz)$ implying motion in the z-direction only. In order to enable the atom to capture this particle so as to form a new combined system, it is necessary that there be something to carry away the energy that is given off. As happened in the case of ionisation, this will be done by an additional particle with coordinate $\boldsymbol{y}$.

Let us suppose there are further particles in the atom with coordinates designated by ξ_i. The initial state of the atom will then be represented by $\phi_0(\boldsymbol{y}, \xi_i) \exp[-iE_0 t]$. The combined initial wave function for the entire system is therefore

$$\Psi_0 = (\boldsymbol{x}, \boldsymbol{y}, \xi_i, t) = \phi_0(\boldsymbol{y}, \xi_i) \exp[ikz] \exp\left[-i\left(E_0 - \frac{k^2}{2m}\right)t\right]. \qquad (5.22)$$

Because of the interaction between these particles, this combined wave function will change. We shall assume that the main possibility for change is the introduction of a ground state, in which the $\boldsymbol{x}$-particle has been captured, while the $\boldsymbol{y}$-particle has been released. This wave function will thus be $\phi_f(\boldsymbol{x}, \xi_i) \exp[-iE_f t]$. The $\boldsymbol{y}$-particle will then have a wave function whose typical component is $\exp[i\boldsymbol{p} \cdot \boldsymbol{y} - (ip^2/2m)t]$. The net change of the overall wave function can then be written as

$$\Delta\psi = \sum_p C_p(t) \exp i\left(\boldsymbol{p} \cdot \boldsymbol{y} - \frac{p^2}{2m}t\right) \phi_f(\boldsymbol{x}, \xi_i) \exp[-iE_f t]. \qquad (5.23)$$

We assume that this process can be treated by perturbation theory in a way similar to what was done with ionisation in the previous section. Of course the details will be considerably more complex because the outgoing wave cannot in this case be restricted to a spherically symmetric form. We shall however give only the general results here. The essential point is that after some time, the outgoing wave takes the form

$$\Delta\psi = f(\theta_B) \frac{\exp[i\boldsymbol{p} \cdot \boldsymbol{r}_B]}{\boldsymbol{r}_B} \phi_f(\boldsymbol{x}, \xi_i) \exp\left[-iE_f t - \frac{ip^2 t}{2m}\right], \qquad (5.24)$$

where θ_B is the azimuthal angle of the outgoing particle relative to the direction of the incident particle. If the initial conditions are such that the particle enters this outgoing wave, then there is no overlap with the incident wave function. The quantum potential acting on the particle will then be the same as if the incident wave function were not present and $\Delta\psi$ were the whole wave function. It is clear that in this case the incident particle will have been captured into a 'state' with wave function $\phi_f(\boldsymbol{x}, \xi_i, t)$

corresponding to a stable limit cycle of the kind discussed in the previous section. At the same time the $\boldsymbol{y}$-particle will be moving outward carrying away the excess energy. The probability of capture can be calculated by averaging with the probability density $P = |\psi|^2$ and this will give the same result as is obtained in the usual interpretation.

Initially the combined system has a product wave function implying, as we have seen in chapter 4, that the two subsystems represented by $\boldsymbol{x}$ and $(\boldsymbol{y}, \xi_i)$ respectively behave independently. If the initial conditions were such that the $\boldsymbol{y}$-particle did not enter the outgoing wave, the effective wave function would have ended up as essentially unchanged so that the subsystems would remain independent after collision. However, if the $\boldsymbol{y}$-particle did enter the outgoing wave, then we will have formed a new whole with wave function $\phi_f(\boldsymbol{x}, \xi_i)\exp[-iE_f t]$ along with the $\boldsymbol{y}$-particle, which is now a new independent part. So what happened is that we began with one whole which included the ξ_i-particle and the $\boldsymbol{y}$-particle, and along with it was an independent particle represented by $\boldsymbol{x}$. This whole broke up, but at the same time a new whole was formed with different constituents. This process could be described picturesquely by saying that the original whole had undergone 'fission', while the product atom had undergone 'fusion' with the $\boldsymbol{x}$- and ξ_i-particles. Through the fission-fusion process we have been able to describe objective changes in the wholes that are present [5].

To come back to our description in terms of active information, we could also say that the original atom was characterised by a certain common pool of such information which broke up and reformed into a new pool that included the $\boldsymbol{x}$-particle instead of the $\boldsymbol{y}$-particle. We emphasise once again that this process of forming and dissolving wholes is essentially quantum mechanical. For it is only through the existence of such pools of information which are not expressible solely in terms of relationships of the actual particles that the notion of an objective whole can be given meaning. In contrast, as we have said earlier in classical physics there is no such objective significance to the whole, the latter can at best be regarded as a convenient way of thinking about what is considered to be in reality a set of parts. Quantum mechanics thus implies a new kind of process; i.e. the collection and dissolution of wholes. The language of ordinary quantum mechanics already tends to suggest this, in so far as it treats the quantum state as an unanalysable whole. However, as we have already suggested in earlier chapters, in our interpretation this wholeness becomes more intuitively intelligible because this interpretation highlights the new features of the quantum theory that bring it about.

5.6 Summary and conclusions

The essential point of this chapter was to demonstrate with the aid of a number of typical examples that the transition process can be treated ontologically, i.e. as having no essential relationship to measurement or observation. We showed that in such a transition the wave function effectively splits so as to define a number of separate channels. The quantum potential was determined only by the channel that the particle actually enters. The channel entered by the particle depends on the initial conditions which are distributed statistically in an ensemble with $P = |\psi|^2$. From this it follows that the probability of obtaining a certain result is the same as in the usual interpretation.

In effect everything happens as if the wave function 'collapses' to the final result, while no collapse actually takes place. The channels not occupied by the particles correspond to inactive information. We showed that if the particles make collisions with additional systems, there is a constant narrowing down of the domain in which the wave function can significantly affect the quantum potential at the location of the particles. From this it follows that the channel entered by the particles eventually becomes irreversibly fixed, so that it cannot be undone.

An analysis of the behaviour of the particle orbits shows that there are bifurcation points dividing those orbits entering one channel from those entering another. Near these points, the motion is highly unstable and, indeed, chaotic in the sense of modern chaos theory. Stationary states then correspond to an extension of the concept of stable limit cycles which is found in the theory of non-linear differential equations.

Finally we discuss the process of formation and dissolution of wholes (fission-fusion). This brings out the possibility of an objective ontological wholeness and of a distinction between states of such wholeness and states in which the parts or sub-wholes behave independently. We understand this through the formation of common pools of information. This information brings about nonlocal interaction, but quantum wholeness implies even more than this. For it arises out of the quantum field which cannot be understood solely in terms of preassigned properties and interrelationships of the particles alone. Rather the whole is presupposed in the quantum wave function and it is the active information in this wave function that forms and dissolves wholes.

5.7 References

1. D. Bohm, *Quantum Theory*, chapter 11, section 11.5, Prentice-Hall,

Englewood Cliffs, New Jersey, 1951.

2. C. Dewdney and B. J. Hiley, *Found. Phys.* **12**, 27–48 (1982).
3. P. Cvitanovic, *Universality in Chaos*, Adam Hilger, Bristol, 1984.
4. W. Heitler, *The Quantum Theory of Radiation*, Dover, New York, 1984, 181.
5. A. Baracca, D. Bohm, B. J. Hiley and A. E. G. Stuart, *Nuovo Cimento* **28B**, 453–466 (1975).

Chapter 6
Measurement as a special case of quantum process

As pointed out in chapter 2, in the conventional interpretation of the quantum theory (which is basically epistemological in nature) measurement plays a key role in the sense that without it the mathematical equations would have no physical meaning. In our ontological interpretation, however, we have started with a treatment of the individual, actual process, e.g. a particle penetrating a barrier, undergoing transition between stationary states etc. Evidently it is necessary to deal with the measurement process in essentially the same way; i.e. as a special case of quantum processes in general. What is particularly significant in a measurement process is that from a large scale result that is observable in a piece of measuring apparatus, one can infer the state of the observed system, or at least the state in which it has been left after the measurement process is over.

In this chapter we shall treat the measurement process in some detail and show how it is to be understood ontologically. We shall first discuss the general principles of this process. We shall show that the essential new feature of quantum measurement is that there is mutual and irreducible participation of the measuring instrument and the observed object in each other. As a result, any attempt to discuss this process as measuring 'a property of the observed object alone' will not be consistent with our interpretation. Rather we say that the result of a measurement is a potentiality of the combined system and can be determined only in terms of the properties of the particles, along with the wave function of the combined system as a whole. On the basis of this notion, we shall go into several examples that help illustrate the principles involved and clarify some of the puzzles to which these examples give rise in the conventional treatment.

6.1 Brief treatment of the measurement process

In treating measurements ontologically we shall for the sake of convenience divide the overall process into two stages. In the first stage the measuring apparatus and the observed system interact in such a way that the wave function of the combined system breaks into a sum of non-overlapping packets, each corresponding to a possible distinct result of the measurement. In the second stage this distinction is magnified by some detecting device, so that it is directly observable at the large scale level.

In the first of these stages, the wave function can be treated in the same way as was done in the discussion of von Neumann's approach given in chapter 2, section 2.4. As was done there, we let the initial wave function be represented by

$$\psi_i'(\boldsymbol{x}) = \sum_n C_n \psi_n(\boldsymbol{x}), \tag{6.1}$$

where $\psi_n(\boldsymbol{x})$ are the possible eigenfunctions of the operator O that is being measured. We also let the initial wave function of the measuring apparatus be represented by $\phi_0(\boldsymbol{y})$ which corresponds to a suitable wave packet (describing, for example, the position of a pointer). The initial wave function of the combined system is then

$$\Psi_i(\boldsymbol{x}, \boldsymbol{y}) = \phi_0(\boldsymbol{y}) \sum_n C_n \psi_n(\boldsymbol{x}). \tag{6.2}$$

As a result of the interactions, this wave function is modified and it becomes

$$\Psi(\boldsymbol{x}, \boldsymbol{y}, t) = \sum_n C_n \psi_n(\boldsymbol{x}) \phi_0(\boldsymbol{y} - \lambda O_n t), \tag{6.3}$$

where the O_n are the eigenvalues of the operator O. After the time Δt when the interaction is over, we obtain

$$\Psi_f = \sum_n C_n \psi_n(\boldsymbol{x}) \phi_0(\boldsymbol{y} - \lambda O_n \Delta t). \tag{6.4}$$

If the interaction is strong enough to satisfy equation (2.5), then the $\phi_n = \phi_0(\boldsymbol{y} - \lambda O_n \Delta t)$ will represent distinct and non-overlapping packets that correspond to the various possible results of the measurement.

It is clear from (6.3) that during the period of interaction, the various components of the wave function will overlap and interfere. The amplitude of the wave function will then become a very complex and rapidly fluctuating function of $\boldsymbol{x}$ and $\boldsymbol{y}$ and, of course, the time as well. The quantum potential will evidently be correspondingly complicated so that the motion of the particle becomes difficult to predict in much the same way as happened in the case of transitions between stationary states discussed in chapter 5. However, when the wave packets have definitely separated, the apparatus particle $\boldsymbol{y}$ must have entered one of them, say m, in which it remains indefinitely, since the probability of being between packets is zero. From then on, the action of the quantum potential (and, of course, the guidance condition) on the particles will be determined only by the packet $\psi_m(\boldsymbol{x})\phi_m(\boldsymbol{y})$, because all the other packets (which do not overlap this one) will not contribute to it. So at least as far as the particles are concerned, we may ignore all the other packets and regard them as constituting inactive or physically ineffective information in the sense discussed in chapter 5. Here it must be emphasised that, as was shown for the transition processes for example, this will still happen even when there is some spatial overlap between $\psi_m(\boldsymbol{x})$ and the remaining packets $\psi_n(\boldsymbol{x})$. This is because of the multidimensional nature of the wave function of the combined system, which implies that the product, $\psi_m(\boldsymbol{x})\phi_m(\boldsymbol{y})$, and any other product, say $\psi_s(\boldsymbol{x})\phi_s(\boldsymbol{y})$, will fail to overlap as long as *one* of its factors fails to overlap, even though the other factor will still have some overlap.

It is clear that measurement has in effect been treated as a particular case of quantum transition even though one of the systems concerned, i.e. the apparatus, is considered to be macroscopic. Indeed the combined system makes a transition from an initial state $\phi_0(\boldsymbol{y})\sum_n C_n\psi_n(\boldsymbol{x})$ to one of a set of final states, $\phi_m(\boldsymbol{y})\psi_m(\boldsymbol{x})$. A simple calculation will show that the probability of this transition is just $|C_n|^2$ (as it would be also in the conventional interpretations). We may indeed use the language of chapter 5 and say that each of the possibilities $\phi_m(\boldsymbol{y})\psi_m(\boldsymbol{x})$ constitutes a kind of channel. During the period of interaction the quantum potential develops a structure of bifurcation points, such that apparatus particles initially on trajectories leading to one side of these points enter, for example, the mth channel, while the others do not. Eventually each particle enters one of the channels to the exclusion of all the others and thereafter stays in this channel. When this has happened the 'observed particle' will behave from then on as if its wave function were just $\psi_m(\boldsymbol{x})$, even if $\psi_m(\boldsymbol{x})$ and the rest of the $\psi_n(\boldsymbol{x})$ should still overlap. The fact that the apparatus particle must enter one of the possible channels and stay there is thus what is behind the possibility of a set of clearly distinct results of a quantum measurement.

Nevertheless, as was brought out in chapter 5, the 'inactive' packets not containing the particles need not at this stage remain permanently inactive. Rather it is still possible, in principle, to bring the apparatus packet, say $\phi_m(\boldsymbol{y})$ and $\phi_n(\boldsymbol{y})$, together so they will interfere. In this case the quantum potential may now be affected by the previously inactive packets and so the production of distinct and separate results of the measurement process is not yet irrevocable.

As shown in chapter 5, the irrevocability of such a process is established only when the apparatus particle interacts with a macroscopic system of particles. In the case of a measurement, this will be the detecting device which amplifies the results to a scale large enough to be directly observed. Such a detecting device will of course contain a very large number, N, of particles with coordinates z_i. When this interacts with the apparatus particle, $\boldsymbol{y}$, its wave function $\lambda(z_1 \ldots z_N)$ will change. To each distinct state, $\boldsymbol{n}$, of the apparatus variable, there will be a corresponding state $\chi_n(\boldsymbol{y}, z_1 \ldots z_N)$ of the combined system of the apparatus and the detector. This wave function will then be

$$\psi = \sum_n C_n \psi_n(\boldsymbol{x}) \chi_n(\boldsymbol{y}, z_1 \ldots z_N).$$

We have assumed that the effect of this interaction on the observed particle is negligible. In general however, some correlation of the $\boldsymbol{y}$ and the set $\{z_i\}$ is to be expected. But within the context under discussion this will have no appreciable effect on the meaning of the experimental results. For most purposes we can therefore simplify the treatment by writing

$$\chi_n(\boldsymbol{y}, z_1 \ldots z_N) = \phi'_n(y) \lambda_n(z_1 \ldots z_N)$$

where $\lambda_n(z_1 \ldots z_N)$ represents the wave function of the detector corresponding to $\phi'_n(\boldsymbol{y})$ for the apparatus variable.

In the above we have allowed scope for the apparatus packet $\phi_n(\boldsymbol{y})$ to change to some other form $\phi'_n(\boldsymbol{y})$ as a result of interaction with the detector. In general, conditions can evidently be arranged so that this change will not produce any overlap between the $\phi'_n(\boldsymbol{y})$ corresponding to different values of $\boldsymbol{n}$. Even if it were possible, either naturally or by some special experimental arrangement for these to come to overlap, any given λ'_n will not overlap with the other λ_n in the configuration space of the detecting device. Therefore such overlap of the ϕ'_n will not alter the quantum potential acting on the $\boldsymbol{x}$- and $\boldsymbol{y}$-particles. This is because the particles of the detecting device will now be in distinct channels in the overall configuration space of the detector. Moreover since these channels will be so complex and multiple in nature, it will not be possible to arrange them to overlap in the future, while the probability for this to happen spontaneously will be negligible. The channels corresponding to distinct possible results of a measurement are in this way established essentially irrevocably.

The above treatment discusses measurement in terms of von Neumann's analysis in which one introduces an interaction between measuring apparatus and observed system that correlates the value of the observed result directly to the position of a suitable apparatus variable. However, there is a wide class of measurement in which the observed result is, in the first instant, correlated to the *momentum* of such an apparatus variable. With the passage of time the resulting motion of the apparatus variable then leads to a separation of the apparatus packets corresponding to different values of the final state of the observed system.

A typical example of this sort of measurement is the Stern-Gerlach experiment. Suppose we have a beam of atoms with coordinate $\boldsymbol{y}$, each containing an electron with coordinate $\boldsymbol{x}$. Let the orbital angular momentum of the atoms be $\hbar$. Relative to any convenient z-axis there are three possible quantum fields $\psi_j(\boldsymbol{x})$ where $j = 0, 1, -1$ corresponding to the different values of the z-component of the angular momentum. The first stage of the measurement is achieved by letting the atoms pass through an inhomogeneous magnetic field in the z-direction. Before the atoms reach this field, the wave function of the combined system is

$$\Psi_i = \phi_0(\boldsymbol{y}) \sum_j C_j \psi_j(\boldsymbol{x}), \tag{6.5}$$

where $\phi_0(\boldsymbol{y})$ represents the initial packet corresponding to the location of the atom.

While the atoms are passing through the magnetic field, the interaction Hamiltonian is going to be approximately

$$H_I = \left(\frac{\partial B_z}{\partial z}\right)_0 L_z \tag{6.6}$$

where $(\partial B/\partial z)_0$ is the value of the gradient of the magnetic field at the centre of the beam and L_z is the z-component of the angular momentum operator. Recalling that the interaction is impulsive so that all changes of the wave function due to the rest of the Hamiltonian can be neglected during the interaction, we obtain for the wave function immediately after passing through the inhomogeneous magnetic field,

$$\Psi_{f_1} = \sum_j C_j \psi_j(\boldsymbol{x}) \phi_0(\boldsymbol{y}) e^{ij(\partial B/\partial Z)_0 \Delta t \boldsymbol{y}} \tag{6.7}$$

where Δt is the time spent in this field. Thus the apparatus packet obtains a momentum proportional to the angular momentum of the atom. To make a measurement possible, the momentum transferred to the atom, $(\partial B/\partial z)_0 \Delta t$ must evidently be much greater than the width δp of the packet.

After this interaction takes place, the packets will begin to separate. With the approximations made here, the spread of the packet can be neglected. We then obtain, after a suitable time t,

$$\Psi_{f_2}(x, y, t) = \sum_j C_j \psi_j(x) e^{ijp_0 y} \phi_0 \left(y - \frac{jp_0}{m} t \right), \tag{6.8}$$

where $p_0 = (\partial B/\partial z)_0 \, \Delta t$. Clearly the three packets will eventually separate after which they may be detected, for example, at a screen.

From here on the treatment goes through in essentially the same way as with the von Neumann type of measurement interaction. Thus, while the packets are separating but still overlap, there will be a complicated and rapidly varying quantum potential containing bifurcation points of more or less the same kind that has been discussed in earlier examples. According to the initial position of the x- and y-particles, the system will enter one of these packets and, as explained earlier, this will determine a channel in which it remains from then on. The other two channels will then have no effect on the quantum potential (or on the guidance conditions) acting on the particle. They can therefore be ignored. However, as we have already pointed out in several contexts, it is still possible for the atomic quantum fields $\phi_j(y)$ to be made to overlap, for example, with the aid of suitable magnetic fields. The corresponding channels are thus not yet irrevocably determined.

The next stage is to allow the atoms to enter a detecting device. For example, this may be done by having each beam enter an ion chamber. As the atom goes through the chamber it will ionise molecules of the gas that are present there. Let the relevant coordinates of these molecules be denoted by $z_1^n \dots z_N^n$ (where z_s^n represents the s molecule in the nth chamber). If an electron is liberated, it will be accelerated by the applied field of the ion chamber as shown in figure 6.1. This will initiate a cascade giving rise ultimately to a pulse containing so many electrons that it can be observed by ordinary means (e.g. a perceptible deflection of a galvanometer).

To discuss the behaviour of the wave function Ψ of the total system, let $\chi_0 = \prod_n \lambda_0^n(z_1^n \dots z_N^n)$ represent the state in which all the counters are unfired and let $\chi_j^n = \lambda_j^n(z_1^n \dots z_N^n)$ represent the state in which the nth counter has fired and the others have not. The total wave function just before the particle enters a counter is

$$\chi_0(z_1 \dots z_N) \sum_j C_j \psi_j(x) \phi_j'(y),$$

where

$$\phi_j'(y) = \phi_0 \left(y - \frac{jp_0}{m} t \right) \exp(ijp_0 y).$$

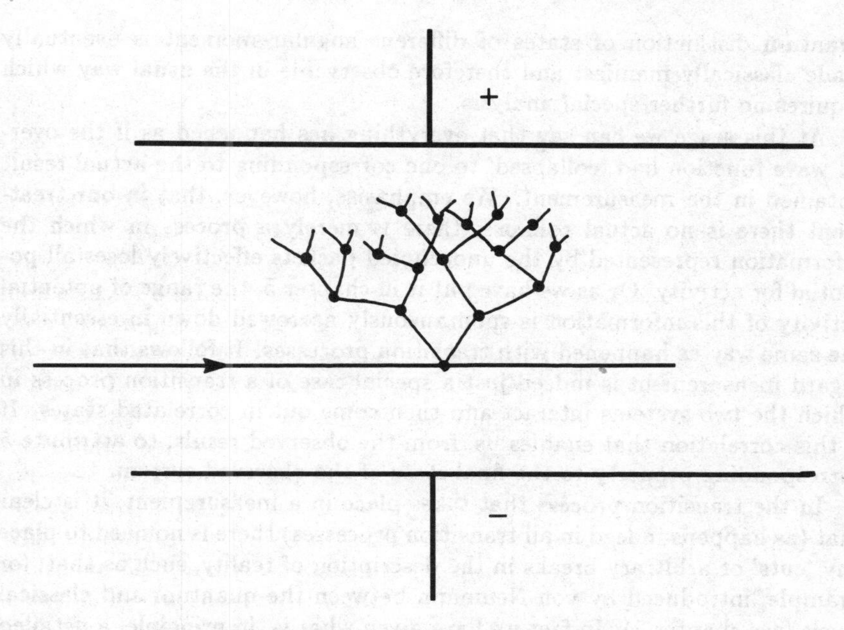

Figure 6.1: Discharging ion chamber

After one of the counters has functioned this becomes

$$\sum_j C_j \psi_j(\boldsymbol{x}) \phi'_j(\boldsymbol{y}) \chi^j_f(z^j_1 \ldots z^j_N).$$

It is clear that in order to obtain interference of the various states of the $\boldsymbol{x}$-particle, it will be necessary that the quantum fields of the total system of $\boldsymbol{z}$-particles in the fired state overlap those in the unfired state; even while the cascade is developing, it is evident that there will be no such overlap. And after all the cascade particles have reached the electrodes and flowed along the wire to a very different place, the possibility of establishing an overlap becomes even less likely than during the cascade.

Along the lines that we have described above, we can see that it is in principle possible to follow the measurement process from the original atom, through the cascade, and to show how what was an essentially quantum mechanical process gradually turns into a directly observable pulse. We may here anticipate the results of chapter 8, where it is shown that large scale movements such as that of the pulse which reaches the electrodes of the ion chamber can be described by collective coordinates for which the WKB approximation is valid. This is what justifies us in saying a quantum distinction of states of different angular moment is eventually made classically manifest and therefore observable in the usual way which requires no further special analysis.

At this stage we can say that everything has happened as if the overall wave function had 'collapsed' to one corresponding to the actual result obtained in the measurement. We emphasise, however, that in our treatment there is no actual collapse; there is merely a process in which the information represented by the unoccupied packets effectively loses all potential for activity. Or as we have put it in chapter 5, the range of potential activity of this information is spontaneously narrowed down in essentially the same way as happened with transition processes. It follows that in this regard measurement is indeed just a special case of a transition process in which the two systems interact and then come out in correlated states. It is this correlation that enables us, from the observed result, to attribute a corresponding property to the final state of the observed system.

In the transition process that takes place in a measurement, it is clear that (as happens indeed in all transition processes) there is no need to place any 'cuts' or arbitrary breaks in the description of reality, such as that, for example, introduced by von Neumann between the quantum and classical levels (see chapter 2). In fact we have given what is, in principle, a detailed account of the whole process of measurement which is not restricted to the phenomena alone as Bohr requires (while nevertheless retaining a wholeness similar to that of Bohr in certain ways which will be discussed in more detail later).

6.2 Loss of information in the unoccupied wave packets

For those who are accustomed to the conventional interpretation of the quantum theory, the notion of empty wave packets arising naturally in transition processes and artificially in measurement processes may not seem to be acceptable, even though we have shown that they constitute inactive information that will never have any subsequent effect on the particles.

This objection probably arises mainly because the conventional interpretation assumes either tacitly or explicitly that the wave function corresponds to the actual state of the system and as such constitutes the basic description of reality, at least in so far as this latter can be described in any way at all. Therefore it could be felt that the empty packets, which also satisfy Schrödinger's equation, constitute a vast mass of 'bits of reality' that are, as it were, 'floating around' interpenetrating that part of reality which corresponds to the occupied packets.

In our interpretation, however, we do not assume that the basic reality is thus described primarily by the wave function. Rather, as we have stated in chapter 3, we begin with the assumption that there are *particles* following definite trajectories (and as we shall see in chapter 11 we can also incorporate ordinary fields such as the electromagnetic in this general approach). We then assume that the wave function, ψ, describes a qualitatively new kind of quantum field which determines the guidance conditions and the quantum potential acting on the particle. We are not denying the reality of this field, but we are saying that its significance is relatively subtle in the sense that it contains active information that 'guides' the particle in its self-movement under its own energy. The behaviour of the particles is profoundly affected by this field so that it has fundamentally new features relative to those of classical physics. The effects of the quantum field will, however, become manifest only at the classical level in the collective movement of large numbers of particles. In the latter, the WKB approximation will generally hold and the quantum fields can be neglected. So ultimately all manifestations of the quantum fields are through the particles.

We may make an analogy here to human relationships in society. The most immediate and concrete reality is the collection of individual human beings. In so far as these are related by pools of information, this latter will become manifest in the behaviour of the human beings. The behaviour both of the individual and of the society depend crucially on this information (rather as happens with the particles of physics). The information itself is held at some very subtle level which does not show directly and which has negligible energy compared with that involved in the physical movements of the people. To complete the analogy we might surmise that perhaps the information in the wave function is likewise contained at a more subtle level of negligible energy in a way that has not shown yet and that we have not thus far been able to study.

In this analogy we can see that information is constantly losing its potential for activity. For example, people may see that some of it is irrelevant or wrong, while a great deal may simply be forgotten. Even if the information is preserved in books or on discs, most of these latter eventually become more and more difficult to access, for example, by being sent to depositories and ultimately by being shredded. Other information is simply lost by dispersal, for example, by the spread of sound waves and radio waves. In so far as this information can no longer affect human beings (or their proxies in the form of computers) it has lost its potential for activity.

It is clear then that nothing is more common in ordinary experience than for information to lose its potential for activity. If we accept that the particle is what responds to information at the quantum mechanical level, it is clear that there is no serious problem with the empty wave packets. The essential point is that information is immensely more subtle and less substantial than the systems on which it acts. If we give up the idea that the wave function is what comes closest to describing that which may be called the substance of reality, then the loss of potential for activity will be quite a natural and, indeed, even an expected feature.

Another reason why many may have found it difficult to accept our ontological interpretation is that the wave function is all that is used in statistical calculations, which indeed makes no mention of the particles, except to say that the statistics yield the frequency with which they may be registered in specified experiments. Since particles seem to play no part in the calculations, and since the latter are what occupy the major part of the attention of physicists, it may be felt that the wave function (which corresponds to the quantum state) is in essence what the theory is all about. Therefore to have inactive parts of the wave function may seem unacceptable.

However, if we think further about our analogy to the behaviour of human beings we may note, for example, that in the statistical behaviour of people moving on roads with signs that inform them, their behaviour could in certain cases be calculated from the information implied by the signs without the need for a detailed description of the movement and activity of the human being as a whole, e.g. including his muscles, his senses, his brain and nervous system etc. Similarly it need not be surprising to discover that the statistical behaviour of the particles of physics can be calculated from the information in the wave function without knowledge of the detailed constitution of the particles themselves.

In both cases the fact that a quantity provides a complete basis for statistical calculations does not imply that it also provides a complete basis for an ontology.

6.3 Quantum properties not generally attributable to the observed system alone

The very notion of a measurement implies, as has been explained earlier, that we are using the macroscopic observable results of the operation of a measuring instrument to attribute some property to the object itself. However it can easily be seen that in what is called a quantum measurement, such an attribution is not in general possible. For during the interaction between object and measuring instrument, there is the linear combination (6.4) of wave functions. It follows that through this period, Δt, the two systems are 'guided' by a common pool of information implying a quantum potential that connects them in a nonlocal way. As pointed out in section 6.2, the orbits of both particles are then strongly correlated and yet highly unstable. Given that we always start with a statistical ensemble $P = |\psi|^2$ which fluctuates at random from one experiment to the next, there will evidently be no way to predict or control in each individual case which channel the particle will enter (even though this motion is determinate in itself). When the interaction is over, the channels will be separate and the information in the unoccupied channel will, as we have already seen, become inactive, leaving only the occupied channel with its quantum potential. The wave function, and therefore the effective quantum potential, will then be profoundly altered (e.g. from a z-component of angular momentum zero to a z-component of $\hbar$). The apparatus particles will also have been correspondingly altered (e.g. one of the counters going from the unfired to fired state). It is clear then that we are not 'measuring' a state that has already been in existence. Rather the apparatus and the observed system have *participated* in each other, and in this process they have deeply affected each other. After the interaction is over we are left, as already pointed out in the previous section, with a situation in which the states of the two are correlated, in accordance with the channel that the particles have actually entered.

In certain ways the above behaviour is strongly reminiscent of Heisenberg's notion of quantum properties as potentialities belonging both to the observed system and the observing apparatus, rather than being just intrinsic properties of the observed system alone. Which potentiality is actualised depends on the observing apparatus as well as on the observed system. For example if the Stern-Gerlach apparatus is oriented in the z-direction, we actualise the potentiality of the atom to come into a state of well-defined angular momentum in this direction. On the other hand if the apparatus is oriented in the x-direction we actualise a different set of potentialities for the angular momentum now to be well defined in the x-direction. Evidently the simultaneous actualisation of the two sets of potentialities is mathematically impossible and this corresponds to the physical impossibility of simultaneously orienting the Stern-Gerlach apparatus in both of these directions.

One can here make use of the analogy of the seed discussed in chapter 2, section 2.3. The different possible orientations of the apparatus correspond to different environments of the seed, which in turn lead to different forms of the plant. In the case of particles, we can see how changing the orientation of the apparatus changes the quantum potential and how this in turn changes the possible states in which the system can be actualised. In this way we obtain an explanation of Bohr's principle of complementarity, which states that the experimental conditions needed for measuring a given physical quantity are incompatible with those needed to measure another one, when the operators corresponding to these quantities do not commute.

What this all means is that quantum properties cannot be said to belong to the observed system alone and, more generally, that such properties have no meaning apart from the total context which is relevant in any particular situation. (In this case, this includes the overall experimental arrangement so that we can say that measurement is context dependent.) The above is indeed a consequence of the fact that quantum processes are irreducibly participatory in the way we have explained here.

Even in classical physics, measurement may strongly disturb the observed system and introduce large changes in its conditions. Moreover it is possible to have situations in which this disturbance is unpredictable and uncontrollable, for example, if the interactions are such as to bring about chaotically unstable motions. Since the initial conditions of the observing apparatus can never be determined exactly we may have an unpredictable and uncontrollable disturbance of a significant magnitude even in these classical situations. Therefore we can evidently have classical conditions in which faithful measurements are not possible because there is no way to correct for, or otherwise take into account, the effects of the disturbance arising in the measurement process.

However, at the quantum level something much more radical is involved in this. For the interactions which now depend on the quantum potential can introduce large nonlocal connections between all the constituents of the total system which are not preassigned functions of the properties of this system. This is as if the basic law of interaction of all the parts were changing in the process in which the observed system and observing apparatus come into contact. The result is therefore, in effect, a transformation of the very nature of the system as a whole and of all its constituents. In the quantum case, the participation therefore is of a deeper and more fundamental nature than it is classically. Indeed in classical physics no matter how much the observed particle is disturbed fundamentally, its basic laws of interaction remain the same (being expressible, for example, as preassigned functions of all the particle variables). Therefore the basic properties of the particles may be regarded as intrinsic, no matter how strong the interaction may be. But in the quantum theory such properties can, as we have already pointed out, have meaning only in the total relevant context.

The nearest classical analogy to the above situation arises in connection with the free energy of a thermodynamic system

$$F = E - TS.$$

In the above, E is the intrinsic energy of the system, but the actual energy available in an isothermal process is $\delta F = \delta E - T\delta S$, where δS is the charge of entropy. But $T\delta S$ is equal to the heat which flows into the bath in an isothermal process. The free energy is therefore not an intrinsic property of the system, but one that has meaning only in the context of a thermal bath with which the system interacts. The analogy to the quantum theory is clear, but it must be emphasised that the context dependence of properties in quantum mechanics is much more radical and far reaching than it is in thermodynamics.

However, apart from the field of thermodynamics, the notion of properties that are not intrinsic but are inherently dependent on a total context has actually been suggested to play a fundamental role in cosmology through the Mach principle. This states that basic properties of a particle, such as mass, depend on some form of collective feature of the general distribution of matter which includes indefinitely far off systems. And as we have seen, the quantum properties imply something similar, but far more widespread and pervasive, in the sense that measured properties are not intrinsic but are inseparably related to the apparatus. It follows then that the customary language that attributes the results of measurements, e.g. momentum and angular momentum, to the observed system alone can cause confusion, unless it is understood that these properties are actually dependent on the total relevant context.

For example, the actual momentum of the particle is given by $\boldsymbol{p} = \nabla S$, but this evidently depends on the wave function. And as we have seen earlier, the latter may either correspond to a pool of information common to many particles or may depend on distant features of the environment such as a slit system. It is clear then that, as we shall bring out in more detail presently, the momentum of the nth particle $p_{in} = \partial S/\partial x_{in}$ is not an intrinsic property of that particle alone, but can only be understood in relationship to other particles and to all the relevant features of the environment.

In this interpretation there is one property that is intrinsic and not inherently dependent in this way on the overall context. This is the particle position, $\boldsymbol{x}$. It is evidently intrinsic because we have defined it in a way that is conceptually independent of the wave function. Moreover it is intrinsic also in the sense that it can be measured without being altered. This can be done, for example, by introducing an interaction between the particle and the observing apparatus in which the interaction Hamiltonian is a function that is proportional to the position, x_i, itself. In equation (2.3), we may therefore write

$$\boldsymbol{O} = \boldsymbol{x}.$$

Because $\boldsymbol{x}$ is a continuous variable, we can write the initial wave function as

$$\Psi_i = \int C(\xi)\delta(\boldsymbol{x} - \xi)\phi_0(\boldsymbol{y})\,d\xi. \qquad (6.9)$$

The final wave function is

$$\Psi_f = \int C(\xi)\delta(x - \xi)\phi_0(y - \lambda_1 x\Delta t)\,d\xi. \tag{6.10}$$

If δy is the width of the packet ϕ_0, this measurement will make possible an observable distinction between values of ξ differing by $\delta x = \delta y/\lambda_1\Delta t$. Recalling that this is an impulsive measurement, the change of x resulting from the original momentum possessed by the particle can be neglected. Moreover the wave equation during the interaction is

$$\frac{\partial\psi}{\partial t} = \lambda_1 x\frac{\partial\psi}{\partial y} \tag{6.11}$$

implying that the only change is to shift $\phi_0(y)$ to $\phi_0(y - \lambda_1 x\Delta t)$. It may be assumed with sufficient generality that ϕ_0 is real so that this term will not introduce any change in the momentum of the x-particle, $p_i = \partial S/\partial x_i$. It follows that the position of the x-particle does not change in this measurement. It is therefore possible to attribute this position variable to the particle itself.

However, while the position has not been changed in the measurement, the wave function has in general been changed because the interaction with the measuring apparatus can bring about a new wave function that can be much more localised in x-space than the original one was. Since all other properties of the electron depend as much on the wave function as they do on the position of the particle, the measurement is still participating in an essential way. Therefore even though the particle position has been determined 'faithfully', that is without changing it, this is not true of the other properties. With regard to the latter we still cannot in general talk of a measurement as something which simply establishes what the state of the observed system is. Therefore while the position of the particle considered as an abstract concept is an intrinsic property, the other properties are, in general, still context dependent.

If we try to measure properties other than position, we find that, as we have already pointed out, the result is affected by the process of interaction in a way that depends, not only on the total wave function, but also on the details of the initial conditions of both the particle and the apparatus. To illustrate this we consider a measurement of the momentum p. In equation (2.3) we introduce the operator $O = p$. We then write the initial wave function as

$$\Psi_i = \int C(p)e^{ipx}\,dp\,\phi_0(y). \tag{6.12}$$

The final wave function is

$$\Psi_f = \int C(p)e^{ipx}\phi_0(y - \lambda_2 p\Delta t)\,dp. \tag{6.13}$$

This measurement will make possible an observable distinction between values of $\boldsymbol{p}$ differing by $\delta \boldsymbol{p} = \Delta \boldsymbol{y}/\lambda_2 \Delta t$. In effect this process turns the original wave function into wave packets of width in momentum space $\delta \boldsymbol{p}$ and in position space $\delta \boldsymbol{x} = 1/\delta \boldsymbol{p} = \lambda_2 \Delta t/\Delta \boldsymbol{y}$. These wave packets do not interfere because they correspond to non-overlapping states of the apparatus variable.

It is especially clear in this example how a measurement transforms the whole system in a radical way. To illustrate this in more detail, suppose we originally had a narrow wave packet in $\boldsymbol{x}$-space of width $\delta \boldsymbol{x}_0$. After interacting with the momentum measuring device we would have a set of wave packets of width $\delta \boldsymbol{x} = 1/\delta \boldsymbol{p}$, where $\delta \boldsymbol{p}$ is the accuracy of the momentum measurement. If $\delta \boldsymbol{p}$ is very small it is clear that the effect of the measurement is to cause the original packet to spread a great deal in position space, in the same operation in which it narrows in momentum space. On the other hand if we had originally a broad packet in $\boldsymbol{x}$-space and then measured $\boldsymbol{x}$ to a great accuracy, the effect would be, of course, to cause the packet to become very narrow in position space, while it spread out over momentum space.

To bring out yet further how the actual particle momentum is generally changed in a momentum measurement, let us consider an initial state in which $C(p)$ is chosen so that the overall wave function is real (i.e. $C(\boldsymbol{p}) = C^*(-\boldsymbol{p})$). The original particle momentum is $\boldsymbol{p} = \nabla S = 0$. After the measurement the particle will end up in one of the packets corresponding to the mean value of the momentum, $\boldsymbol{p}_m$. The momentum has thus changed from zero to something close to $\boldsymbol{p}_m$.

In the ordinary quantum mechanical language of the conventional interpretation, the momentum operator commutes with the Hamiltonian for the measurement process, so that one would have expected it to be a constant of the motion. How then is a change in this momentum possible? The answer is that, in our interpretation, the momentum operator is definitely related to the particle momentum only when the wave function is an eigenfunction, $\exp[i\boldsymbol{p} \cdot \boldsymbol{x}]$, of this operator. More generally, as has already been indicated in chapter 3, the actual particle momentum and the corresponding operator are not necessarily so closely connected. In such a measurement the interaction can be described by replacing Δt by t in equation (6.13). The wave function is then

$$\Psi_f = \int C(\boldsymbol{p}) \exp[i\boldsymbol{p} \cdot \boldsymbol{x}] \phi_0(\boldsymbol{y} - \lambda_2 \boldsymbol{p} t) \, \mathrm{d}\boldsymbol{p}. \tag{6.14}$$

The coefficient of $\exp[i\boldsymbol{p} \cdot \boldsymbol{x}]$ becomes $C(\boldsymbol{p})\phi_0(\boldsymbol{y} - \lambda_2 \boldsymbol{p} t)$, with $C(\boldsymbol{p}) = C^*(-\boldsymbol{p})$.

The reality condition is no longer satisfied because

$$\begin{aligned}[C(\boldsymbol{p})\phi_0(\boldsymbol{y}-\lambda_2\boldsymbol{p}t)]^* &= C^*(\boldsymbol{p})\phi_0(\boldsymbol{y}-\lambda_2\boldsymbol{p}t) \\ &\neq C(-\boldsymbol{p})\phi_0(\boldsymbol{y}+\lambda_2\boldsymbol{p}t),\end{aligned} \tag{6.15}$$

therefore $\boldsymbol{p} = \nabla S$ is no longer zero. As t changes, the phase will change and this implies that the particle is steadily changing momentum, at least until the wave packets cease to overlap, when its momentum will be some constant $\boldsymbol{p}_f$ (according to what packet the apparatus particle actually enters). If we were to work out the quantum potential, we would see that this would explain the change of momentum that takes place during the process of interaction.

To bring this out in yet another way, let us consider the example of a one-dimensional particle in a box of length L. A particular initial wave function of the system may be

$$\begin{aligned}\Psi_i &= A\sin\frac{n\pi x}{L}\phi_0(y) \\ &= \frac{A}{2i}\left[\exp\left[i\frac{n\pi x}{L}\right] - \exp\left[-i\frac{n\pi x}{L}\right]\right]\phi_0(y).\end{aligned} \tag{6.16}$$

The initial momentum of the x-particle is evidently zero. The quantum potential is

$$\begin{aligned}Q &= -\frac{1}{2m}\frac{\partial^2/\partial x^2\sin(n\pi x/L)}{\sin(n\pi x/L)} \\ &= \left(\frac{n\pi}{L}\right)^2\frac{1}{2m}.\end{aligned} \tag{6.17}$$

As pointed out in chapter 3, section 3.5, the quantum potential now contains all the energy which, in the conventional interpretation, is attributed to the kinetic energy of the particle. In the conventional interpretation, the latter is supposed to be moving back and forth with equal probability of being in either direction. Yet if this were actually the case, we would not understand why a particle with constant speed, p/m, could cross the regions where the probability is zero. Evidently we should not take this picture too literally, but it is better to think of it as a figure of speech that people have become accustomed to use. On the other hand in our interpretation we have a consistent picture because the particle is at rest and there is no force on it because the quantum potential is constant.

After interacting with the measuring apparatus, this wave function splits into two non-overlapping parts, which may be written as

$$\Psi_f = \frac{A}{2i}\exp[ipx]\phi_0(y-\lambda_2 p\Delta t)$$

$$+\frac{A}{2i}\exp[-ipx]\phi_0(y+\lambda_2 p\Delta t). \qquad (6.18)$$

But during the interaction there is a period throughout which these two parts overlap. In this period, the quantum potential will not be a constant and we see that the particle will be accelerated. Ultimately we end up with the particle in one of these packets with equal probabilities for both possibilities as in the conventional interpretation.

However, there is a key difference in the language that we use to describe this process. In the conventional interpretation it is stated that we have measured the momentum without changing it, but in our interpretation, the particle's momentum has been changed. This emphasises that, in a measurement process, we do not in general find an intrinsic property of the observed system that was already there. Rather we have an irreducible participation of each system in the other. Therefore the momentum is not, as we have already said, an intrinsic property. This will be true for all properties other than the position.

We can bring out this character of the momentum further by considering the formulae for the mean values of p and of p^2. These are

$$\begin{aligned}\bar{p} &= \frac{\hbar}{i}\int \psi^*(x)\frac{\partial}{\partial x}\psi(x)\,\mathrm{d}x \\ &= \int \psi^*(x)\psi(x)\frac{\partial S}{\partial x}\,\mathrm{d}x \\ &= \int P(x)\frac{\partial S}{\partial x}\,\mathrm{d}x\end{aligned}$$

and

$$\begin{aligned}\bar{p^2} &= -\hbar^2\int \psi^*(x)\frac{\partial^2}{\partial x^2}\psi(x)\,\mathrm{d}x \\ &= \hbar^2\int \frac{\partial\psi^*(x)}{\partial x}\frac{\partial\psi(x)}{\partial x}\,\mathrm{d}x \\ &= \int P(x)\left[\left(\frac{\partial S}{\partial x}\right)^2+\frac{(\partial R/\partial x)^2}{R^2}\right]\mathrm{d}x.\end{aligned}$$

$\bar{p}$ is obtained simply by averaging the momentum of the particle $\partial S/\partial x$ over the probability distribution $P(x)$. But to obtain $\bar{p^2}$ it is not enough to consider the average of $(\partial S/\partial x)^2$. We have also to add a term $(\partial R/\partial x)^2/R^2$. This shows that the average value of an operator as computed quantum mechanically cannot in general be obtained by averaging the corresponding physical property of the particle over the position distribution. We understand this as a consequence of the fact that the momentum is not an intrinsic property, but rather is inherently context dependent so that it involves the participation of the measuring apparatus and the quantum field as a whole.

Quite generally all powers of the momentum operator higher than the first are to be understood in a similar way. (Corresponding results can be obtained for the angular momentum operator, $\hbar/i\ \partial/\partial\phi$.)

6.4 The meaning of the uncertainty principle

Since our interpretation gives the same statistical distribution of experimental results as does the conventional one, it follows that the proofs usually given for the uncertainty principle will also be valid in our approach. The meaning of the uncertainty principle is, of course, that there is a limit to the precision with which properties can be attributed to the observed object on the basis of observed experimental results. Or as Bohr [1] has put it, there is an irreducible ambiguity in the meaning of these results as referring to intrinsic properties of the observed object.

Because of this ambiguity, it becomes impossible to know from a measurement exactly what the state of the particle is and, of course, this implies that we do not know exactly how it is going to behave after the measurement. In general, the ambiguities of the actual particles' positions and momenta will be of the order of those present in the spread of the wave packets (as we have explained in more detail in chapter 3). But in particular cases this may not be so. For example, in a stationary state, the ambiguity in the momentum of the particle $\boldsymbol{p} = \nabla S$ was seen to be zero. So if $\boldsymbol{p}$ represents the *actual* momentum of a particle as it moves (or is at rest) in its trajectory, Heisenberg's principle can in certain cases be violated. In our interpretation, however, what Heisenberg's principle refers to is not the actual momentum of the particle itself, but the value of the momentum that can be attributed to the particle after what is commonly called a measurement of the momentum. And because measurements are actually participatory, these two can differ. Indeed it is misleading and even confusing to describe what happens as a measurement (rather as if in the observation of the mature plant, we were said to be 'measuring' the properties of the seed). Perhaps as Bell [2] has suggested, we should call the whole process an experiment or even a participatory experiment revealing the potentialities of the system in question. (Indeed this participatory feature has already been noted by Wheeler [3] in a different context in which, however, there is no way to explain how it comes about.)

Another way of looking at the participatory nature of quantum experiments in which something is said to be 'measured' can be seen by considering equations (6.2) and (6.3). These describe a process in which the wave function is eventually transformed into a set of eigenfunctions of the operator that is being 'measured', and each eigenfunction is multiplied by a corresponding apparatus wave function which does not overlap the others. The function actually containing the particles (of both the apparatus and the observed system) then selects what will be the effective wave function realising one of the potentialities of the overall experimental arrangement, while the empty channels can be neglected.

What we learn from the measurement is not only the value of the operator but also the final effective wave function of the observed system. From this we can locate the observed particle somewhere in the region in which the wave function of the particle is appreciable. The role of the particle is crucial here and it makes it possible for the system to select one of the eigenfunctions which will be effective without the need for a collapse.

If in the quantum domain experiments are thus mutually participatory transformations how do we understand the ordinary classical experience in which we definitely obtain measured results that can be unambiguously attributed to the observed system without any question of mutual participation? The answer is, of course, that for measurements carried out in the classical limit such participation is negligible, and the transformation of the wave function is unimportant because the quantum potential is negligible.

We may bring out this point in more detail by considering a system with wave function $\psi(x)$ whose Fourier coefficient is $\phi(p)$. Let Δx be the width of the wave packet in x-space and Δp be its width in p-space. We can then show that in measurements of x and p carried out to respective accuracies δx and δp, the change of the wave function will be negligible when $\delta x \gg \Delta x$ and $\delta p \gg \Delta p$. In this case the participatory character of the interaction will not be significant so that the net result is just to provide information (of limited accuracy) about the system.

Along the lines leading to equation (6.3) we can say that in a measurement of x, an initial wave function

$$\psi_i(x) = \int \psi(\xi)\delta(x-\xi)\,\mathrm{d}\xi \tag{6.19}$$

goes over into

$$\psi_f(x,y) = \int \psi(\xi)\delta(x-\xi)\phi_0(y-\lambda_1 x\Delta t)\,\mathrm{d}\xi, \tag{6.20}$$

with the accuracy of measurement given by $\delta x = \delta y/\lambda_1\Delta t$ where δy is the width of the apparatus wave packet. If we choose $\delta x \gg \Delta x$, the change of $\phi_0(y-\lambda_1 x\Delta t)$ over the region in which $\psi(\xi)$ is appreciable can be neglected and so we obtain to a good approximation

$$\psi_f = \int \psi(\xi)\delta(\boldsymbol{x} - \xi)\phi_0(\boldsymbol{y} - \lambda_1\bar{\boldsymbol{x}}\Delta t)\,\mathrm{d}\xi, \qquad (6.21)$$

where $\bar{\boldsymbol{x}}$ is the value of $\boldsymbol{x}$ at the centre of the particle's wave packet. This gives

$$\psi_f = \psi_i\phi_0(\boldsymbol{y} - \lambda_1\bar{\boldsymbol{x}}\Delta t). \qquad (6.22)$$

If the value of $\boldsymbol{y}$ is observed as $\boldsymbol{y}_0$, the value $\bar{\boldsymbol{x}}$ can be inferred as $\bar{\boldsymbol{x}} = \boldsymbol{y}_0/\lambda_1\Delta t$. Thus by observing the value of $\boldsymbol{y}$ we are able to measure $\bar{\boldsymbol{x}}$ to an accuracy $\delta\boldsymbol{x} = \delta\boldsymbol{y}/\lambda_1\Delta t$ without making any significant changes in the observed system.

A similar argument can be made for the measurement of momentum, showing that it is also possible to measure the latter without significant participation provided that $\delta\boldsymbol{p} \gg \Delta\boldsymbol{p}$. If we then consider the two results together, we obtain

$$\delta\boldsymbol{x}\delta\boldsymbol{p} \gg \Delta\boldsymbol{x}\Delta\boldsymbol{p} \geq \hbar.$$

We conclude then that the ordinary idea of non-participatory measurement holds only when the measurements are much less accurate than the limits set by the uncertainty principle. But the conditions for this are the same as those for the classical limit. So it follows from our interpretation that in this limit we can ignore the participatory implications of the quantum theory in discussing the meaning of measurements.

6.5 On proofs of the impossibility of hidden variables in the quantum theory

Historically the question of an ontological interpretation of the quantum theory was first put in the form of asking whether there could be 'hidden variables'. These, along with the wave functions, would determine the results of individual experiments and when treated as a statistical ensemble, they would also explain the probability distribution of the corresponding ensemble of experiments. (A detailed discussion of the work up to 1973 can be found in Belinfante [4].)

Rather early in the development of the theory, work along these lines was strongly discouraged by a theorem of von Neumann [5] claiming to show that no distribution of hidden variables could account for the statistical predictions of the quantum theory. We shall not reproduce his argument in detail here, but we shall merely focus on what we regard as the essential point at issue.

According to the uncertainty principle, the quantum theory implies that there is no state in which the values of non-commuting operators are well defined. We can ask however whether there are not some additional 'hidden' parameters that would select sub-ensembles in which such operators could be defined within limits that are narrower than the uncertainty principle. An extreme case would be what von Neumann called a dispersionless ensemble in which these operators would all have definite values.

Von Neumann [5] gave an argument aiming to prove that if there are such dispersionless ensembles then quantum mechanics must be 'objectively false'. A basic assumption underlying his proof is that if A and B are any two operators there is an operator $A+B$ such that

$$\langle A\rangle + \langle B\rangle = \langle A+B\rangle, \tag{6.23}$$

where $\langle A\rangle$ etc. denote the expectation values of the operator. This relationship does in fact hold for the ensembles treated in the quantum theory. But von Neumann assumes that it also holds for sub-ensembles. In particular it would have to hold for a dispersion-free ensemble in which the values $V(A), V(B), \ldots$ of the quantities associated with the operators $A, B, \ldots$ were well defined and therefore equal to the corresponding eigenvalues of these operators.

It is easy to see that this requirement cannot be satisfied. We shall discuss this question with the aid of an example given by Bell [6], i.e. a particle of angular momentum with the eigenvalues of the components L_x, L_y, L_z evidently restricted to $\hbar, 0, -\hbar$. We begin with the two operators $A = L_x$ and $B = L_y$, then the angular momentum L' at an angle of 45° between these directions is

$$L' = \frac{1}{\sqrt{2}}(L_x + L_y). \tag{6.24}$$

In the dispersion-free case this becomes

$$V(L') = \frac{1}{\sqrt{2}}\left[V(L_x) + V(L_y)\right].$$

But there is no way to satisfy this if $V(L') = \pm\hbar$ and so one might conclude with von Neumann that there can be no dispersion-free state.

Bell has given an answer to this. As is well known (and as von Neumann agrees) there is really no meaning to combining the results of non-commuting operators such as L_x, L_y and L'. For example to measure L_x we need a Stern-Gerlach magnet with a field in the x-direction, while L_y and L' would require fields in their corresponding directions. These measurements are incompatible and mutually exclusive. Therefore we cannot arrive at an inconsistency in this way. However, the interesting point is that in spite of this incompatibility, equation (6.23) is still true for quantum mechanical averages over the three separate series of measurements of L_x, L_y and L' respectively. As we shall make clear later, the essential point is that von Neumann had in mind hidden parameters that belonged only to the observed system itself and were not affected by the apparatus. The three different pieces of apparatuses discussed above constitute three different and incompatible contexts. Therefore for an individual experiment there is no reason why L_x, L_y and L' should be related in the way that operators are related. However, as we have already remarked, equation (6.23) is still valid for averages over statistical ensembles of experiments even though it has nothing to do with individual experiments.

The above criticism of von Neumann applies to non-commuting operators. But if the operators commute then it might seem that we can reasonably expect that even for individual cases the values of operators will have the same functional relationships as the operators themselves. There has been a continuing effort to improve on von Neumann's treatment by considering sets of commuting operators. Among the earlier discussions of this problem we may mention Gleason's work [7] on the basis of which Jauch and Piron [8] thought that they had proved the impossibility of hidden variables. Bell [6] later showed that they had merely ruled out a limited class of such variables, i.e. those that are context independent. Bell did however point out that this proof did not apply to our interpretation because the latter implied context dependent hidden variables.

This latter point does not seem to have been taken seriously by a number of workers in the field. Thus Kochen and Specker [9] gave another independent proof which was similar in content to that of Jauch and Piron [8], but which had the advantage of using operators that could be measured by known physical procedures. However, their proof was rather complicated and later Peres [10] gave a much simpler proof for the case of two particles of spin one-half. Meanwhile Greenberger, Horn, Shimony and Zeilinger [11] have also developed a simple proof along similar lines for the case of three or more particles of spin one-half. Mermin [12] summarised these in a very concise way. (For a helpful treatment of this phase of the work see Brown and Svetlichny [13].) All these authors use the example of commuting operators and are therefore basically similar in their fundamental assumptions. We shall therefore carry out the further discussion in terms of Mermin's presentation which is the simplest of these.

Mermin begins with a set of commuting operators $A, B, C, \ldots$. He then points out that if some fundamental relation

$$f(A, B, C, \ldots) = 0 \tag{6.25}$$

σ_{1x}	σ_{2x}	$\sigma_{1x}\sigma_{2x}$	1
σ_{2y}	σ_{1y}	$\sigma_{1y}\sigma_{2y}$	1
$\sigma_{1x}\sigma_{2y}$	$\sigma_{1y}\sigma_{2x}$	$\sigma_{1z}\sigma_{2z}$	1
1	1	-1	

Table 6.1: Spin operators

holds as an operator identity, then since the results of the simultaneous measurements of $A, B, C, \ldots$ will be one of the sets of simultaneous eigenvalues $a, b, c, \ldots$ of the operators in question, these results must also satisfy

$$f(a, b, c, \ldots) = 0$$

independently of the state of the system prior to measurement.

He then assigns to the observables $A, B, C, \ldots$ for an individual system, the values $V(A), V(B), V(C), \ldots$. These values are assumed simply to be revealed by measurement of the corresponding operator. Since any measurement must give rise to an eigenvalue of the corresponding operator, the value assigned to that operator must be one of its eigenvalues. And since any commuting sub-set of the full set of operators can be measured simultaneously if the values are to agree with the predictions of quantum mechanics, they must be constrained by the relation

$$f(V(A), V(B), V(C), \ldots) = 0. \tag{6.26}$$

We now go on to consider a system of two particles, each of spin-half, and the nine operators shown in table 6.1.

It is easy to see that the product of the three operators in each row is 1 as is indicated in the table. The product of the operators in the first two columns is also 1, but the product of the operators in the third column is -1. Moreover it is also easily shown that the operators in each row commute, as do those in each column. Therefore the values of the operators must satisfy the same constraints as do the operators themselves. Thus we have

$$\begin{aligned}
V(\sigma_{1x}\sigma_{2x})V(\sigma_{2x})V(\sigma_{1x}) &= 1\\
V(\sigma_{1y}\sigma_{2y})V(\sigma_{1y})V(\sigma_{2y}) &= 1\\
V(\sigma_{1x}\sigma_{2y})V(\sigma_{1y}\sigma_{2x})V(\sigma_{1z}\sigma_{2z}) &= 1\\
V(\sigma_{1x}\sigma_{2y})V(\sigma_{2y})V(\sigma_{1x}) &= 1\\
V(\sigma_{1y}\sigma_{2x})V(\sigma_{1y})V(\sigma_{2x}) &= 1\\
V(\sigma_{1z}\sigma_{2z})V(\sigma_{1y}\sigma_{2y})V(\sigma_{1x}\sigma_{2x}) &= -1.
\end{aligned}$$

In the above set of equations each value appears twice, once as a member of a row and once as a member of a column. The values of all nine operators are ± 1. Therefore if we multiply all these equations together, the product of the left hand side is 1. On the other hand, the product of the right hand side is -1. This contradiction implies that there is no consistent way to assign values to all the operators.

This contradiction rules out the model of quantum theory that is proposed by Mermin (and, of course, all the other authors who use essentially the same model). The basic assumption in this model is that the values $V(A), V(B), V(C), \ldots$ of the set of variables specifying the individual system are equal to the results $R(A), R(B), R(C), \ldots$ of measurements of the corresponding operators. The above is equivalent to supposing that the beables are specified by $V(A), V(B), V(C), \ldots$ and that, as we have already said, the measurement simply reveals the values of these beables (as would happen in classical physics).

In our interpretation we do not assign values such as $V(A), V(B), V(C), \cdot$ to the operators. For these operators do not correspond to beables in our approach. Rather the beables are the overall wave function together with the coordinates of the particles, both of the observed system, $\mathbf{x}$, and the of the observing apparatus, $\mathbf{y}$. These beables determine the results $R(A), R(B), R(C), \ldots$ of each individual measurement *operation*. But these results are not present before the measurement operation has been completed. Rather they are then only the potentialities whose realisation depends not only on $\mathbf{x}$ and $\mathbf{y}$, but also on the overall wave function of the system and on the interaction Hamiltonian H_I between the apparatus and the observed object (the latter being determined in principle by the operator O that has to be measured). Therefore as we have already pointed out earlier, there is no *pre-existing* quantity that is actually revealed in this process.

The above may be said to give the full meaning of the statement that the results $R(A), R(B), R(C), \ldots$ are context dependent. On the other hand in the model used by Mermin, the results are assumed to be determined solely by the values $V(A), V(B), V(C), \ldots$ and are therefore context independent.

Let us now return to a consideration of Mermin's treatment. Since we do not assign values to operators, there is, of course, no unambiguous way to apply Mermin's argument to our model. What we do instead is to show that the measurement of the operators given by Mermin do not lead to *contradictory* results. This is, of course, sufficient to demonstrate that our approach is consistent.

Strictly speaking we ought to describe by means of some physical arrangement how each of the operators above could give rise to the results $R(A), R(B), R(C), \ldots$ of the corresponding measurement process. With operators such as σ_{1x} and σ_{2y}, this can be done fairly easily by assuming that the particles are far apart and that the first particle undergoes a

Stern-Gerlach process with the magnetic field in the x-direction, while the second particle undergoes a process with the field in the y-direction (see chapter 10). However, with product operators like $\sigma_{1x}\sigma_{2y}$ no one has yet suggested any physical process that could lead to results corresponding to their measurements (which would have to be of a nonlocal kind). Until such a procedure is suggested, we cannot demonstrate in detail how our approach leads to no contradiction with regard to operators of this kind. However, if we accept Mermin's assumption that such a measurement is somehow possible, then we can show that its results will not lead to a contradiction with those operators that may be measured.

Let us first recall with the aid of equation (6.4) and the paragraphs that follow it, that any operator O which can be measured will lead to a linear supposition of wave functions corresponding to its different eigenvalues O_n. The active part of the wave function then effectively reduces to one of these eigenfunctions determined by the channel that is actually entered by the particles concerned. In this way we show that for a general operator O, the result of measurement is one of its eigenvalues O_n, and that the observed system is left with the corresponding effective wave function ψ_n.

We now apply these ideas to the operators in table 6.1. It will be sufficient for our purposes to consider the operators in the first column and in the last row. It will also be convenient to write

$$\sigma_{1x} = A \qquad \sigma_{1y}\sigma_{2x} = B'$$
$$\sigma_{2y} = B \qquad \sigma_{1z}\sigma_{2z} = A'$$
$$\sigma_{1x}\sigma_{2y} = C$$

together with the identities

$$C = AB = A'B'.$$

It is crucial to our argument to note that although A, B, C commute as do A', B' and C, A and B do not commute with A' and B'. If we measure A, the result will be an eigenvalue a_n and the effective wave function will be the corresponding eigenfunction ψ_{a_n}. Because A and B commute, if we then measure B the result will be a simultaneous eigenfunction of both operators, $\psi_{a_n b_n}$. It follows trivially that $\psi_{a_n b_n}$ is an eigenfunction of C corresponding to eigenvalue $a_n b_n$. So if we measure C, the result will be the product $a_n b_n$. Therefore the operator identity $C = AB$ implies the corresponding identity of the results $R(C) = R(A)R(B)$.

A similar argument can be made for the commuting set of operators A', B' and C. However, it is important to keep in mind that A' and B' do not commute with A and B. Therefore a different and incompatible measuring apparatus is needed to measure A' and B' which excludes the apparatus necessary to measure A and B.

Let us recall here that the results depend as much on the wave function as on the apparatus variables **y** as it does on the variables of the observed system, **x**. This implies that the results obtained for C when the apparatus used to 'measure' it can also be used to 'measure' A and B simultaneously, will not in general agree with the result obtained for C using an apparatus that would allow A' and B' to be 'measured' simultaneously instead. It is this point that enables us to avoid the contradiction produced by Mermin. This contradiction depends on assuming that the result of C, $R(C)$, is the same in both cases.

As we have already pointed out it may be said that a measurement of C is context dependent in the sense that the result obtained when A and B can be measured along with it has no necessary relationship to the result that would be obtained for C when A' and B' could be measured simultaneously. Because A' and B' do not commute with A and B, the corresponding pieces of measuring apparatus provide contexts that are mutually incompatible. It follows then that Mermin's argument makes assumptions that have no place in our interpretation, so that the contradiction to which it leads is not relevant to our approach.

Kochen and Specker [9] have argued that the context dependence described above does not invalidate the kind of proof of the impossibility of hidden variables that we describe here. This is because the results of the quantum mechanical observations should be determined by the hidden variables of the combined apparatus and observed system so that the net result for the combined system should be predictable in each case. Therefore the argument against hidden variables should still go through when applied to the combined system. Our answer to this is that when non-commuting operators are measured, different and incompatible pieces of apparatus are required. Therefore we have as many combined systems as there are relevant sets of non-commuting operators (so that there is no single 'phase space' for the combined system). This means that one of the basic assumptions underlying the proof of the impossibility of hidden variables is not satisfied in our approach.

The context dependence of results of measurements is a further indication of how our interpretation does not imply a simple return to the basic principles of classical physics. It also embodies, in a certain sense, Bohr's notion of the indivisibility of the combined system of observing apparatus and observed object. Indeed it may be said that our approach provides a kind of intuitive understanding of what Bohr [16] was saying. He described a 'measurement' as a whole phenomenon not further analysable. The description of this phenomenon includes a specification of the experimental conditions (which are equivalent to the context that we have been discussing), along with a statement of the results. From the results and the conditions, one infers the 'values' of the 'measured' qualities, but as in our approach, it is not implied in Bohr's treatment that these values correspond to 'beables' that exist independently of the overall experimental context.

6.6 No measurement is a measurement

In the conventional interpretation of the quantum theory one obtains a certain paradoxical quality for the case of what is called 'negative measurement', i.e. measurements which establish a given result by showing that the only possible alternative is not present. To illustrate this sort of paradox, suppose that by means of a Stern-Gerlach apparatus, we select a beam of atoms with a definite value, zero, for the z-component of the orbital angular momentum. We then do a further Stern-Gerlach experiment with the apparatus oriented in the x-direction. The beam splits into two parts corresponding to values of the x-component of $+\hbar$ and $-\hbar$. The wave function of the atom is then

$$\Psi(x, y) = \frac{1}{\sqrt{2}} [\phi_0(y - at)\psi_+(x) + \phi_0(y + at)\psi_-(x)], \tag{6.27}$$

where, as shown in equation (6.4), ϕ_0 represents the packet defining the coordinates of the atom as a whole, while x is the coordinate of the orbital electron, and a is a factor proportional to the strength of the interaction and to the time of its duration.

Let us now suppose that a detector is placed in the beam corresponding to the $-\hbar$ component of the angular momentum. In terms of the von Neumann interpretation of the quantum theory, we can say that if this detector fires, the wave function 'collapses' from its original form (6.27) to

$$\Psi_{f_1}(x, z) = \phi_0(y + at)\psi_-(x). \tag{6.28}$$

And as pointed out earlier, in the conventional interpretation one can regard this 'collapse' as produced by the interaction with the apparatus.

But suppose that in a certain individual case the detector does not fire. The wave function then 'collapses' to

$$\Psi_{f_2}(x, z) = \phi_0(y - at)\psi_+(x). \tag{6.29}$$

It seems difficult to accept that the *failure* of this apparatus to operate will produce a 'collapse' of the other beam which, in principle, could be very far away. Or as it has been put, it seems that "no measurement is a measurement".

Thus the paradox arises because a 'non-event' (i.e. the failure of the counter to fire) precipitates an event (i.e. the collapse of the wave function to a state in which the atom never even contacts the counter). It is clear that the assumption that the wave function actually collapses is the root of the difficulty.

In the conventional interpretation this is a rather strange behaviour. However, what actually happens in this experiment can be explained fairly directly in terms of our interpretation. We begin by introducing the relevant coordinates of the detector, which may be represented by $z_1 \ldots z_n$. We have to take into account, not only the wave function $\lambda_f(z_i)$ representing the detector that has fired, but also the wave function $\lambda_0(z_i)$ representing the detector that has not fired. Before interaction, the wave function for the whole system is

$$\Psi_0(x,y,z) = \frac{1}{\sqrt{2}}\left[\phi_0(y-at)\psi_+(x) + \phi_0(y+at)\psi_-(x)\right]\lambda_0(z_i). \qquad (6.30)$$

After the interaction with the detector the wave function is

$$\Psi_f(x,y,z) = \frac{1}{\sqrt{2}}\left[\phi_0'(y-at)\psi_+(x)\lambda_0(z_i) + \phi_0'(y+at)\psi_-(x)\lambda_f(z_i)\right] \qquad (6.31)$$

where ϕ_0' represents the state of the particle as altered by the detector.

It is clear from the above that we cannot leave out $\lambda_0(z_i)$ in equation (6.31) just because it describes a situation in which the apparatus has not functioned. Rather, it is evident that each part of the wave function contains some component corresponding to the detector, whether the detector has acted or not.

In the above wave function not only is there no overlap between $\phi_0'(y-at)$ and $\phi_0'(y+at)$, but also between $\lambda_0(z_i)$ and $\lambda_f(z_i)$. For example, if the detector is an ion chamber, then as we have brought out in section 6.1, $\lambda_0(z_i)$ corresponds to electrons attached to the atoms of the gas, while$\lambda_f(z_i)$ corresponds to electrons that have cascaded and have been collected on the electrode to give rise to a current in a wire leading away from the electrode. Clearly the wave function for the electrons in these two cases do not overlap.

Taking the above into account, we can see that there are only two mutually exclusive possible configurations of the whole set of particles for which the total wave function is not zero. These are:

1. The atom has angular momentum $+h$ and the counter has not fired.

2. The atom has angular momentum $-h$ and the counter has fired.

It follows then that if the counter has not fired, the atom must be in the channel corresponding to angular momentum $+h$, while if it has fired, it must be in the channel corresponding to angular momentum $-h$. In this way a negative result implies that a measurement has nevertheless been made. What happens is that if the atom enters the channel that passes through the counter, the latter must fire, while if it enters the other channel, it does not. This is the sort of thing that would also happen in classical physics. We emphasise again that we do not assume the collapse of the wave function. The particle enters one channel and is only affected by the active information of that channel. Thus there is no paradox in this conclusion.

As in many other examples treated in this and other chapters, we avoid a paradox in our interpretation by giving up the assumption that the wave function is a complete description of reality (which is, incidently, what makes the further assumption of collapse necessary). By introducing the particle's position as essential for a more complete description, we turn what was a paradoxical situation in the conventional interpretation into one that is quite simple and straightforward.

6.7 The Schrödinger cat paradox

Most of the further paradoxes of the conventional interpretations of the quantum theory arise from the above described assumption, made especially sharply by von Neumann, that the wave function provides the most complete description of reality that is possible. For when the wave function is constituted of a linear combination of components, the actual state of the system is ambiguous in the sense that there is no way to define it as being in one of the components or the other. The process by which the system 'collapses' is then mysterious and often ultimately paradoxical.

In a long article devoted to this whole subject, Schrödinger [14] criticised such assumptions with the aid of his now famous cat paradox. To formulate this paradox we begin by assuming a cat isolated in a box with a gun pointed at the cat. The firing mechanism for the gun is activated when an individual electron strikes it. We further suppose a source of the individual electrons which pass through a beam splitter. The effect of this splitter is to divide the wave function into two coherent parts $\psi_1(\boldsymbol{x})$ and $\psi_2(\boldsymbol{x})$. The first part simply goes off harmlessly and is absorbed. The second part is aimed at the device that fires the gun. Classically one would expect from this arrangement that if a single electron enters the system, there would be equal probability after the experiment was over for the cat to be alive or dead. This statement could be made meaningfully even before anyone had looked into the box to see what had happened.

Quantum mechanically, however, we must treat this situation in terms of the wave function. Let $z_1 \ldots z_N$ represent the particles constituting the cat and let $\psi_L(z_1 \ldots z_N)$ represent a living cat while $\psi_D(z_1 \ldots z_N)$ represents a dead cat. Furthermore let $y_1 \ldots y_M$ represent the particles of the gun, the bullet, the powder and the firing device. Let $\psi_U(y_1 \ldots y_M)$ represent the unfired state of this system and $\psi_F(y_1 \ldots y_M)$ represent the fired state.

Initially (just after the electron passes through the beam splitter) the wave function of the total system is

$$\Psi(x, y, z) = \frac{1}{\sqrt{2}} (\psi_1(x) + \psi_2(x))\, \psi_U(y_1 \ldots y_M) \psi_L(z_1 \ldots z_N).$$

After the entire process is finished, the wave function is

$$\begin{aligned} \Psi \;=\; & \frac{1}{\sqrt{2}} \psi_1(x) \psi_U(y_1 \ldots y_M) \psi_L(z_1 \ldots z_N) \\ & + \frac{1}{\sqrt{2}} \psi_2(x) \psi_F(y_1 \ldots y_M) \psi_D(z_1 \ldots z_N). \end{aligned}$$

Since the two components of this wave function are coherent, it then follows in terms of the von Neumann interpretation, that the state of the total system cannot be one or the other. It therefore seems that the cat is in a state that is neither alive nor dead. This ambiguity could be removed only when someone looked into the box and saw whether the cat was dead or alive. The question immediately arises then of whether in the case that the cat was found to be dead, the observer had not actually participated in 'killing the cat'.

Schrödinger regarded this whole situation as absurd. Firstly it did not make sense to him that observation was needed to answer the question of whether the cat was alive or dead. In addition, even before observation, one might well think that the cat should at least know more than this about its own state. Or if one does not wish to attribute consciousness to the cat, suppose instead that it had been a human being in the box. To avoid violence we could replace the gun by a device that would prick his skin. Would this person find himself in an ambiguous state in which his skin was neither clearly untouched nor clearly pricked by the device?

In our approach this sort of paradox does not arise because we go beyond the assumption that the wave function provides the most complete possible description of reality. To define the actual state of being of the cat, we have to consider in addition the particles that constitute it. It is evident that when the cat is alive, many of these particles will be in quite different places and will move quite differently than they would if the cat were dead. Indeed one can easily see that in the configuration space of the particles of the cat, there will be no overlap of $\psi_L(z_1 \ldots z_N)$ and $\psi_D(z_1 \ldots z_N)$. Therefore it is clear that at the end of the experiment the cat has to be either alive or dead. For the wave function corresponding to the live cat has no effect on the quantum potential acting on the dead cat or vice versa.

Indeed this sort of nonoverlap is present also in the conventional interpretations. However, in these latter, there is no way to remove the ambiguity in the state of being in a coherent wave function even when the components do not overlap. But in our interpretation the state of being also depends on the positions of the particles that constitute it and it is this which enables us to treat this situation non-paradoxically as in essence a simple case of pairs of alternative and mutually exclusive states.

As a matter of fact it is not even necessary to go as far as the cat to resolve this paradox. Thus if we consider the system of firing device, plus gun, bullet and powder, it is clear that there is no overlap between the wave function $\psi_U(y_1 \ldots y_M)$ and $\psi_F(z_1 \ldots z_M)$ (i.e. because after firing, the powder will become gas and the bullet will move so that their particles will be in very different positions). Therefore the particles constituting this system will either be in the state corresponding to the firing or non-firing of the gun. If the gun has actually fired, the bullet will be speeding towards the cat (and will *not* be in an ambiguous state in which it is neither moving towards the cat nor at rest). The state of the dead cat will then follow only if the bullet is actually moving towards it. In other words, once the electron has in effect been 'detected', everything proceeds in essentially the same way as it does in classical physics.

6.8 Delayed choice experiments

Wheeler [3] has given a detailed description of several possible types of 'delayed choice' experiments. These are designed to show that, according to the quantum theory, the choice to measure one or another of a pair of complementary variables at a given time can apparently affect the physical state of things for considerable periods of time before such a decision is made.

Like Bohr, Wheeler puts a primary emphasis on the phenomenon, in which the experimental arrangement and the experimental result constitute an indivisible whole in the way discussed in chapter 2, section 2.1. From his analysis of the delayed choice type of experiment, he concludes that "no phenomenon is a phenomenon until it is an observed phenomenon", so that

> the universe does not 'exist out there' independently of all acts of observation. It is in some strange sense a participatory universe. The present choice of the mode of observation...should

> influence what we say about the past...The past is undefined and undefinable without the observation.

We can agree with Wheeler that no phenomenon is a phenomenon until it is observed, because, by definition, a phenomenon is what *appears*. Therefore it evidently cannot be a phenomenon unless it is the content of an observation. The key point about an ontological interpretation such as ours is to ask the question as to whether there is an underlying reality that exists independently of observation, but that can appear to an observer when he 'looks' (in physics, with the aid of suitable instruments). We have proposed a model of such a reality in which we say, along with Wheeler, that the universe is essentially participatory in nature. However, unlike Wheeler, we have given an account of this participation, which we show throughout this book to be rational and orderly and in agreement with all the actual predictions of quantum theory. In doing this we assume, as we have been emphasising throughout this chapter, that the underlying reality is not just the wave function, but that it also has to include the particles. As we shall see in this section, when we take this into account there is no need to say that the past is affected by our observation in the present. Nor do we imply even that what we *say about the past* is thus affected. Therefore the need to introduce such ideas is based on the insistence that the wave function provides a complete description of reality. Such insistence has led, as we have seen, to many other strange paradoxical features, e.g. with negative measurements and with the Schrödinger cat paradox. It seems to us preferable to consider extending our notion of reality by including the concept of particles, rather than to go on with an indefinite series of paradoxical consequences of sticking rigidly to the assumptions underlying the conventional interpretation of the quantum theory.

To illustrate what is involved in the delayed choice experiment, we first describe one of the simplest arrangements suggested by Wheeler [3] (the generalisation to more complex cases is straightforward). Wheeler uses a photon interference experiment. To apply our interpretation [15] to this experiment would require a treatment of the electromagnetic field which we do indeed outline in chapter 11. However, to simplify the discussion, we shall consider instead the interference of particles such as electrons or neutrons. The experiment is shown in figure 6.2, respective beams being then reflected by mirrors at E and B. The beams meet at an additional beam splitter F and interference can be detected either in beam G or in beam H. Two particle detectors are mounted so that they can be moved easily and quickly either to C_1 and C_2 or to C_1' and C_2'. Then when the detectors are in the position C_1 and C_2, the relative counting rates will be determined by the phase difference between the two routes reflecting

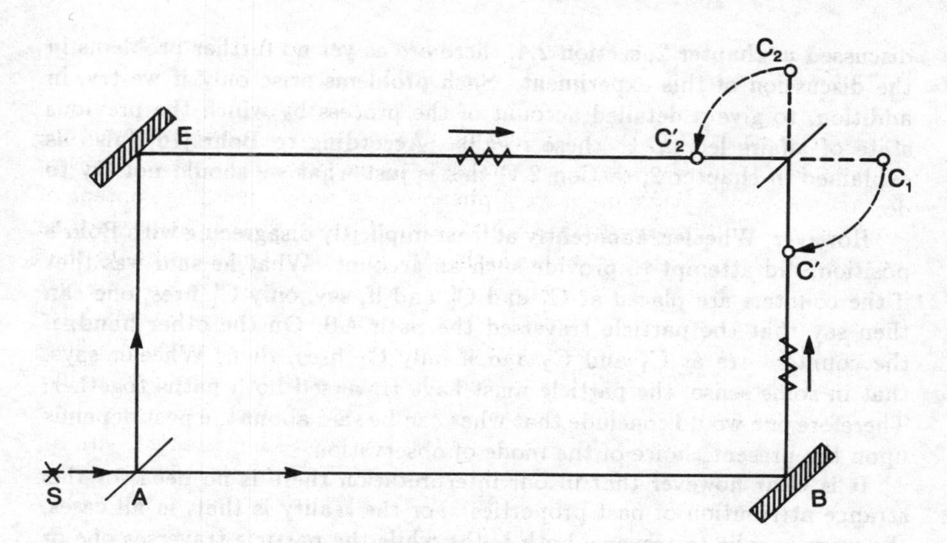

Figure 6.2: Schematic view of the delayed-choice experiment for photons passing through a half-silvered mirror at A. Counters C_1 and C_2 can be moved into positions C_1' and C_2'

respectively from E and from B. But with the counters at C_1' and C_2', we can then tell which route a particle actually takes. But with the counters at C_1 and C_2, there is no way to tell this because the result depends on the interference between the waves that have traversed these two paths.

Suppose now that a single particle enters the system and furthermore suppose that it is represented by a wave packet that is not too broad. We can make the choice on which of these experiments to perform at sometime after the particle has entered the apparatus, but before it reaches the beam splitter F.

In the conventional interpretation of the quantum theory, the wave function is, as we have already pointed out many times, assumed to provide a complete description of reality. Therefore we must say that the wave function collapses to the actual result of the measurement. Thus in the first kind of experiment, it will collapse to a state corresponding either to the firing of C_1' or of C_2', and in the second experiment it will collapse to a state corresponding to the firing of C_1 or of C_2. If the first experiment is done, then clearly the interference will be destroyed before the waves can arrive at the beam splitter F. But if the second experiment is done, interference will take place at the beam splitter F and then the wave function will collapse.

Aside from the difficulties in the von Neumann concept of collapse as discussed in chapter 2, section 2.4, there are as yet no further problems in the discussion of this experiment. Such problems arise only if we try, in addition, to give a detailed account of the process by which the previous state of affairs led up to these results. According to Bohr [16] (also as explained in chapter 2, section 2.1) this is just what we should not try to do.

However, Wheeler, apparently at least implicitly disagreeing with Bohr's position, did attempt to provide such an account. What he said was that if the counters are placed at C'_1 and C'_2 and if, say, only C'_1 fires, one can then say that the particle traversed the path AB. On the other hand, if the counters are at C_1 and C_2 and if only C_1 fires, then, Wheeler says, that in some sense, the particle must have traversed both paths together. Therefore one would conclude that what can be said about the past depends upon the present choice of the mode of observation.

It is clear however that in our interpretation there is no need for this strange attribution of past properties. For the reality is that, in all cases, the wave is split to traverse both paths while the particle traverses one or other in each individual case. But as has been brought out in a number of contexts, the interaction with the measuring apparatus implies mutual participation in which the final state need not directly reveal the earlier properties of the system. If the counter is placed at C'_2 and fires, then, as explained earlier, the system is left with the corresponding wave function and the interference at F is destroyed. But if the counter is placed at C_2 and fires, then the system is left with the wave function corresponding to a particle moving from that point. But in getting there (whether it traverses the path ABF or AEF) it will have been affected by the quantum potential in the region of interference at F. Therefore we are not surprised to discover that the frequency of detection of such particles reflects the difference of phase between the two beams.

It should be recalled, however, that even the conventional interpretation of this experiment does not imply that what we say about the past is affected by our present choice of experiments. In fact the conventional interpretation would say that, in general, we cannot make unambiguous statements about the past on the basis of present observations. As we have seen our interpretation agrees with this in a certain way in that we cannot say from our observations at C_1 which path the particle actually took. But we say that it actually took a definite path even though we cannot say what that path actually was. It is only if with Wheeler we tried to infer this past while holding to the conventional interpretation that we get into this sort of paradoxical way of talking about the past.

6.9 The watchdog effect and Zeno's paradox

In the conventional interpretation, the quantum theory puts, as we have seen, the observing instrument into an essential role, in the sense that without it the equations of quantum theory would have no physical meaning. At first sight one might then think that the more carefully we 'watched' a given process, the more that we would learn about how it takes place. Yet as has been shown [17,18], in the limit when the system is constantly watched, the process itself disappears and one is left with a constant state. While this is not actually a paradox, it does seem to suggest something unsatisfactory about the whole approach in the conventional interpretation. We shall see however that when this question is viewed ontologically, the whole effect can be seen to arise quite naturally in a way that is easily explained.

To discuss this question, let us consider the simplified example of an atom of angular momentum $\hbar$ in a magnetic field B in the z-direction. The possible eigenvalues of energy of the atom in this field will be

$$\frac{eB}{2mc}, 0, -\frac{eB}{2mc}.$$

Suppose that initially the wave function corresponds to an angular momentum of zero. The transition to the state with angular momentum $-\hbar$ can be described more or less along the lines given in chapter 5, section 5.3. To do this we may assume, for example, that the atom in question contains a loosely bound electron which can, in an Auger-like effect, absorb the energy given off in a transition to the state of lower angular momentum $-\hbar$. Again as shown in chapter 5, section 5.3, the probability of transition to such a state will then be proportional to the time t.

Let us now suppose however that the angular momentum is measured by means of a very strong interaction which allows the measurement to take place in a time Δt much less than $1/\Delta E$ where ΔE is the energy given off in this transition.

Without the measurement, the wave function of this atom would be given by (5.18) but only for times much longer than $1/\Delta E$. For shorter times we may write

$$\Psi \propto \chi_0(\boldsymbol{r}_A)\lambda_0(\boldsymbol{r}_B, t) + \chi_{-\hbar}(\boldsymbol{r}_A)\lambda_{-\hbar}(\boldsymbol{r}_B, t),$$

where $\chi_0(\boldsymbol{r}_A)$ represents a wave function corresponding to the zero angular momentum of the atom, while $\chi_{-\hbar}(\boldsymbol{r}_A)$ represents a wave function corresponding to an angular momentum $-\hbar$. And $\lambda_0(\boldsymbol{r}_B, t)$ represents the initial wave function of the Auger particle and$\lambda_{-\hbar}(\boldsymbol{r}_B, t)$ represents the change of the wave function of this particle as it begins to move out of the atom. As pointed out in chapter 5, section 5.3, for sufficiently short times this wave function is proportional to the time.

When the measuring apparatus interacts with the atom, we will get

$$\Psi \propto \chi_0(\boldsymbol{r}_A)\lambda_0(\boldsymbol{r}_B)\phi_0(\boldsymbol{z}) + \chi_{-\hbar}(\boldsymbol{r}_A)\lambda_{-\hbar}(\boldsymbol{r}_B, t)\phi_0(\boldsymbol{z}+\alpha),$$

where $\phi_0(\boldsymbol{z})$ represents the initial wave packet of the apparatus (which will not change for a particle of angular momentum zero) while $\phi_0(\boldsymbol{z}+\alpha)$ represents the apparatus wave packet corresponding to an angular momentum $-\hbar$. As explained earlier in this chapter, $\phi_0(\boldsymbol{z})$ and $\phi_0(\boldsymbol{z}+\alpha)$ do not overlap so that they determine separate channels into one of which the whole system must enter. If the interaction time with the apparatus is very short, then contribution of this term to the normalisation of the wave function will be proportional to t^2 since the perturbed wave function itself is proportional to t.

Suppose then we made a series of measurements each one in a time Δt. In each of these measurements the probability of getting an angular momentum $-\hbar$ is proportional to $(\Delta t)^2$ and the probability that it is zero will be $1 - \gamma(\Delta t)^2$ where γ is a suitable constant. As Δt decreases by a factor of $1/N$, the probability of transition in a given measurement will be proportional to $1/N^2$, and in N measurements it will be proportional to $1/N$. Therefore as N increases without limit, the net probability of transition goes to zero. Or in other words the effect of making repeated measurements of the angular momentum at great speeds is to prevent transitions from taking place.

In our interpretation this behaviour is perfectly comprehensible and indeed quite easy to explain. In order for a transition to take place, it is necessary that the perturbed wave function interfering with the original one shall produce a quantum potential that allows the Auger-like particle to escape. Otherwise it will simply remain stably in its original stationary state. However, if we measure the angular momentum very rapidly, then, as shown in equation (5.18), the perturbed wave function (which is proportional to t for times much less than $1/\Delta E$) will never become large, and furthermore it cannot make a significant contribution to the quantum potential. For this reason no transition can take place.

The behaviour described above brings out once again the essentially participatory nature of quantum interactions. To make measurements in very short times requires very strong interactions, hence very intense participation. The idea of measurement which implies that we are simply watching what is already there and not participating significantly, is seen to be particularly inappropriate in this example.

6.10 References

1. N. Bohr, *Atomic Physics and Human Knowledge*, Science Editions, New York, 1961, 51.
2. J. S. Bell, *Speakable and Unspeakable in Quantum Mechanics*, Cambridge University Press, Cambridge, 1987.
3. J. A. Wheeler, in *Mathematical Foundations of Quantum Mechanics*, ed. A. R. Marlow, Academic Press, New York, 1978, 9–48.
4. F. J. Belinfante, *A Survey of Hidden Variables Theories*, Pergamon Press, Oxford, 1973.
5. J. von Neumann, *Mathematical Foundations of Quantum Mechanics*, chapter 4, Princeton University Press, Princeton, 1955.
6. J. S. Bell, *Speakable and Unspeakable in Quantum Mechanics*, chapter 4, Cambridge University Press, Cambridge, 1987.
7. A. M. Gleason, *J. Math. Mech.* **6**, 885–893 (1957).
8. J. M. Jauch and C. Piron, *Helv. Phys. Acta* **36**, 827 (1963).
9. S. Kochen and E. P. Specker, *J. Math. Mech.* **17**, 59–87 (1967).
10. A. Peres, *Phys. Lett.* **151A**, 107–108 (1990).
11. D. M. Greenberger, M. A. Horn, A. Shimony and A. Zeilinger, *Am. J. Phys.* **58**, 1131–1143 (1990).
12. N. D. Mermin, *Phys. Rev. Lett.* **65**, 3373–3376 (1990).
13. H. R. Brown and G. Svetlichny, *Found. Phys.* **20**, 1379–1387 (1990).
14. E. Schrödinger, *Proc. Am. Phil. Soc.* **124**, 323–338 (1980).
15. D. Bohm, C. Dewdney and B. J. Hiley, *Nature* **315**, 294–297 (1985).
16. N. Bohr, *Atomic Physics and Human Knowledge*, Science Editions, New York, 1961, 50.
17. B. Misra and E. C. G. Sudarshan, *J. Math. Phys.* **18**, 756–783 (1977).
18. W. M. Itano, D. J. Heinzen, J. J. Bollinger and D. J. Wineland, *Phys. Rev.* **41A**, 2295–2300 (1990).

Chapter 7
Nonlocality

7.1 Introduction

We have seen earlier that nonlocality is a basically new feature of the quantum theory, at least in our interpretation. In this chapter we shall go into this topic in more detail, especially with regard to the analysis of the experiment of Einstein, Podolsky and Rosen (EPR) [1].

We shall first review briefly how this question is to be treated in terms of the various interpretations, and we shall show that all but one of them imply some form of nonlocality. However, it has generally been felt that the very concept of nonlocality is unacceptable in a scientific theory and so people have sought an explanation of this behaviour in terms of local hidden variables that would in principle determine the results of each measurement. This leads us then to give a brief discussion of Bell's inequality [2] (which has to be satisfied by any such explanation), and also of the fact that Bell's inequality is not actually satisfied, either by the predictions of the quantum theory or by the experiments that have been done to test this point.

We then go on to discuss how our own interpretation deals consistently with these questions by bringing in a nonlocal connection through the quantum potential. We finally explain the fact that nonlocality is not commonly encountered at the large scale level by showing that it is generally difficult to maintain the wave functions that are needed for this except for certain systems at low temperatures (such as superconductors) and highly isolated systems that have to be produced in a rather artificial way.

7.2 Nonlocality in the conventional interpretations

Nonlocality is usually discussed in terms of the EPR experiment. However, EPR emphasised not only the question of nonlocality, but also that of whether the quantum mechanics was what they called *complete*. By this they did not mean that it should cover everything that could ever be known, but rather that it had all the concepts necessary for its coherence as the basis of an ontology. The result of their analysis was that it was not complete in this sense. We shall give here a brief résumé of their analysis as it would apply to a spin experiment involving a molecule of total spin zero, each of whose atoms have a spin of one-half [3].

Their discussion was based essentially on what they meant by an element of reality. In general terms this signified some feature of the world that is real, independently of its being observed or otherwise known by us. In particular they proposed a sufficient criterion for such an element of reality: if, without disturbing the system in any way, one can predict the values of a given quantity with certainty, i.e. with probability equal to unity, then that quantity corresponds to an element of reality. Clearly this fits in with our general intuitive notions of what is to be meant by an independently existent element of reality.

We now consider an experiment in which the molecule is disintegrated so that the atoms are separated by a large distance. Suppose that the spin of atom A is now measured in the z-direction and that a given result is obtained. We can then predict that the spin of atom B is opposite. Of course, atom A will be disturbed when its spin is measured, so that we cannot conclude that its spin was defined beforehand as an independently existent element of reality. But our predictions of the spin of atom B can now be made without disturbing atom B in any way at all. Because measurements of atom A do not disturb atom B in any way, it follows from the EPR criticism that the z-component of the spin of atom B must have been an element of reality even before atom A was measured and that indeed it was always an element of reality.

But now we could have measured any other component of the spin of atom A instead, and by considering such a measurement it follows that every component of the spin of atom B has always been an element of reality. From this EPR concluded that the quantum theory is conceptually incomplete. For one of its basic principles is that when two observables do not commute, they cannot be defined together. Therefore at most only one of these could be an element of reality at a given time. But if two or more are elements of reality, then this means that the mathematical form of the theory cannot reflect this in any way at all. To take this into account we would need new concepts not contained in the present theory. So EPR would argue that while the results of the quantum theory are statistically correct, this theory does not give an adequate account of the actual reality of the individual system.

Although it is postulated that there is no interaction between the particles through forces of known kinds, one could evidently assume that there was a new and as yet unknown kind of force that connected the two particles. Then when particle A was disturbed in its measurement, this disturbance would communicate itself to particle B in such a way as to bring about a result opposite to that of A for its spin.

If such an interaction was operating, then the criterion of EPR for an element of reality would clearly be irrelevant, since by hypothesis a disturbance of atom A can now bring about a disturbance of atom B. However, EPR did not seem to regard this sort of interaction as worthy of serious consideration. To see why, we note that there are two possibilities:

1. The force is transmitted at some finite speed less than or equal to that of light. In this case the statistical predictions of the current quantum theory would, of course, have to fail for measurements of A and B that are space-like separated. But EPR assumed, as we have pointed out, that these statistical predictions are correct, and thus they did not even envisage such a failure of the current quantum theory.

2. The interaction is transmitted instantaneously. In this case the interaction would have to be nonlocal, i.e. to operate directly and immediately between the two particles with a strong force even at very large distances. Evidently this would violate the special theory of relativity.

However, Einstein felt there was an even more fundamental objection, because independently of relativistic considerations, he thought that all action had to be local, while he regarded nonlocal interactions as 'spooky' and thus in some way unacceptable [4]. Nevertheless he did believe that a suitable local explanation of this sort of experiment could eventually be obtained, perhaps on the basis of a new theory containing additional concepts that would represent new elements of reality. It is clear then that EPR did not make an explicit discussion of the question of nonlocality and focussed instead on that of completeness.

Bohr [5], appreciating the penetrating criticism of EPR, took great care in formulating a response. Essentially this consisted in pointing out that his own approach of complementarity still applied, even when the particles were separated by large distances and when the results of experiments are separated by large intervals of time. The main point was that the attempt to analyse the process in detail, and in doing so to attribute independent reality to the properties of the particle B, for example, was not permissible in the quantum mechanical context. This is because the form of the experimental conditions and the content (meaning) of the experimental results are, in Bohr's view, a whole that is not further analysable. Therefore there is no legitimate way to think about the properties of particle B apart from the experimental context in which they are measured. The context needed to think about the z-component of the spin of atom B is therefore not compatible with that needed to think about its x-component. This signifies that even though we can predict the properties of atom B from those of atom A without disturbing atom B, there is no experimental situation with regard to atom B in which both of the above predictions could have meaning together.

The crucial point here is that Bohr is using a different notion of reality from that of EPR. For Bohr a concept represents reality only in so far as it is in unambiguous correspondence with the whole set of possible phenomena and these phenomena are necessarily such that they have to be described in terms of the concepts of classical physics. For Einstein, however, concepts are a 'free creation of the human mind' and their correspondence with reality is at first assumed and then tested by the phenomena they predict. Therefore there is no problem in assuming the simultaneous reality of all the properties of particle B even though these cannot be simultaneously observed. Bohr regards this as a totally inadmissible way of using concepts in the context of quantum theory in which these have to be ambiguous and mutually exclusive, but nevertheless complementary.

The above constitutes the essence of Bohr's answer to the emphasis of EPR on the question of completeness of the quantum theory. However, from this it also follows that it has no meaning, in his view, to talk about nonlocality. Indeed, according to Bohr nothing can be said about the *detailed* behaviour of individual systems at the quantum level of accuracy. There is only the total unanalysable experimental phenomenon and no way to discuss in detail what this could mean ontologically. Therefore it also follows that there is no meaning to talking about locality either. All that we can do is to use the quantum algorithm to calculate the probabilities of the various experimental results. Nevertheless it still seems that some kind of nonlocality is, at least, implicit in Bohr's approach, because the phenomenon itself, which is spread out over space and time, is considered to be an unanalysable whole.

Recall, however, that the most common approach to quantum theory is along the lines of von Neumann [6] rather than that of Bohr. To show how nonlocality is implied in this approach, let us consider the wave function for a system of two separated particles with total spin zero. This is

$$\Psi = \frac{1}{\sqrt{2}} f(\boldsymbol{r}_A) g(\boldsymbol{r}_B) \left[\psi_+(A)\psi_-(B) - \psi_-(A)\psi_+(B)\right], \tag{7.1}$$

where $f(\boldsymbol{r}_A)$ is the wave function of particle A and $g(\boldsymbol{r}_B)$ is the wave function of particle B and the expression in the square bracket represents the spin part of the combined system. $f(\boldsymbol{r}_A)$ may, for example, represent a situation where particle A is a long way to the left of the original centre of mass of the molecule, while $g(\boldsymbol{r}_B)$ represents a situation in which particle B is equally far to the right. We then assume that the spin of particle A is measured by means of an apparatus whose relevant coordinate is represented by $\boldsymbol{y}$. The theory of measurement that we have given in chapter 2 can readily be extended to the case of spin. Before the interaction with the measuring apparatus, the total wave function is then

$$\Psi_i = \frac{1}{\sqrt{2}} f(\boldsymbol{r}_A) g(\boldsymbol{r}_B) \phi_0(\boldsymbol{y}) \left[\psi_+(A)\psi_-(B) - \psi_-(A)\psi_+(B)\right], \tag{7.2}$$

where $\phi_0(\boldsymbol{y})$ is a wave packet representing the initial state of the measuring apparatus. We then introduce an interaction Hamiltonian which depends upon the spin of particle A in such a way as to separate the wave functions of positive spin from those of negative spin. As shown in the discussion around equation (2.4) and using a similar notation we obtain for the final wave function

$$\begin{aligned} \Psi_f \quad = \quad & \frac{1}{\sqrt{2}} f(\boldsymbol{r}_A) g(\boldsymbol{r}_B) \left[\phi_0(\boldsymbol{y} - \lambda\Delta t)\psi_+(A)\psi_-(B) - \right. \\ & \left. - \phi_0(\boldsymbol{y} + \lambda\Delta t)\psi_-(A)\psi_+(B)\right]. \end{aligned} \tag{7.3}$$

As we have seen in this discussion, $\phi_0(\boldsymbol{y} - \lambda\Delta t)$ and $\phi_0(\boldsymbol{y} + \lambda\Delta t)$ will not overlap after the interaction is over. When an observation is made, the apparatus will be found in one or the other of these states as if the wave function had 'collapsed' to the state in question. We will be left essentially with either

$$\Psi_f^+ = \frac{1}{\sqrt{2}} f(\boldsymbol{r}_A) g(\boldsymbol{r}_B) \phi_0(\boldsymbol{y} - \lambda\Delta t)\psi_+(A)\psi_-(B) \tag{7.4}$$

or

$$\Psi_f^- = \frac{1}{\sqrt{2}} f(\boldsymbol{r}_A) g(\boldsymbol{r}_B) \phi_0(\boldsymbol{y} + \lambda\Delta t)\psi_-(A)\psi_+(B). \tag{7.5}$$

After particle A is measured, particle B is in a well-defined spin state even though there is no measuring apparatus that has interacted with this particle. A similar result would have been obtained if we had measured the component in any direction designated by $\hat{\boldsymbol{n}}$. With regard to atom A, one can say that its interaction with the measuring apparatus was 'responsible' for the 'collapse' on to the eigenfunction of its spin in the direction $\hat{\boldsymbol{n}}$. But if there is no connection between A and B how can the latter 'know' that it too must 'collapse' into an opposite eigenfunction of the spin in the same direction, even though this direction was chosen arbitrarily when someone decided to measure A when the atoms were far apart? Particularly, if this choice is made too rapidly to allow a signal to pass from A to B, it seems very reasonable to suggest that A and B are directly connected, though in a way that is perhaps not yet known. And so one sees the need for some form of quantum nonlocality when we analyse the EPR experiment in terms of the von Neumann approach.

If there is a nonlocal connection between distant particles, what can we say about the possibility of using it to send signals from one of these particles to the other? Within the essentially statistical framework of the current quantum theory, this would require that there be something that we could do, for example, to particle A which would change the statistical result of measurements of the spin of particle B in one or more directions. We shall show that this is, in fact, impossible, from which it follows that EPR correlations cannot make possible the transmission of signals of any kind (including, for example, those that are faster than light).

To show this we use a simplified form of a proof by Eberhard [7]. Let us suppose an external system with coordinate $\boldsymbol{y}$ is allowed to interact with the spin of particle A. The initial wave function for the combined system will be

$$\Psi_0 = \phi_0(\boldsymbol{y})\frac{1}{\sqrt{2}}\left[\psi_+^\alpha(A)\psi_-^\beta(B) - \psi_-^\alpha(A)\psi_+^\beta(B)\right], \qquad (7.6)$$

where α represents the standard spin indices and $\psi_+^\alpha(A)$ represents the wave function for particle A corresponding to a positive spin in any chosen direction, while $\psi_-^\alpha(A)$ represents a negative spin in that direction.

The most general possible result of this interaction will be represented by a unitary transformation on the sub-system consisting of $\boldsymbol{y}$ and A, because, by hypothesis, we are assuming our interaction does not directly disturb B. If it did then this would not constitute sending a signal from A to B, but would just be a direct disturbance of B by its interaction with $\boldsymbol{y}$.

To represent this unitary transformation, we assume that $\phi_0(\boldsymbol{y})$ is the first term in an orthonormal set $\phi_n(\boldsymbol{y})$. The most general unitary transformation can be expressed as the matrix $U_{0\alpha}^{n\alpha'}$ where α and α' represent respectively the initial and final spin indices. In this notation the initial wave function (7.6) can be written as

$$\Psi_0^{\alpha\beta} = \frac{\phi_0(\boldsymbol{y})}{\sqrt{2}}\left[\psi_+^\alpha(A)\psi_-^\beta(B) - \psi_-^\alpha(A)\psi_+^\beta(B)\right]. \qquad (7.7)$$

Under the unitary transformation, this goes over into

$$\Psi_U^{\alpha'\beta'} = U_{0\alpha}^{n\alpha'}\frac{\phi_n(\boldsymbol{y})}{\sqrt{2}}\left[\psi_+^\alpha(A)\psi_-^{\beta'}(B) - \psi_-^\alpha(A)\psi_+^{\beta'}(B)\right]. \qquad (7.8)$$

Let us now consider spin averages for any spin σ_B for particle B. In the undisturbed state this average can be written as

$$\bar{\sigma}_B = \int \Psi_0^{*\alpha\beta} \mathbf{1}_{\alpha\alpha'} \sigma_{B\beta\beta'} \Psi_0^{\alpha'\beta'} \, \mathrm{d}\boldsymbol{y}. \tag{7.9}$$

This will be constituted of four terms, of which two may be called 'diagonal' and the other two called 'off diagonal'. A typical diagonal term will be

$$\int \phi_0^*(\boldsymbol{y})\phi_0(\boldsymbol{y}) \, \mathrm{d}\boldsymbol{y} \, \psi_+^{*\alpha}(B)\psi_+^{\alpha'}(A)\mathbf{1}_{\alpha\alpha'}\psi_-^{*\beta}(B)\sigma_{B\beta\beta'}\psi_-^{\beta'}(B).$$

Since ϕ_0 and $\Psi_+^{\alpha}(A)$ are normalised, this reduces to

$$\tfrac{1}{2}\left[\psi_-^{*\beta}(B)\sigma_{B\beta\beta'}\psi_-^{\beta'}(B)\right]. \tag{7.10}$$

A typical off diagonal term will be

$$\tfrac{1}{2}\int \phi_0^*(\boldsymbol{y})\phi_0(\boldsymbol{y}) \, \mathrm{d}\boldsymbol{y} \, \psi_+^{*\alpha}(A)\psi_+^{\alpha'}(A)\mathbf{1}_{\alpha\alpha'}\psi_-^{*\beta}(B)\sigma_{B\beta\beta'}\psi_-^{\beta'}(B).$$

But because ψ_+^{α} and $\psi_-^{\alpha'}$ are orthogonal this term is zero. We then get the familiar result

$$\bar{\sigma}_B = \tfrac{1}{2}\left[\Psi_+^{*\beta}(B)\sigma_{B\beta\beta'}\Psi_+^{\beta'}(B) + \Psi_-^{*\beta}(B)\sigma_{B\beta\beta'}\Psi_-^{\beta'}(B)\right] \tag{7.11}$$

(which is zero because the second term cancels the first).

It is well known that a unitary transformation will not change the scalar products. It follows then that because this transformation acts only on particle A, the diagonal terms remain the same as before and that the off diagonal terms remain zero. Therefore the average of σ_B will be the same after transformation as before and we conclude then that nothing that we do to particle A will change the average spin properties of particle B.

7.3 Bell's inequalities

We have seen that nonlocality is contained in all the interpretations of the quantum theory that we have discussed so far. As we have already pointed out in chapter 5, our own ontological interpretation also contains nonlocality as a basic feature.

In considering the nonlocal implications of this ontological interpretation, Bell [8] was led to ask whether nonlocality was necessary for all possible ontological explanations of quantum mechanics. He did this by considering the EPR experiment in more detail, representing the states of elements of reality by a set of hidden variables λ. These, together with the combined wave function of the observed system and the observing apparatus, will determine the results of each individual measurement process. In the case of the EPR experiment, we suppose the apparatus that measures the spin of particle A is characterised by an orientation parameter $\boldsymbol{a}$, while the apparatus that measures the spin of particle B is characterised by the orientation parameter $\boldsymbol{b}$ (so that they need not, in general, be in opposite directions). As shown in chapter 6, we may, in general, expect that each individual result of the spin of particle A will depend on hidden variables μ_a associated with the corresponding piece of measuring apparatus, while results for measuring particle B will depend on the hidden variables μ_b of the second piece of apparatus. In addition there will be some further hidden variables. There may be a set of such variables λ_A and λ_B belonging respectively to particle A and B by themselves, and a further set λ which may be associated with the observed system as a whole, rather than just one of the particles.

We then introduce a symbol to represent the result of measurement of the spin of particle A, with $A = +1$ for a positive spin, and $A = -1$ for negative spin, while B is the corresponding symbol for the result of a measurement on particle B. It is clear from what has been said above that, in general, there may be a set of such variables

$$
\begin{aligned}
A &= A(\boldsymbol{a}, \mu_a, \boldsymbol{b}, \mu_b, \lambda_A, \lambda_B, \lambda), \\
B &= B(\boldsymbol{a}, \mu_a, \boldsymbol{b}, \mu_b, \lambda_A, \lambda_B, \lambda).
\end{aligned}
\tag{7.12}
$$

A and B also depend on the wave function of the combined system, but we suppress this for the sake of conciseness.

At this point we have to ask how to take into account the context dependence of the results of quantum mechanical observation that are discussed in chapter 6, section 6.5. It is important not to confuse the question of context dependence with that of nonlocality even though it is difficult to separate the two. To put the question of nonlocality for this experiment, we may assume that only a local context is allowed. In other words, the result of measuring particle A depends only on the context of particle A (i.e. the apparatus that measures its spin) and not on the context of particle B. The interaction between A and B will be local if the result A depends only on μ_a and λ_A and not on μ_b and λ_B (and vice versa for B). Therefore

$$A = A(\boldsymbol{a}, \mu_a, \lambda_A, \lambda) \qquad \text{and} \qquad B = B(\boldsymbol{b}, \mu_b, \lambda_B, \lambda). \tag{7.13}$$

If the above is satisfied (i.e. if each result depends only on a local context) then we may say that we have local hidden variables.

In this connection, however, let us recall that, as indicated in section 7.2, it would be quite possible to conceive of purely local explanations of the EPR experiment by means of forces that are transmitted at the speed of light or less, provided that the predictions of the quantum theory failed when measurements were made with space-like separation and held only when there was sufficient time for signals to be transmitted back and forth between A and B. One might explain such a behaviour by assuming an ambient field that connected both particles. By passing back and forth between A and B a number of times, this field would guarantee that B would 'know' which properties of A were being measured and would behave accordingly. This sort of explanation has always been open to consideration and in fact we have discussed its possibility in some detail in Bohm and Hiley [9]. But Bell's work is based on the assumption that there is no time for a signal that would allow the results of a measurement of one of the particles to depend on the context supplied by the other measuring apparatus. Equation (7.13) then shows that local hidden variables will inevitably satisfy this requirement.

To proceed with the derivation of Bell's inequality we must now consider the distribution of hidden variables. Locality implies that the distributions over μ_a and μ_b are independent. It further implies that when we rotate the apparatus from one orientation to the other, its hidden variables will rotate with it and that such a rotation will not significantly alter the hidden variables, $\lambda_A, \lambda_B, \lambda$, of the two particle system.

This means that the hidden variables, μ_a, can be taken as having the same range independently of the orientation of the apparatus. We may therefore replace the symbol μ_a by μ_A, which denotes the fact that the same set of parameters can be used regardless of the orientation of the apparatus. Similarly we replace μ_b by μ_B. We therefore define three distributions $P_A(\mu_A)$, $P_B(\mu_B)$, and $P(\lambda_A, \lambda_B, \lambda)$. The function P of λ_A, λ_B and λ is able to express the statistical correlations that may be present between the two systems even though the measurement of one of the systems does not affect the other in any way.

We can then define the averages of A and B over the hidden variables μ_A and μ_B respectively. This gives

$$\begin{aligned} \bar{A}(\boldsymbol{a}, \lambda_A, \lambda) &= \int P_A(\mu_A) A(\boldsymbol{a}, \mu_A, \lambda_A, \lambda)\, \mathrm{d}\mu_A, \\ \bar{B}(\boldsymbol{a}, \lambda_A, \lambda) &= \int P_B(\mu_B) B(\boldsymbol{b}, \mu_B, \lambda_B, \lambda)\, \mathrm{d}\mu_B. \end{aligned} \tag{7.14}$$

Since

$$|A(\boldsymbol{a}, \mu_A, \lambda_A, \lambda)| = |B(\boldsymbol{b}, \mu_B, \lambda_B, \lambda)| = 1$$

it follows that

$$|\bar{A}(\boldsymbol{a}, \lambda_A, \lambda)| \leq 1 \qquad |\bar{B}(\boldsymbol{b}, \lambda_B, \lambda)| \leq 1. \tag{7.15}$$

We then go on to consider the correlations of experimental results for A and B with various orientations $\boldsymbol{a}$ and $\boldsymbol{b}$. These are what can be readily measured and they will provide the test for locality. A typical correlation is

$$P(\boldsymbol{a}, \boldsymbol{b}) = \int \rho(\lambda_A, \lambda_B, \lambda)\bar{A}(\boldsymbol{a}, \lambda_A, \lambda)\bar{B}(\boldsymbol{b}, \lambda_B, \lambda)\, \mathrm{d}\lambda_A\, \mathrm{d}\lambda_B\, \mathrm{d}\lambda. \tag{7.16}$$

To obtain the Bell inequality we begin by considering

$$\begin{aligned} P(\boldsymbol{a}, \boldsymbol{b}) - P(\boldsymbol{a}, \boldsymbol{c}) = & \int \rho(\lambda_A, \lambda_B, \lambda) \\ & \times [\bar{A}(\boldsymbol{a}, \lambda_A, \lambda)\bar{B}(\boldsymbol{b}, \lambda_B, \lambda) - \bar{A}(\boldsymbol{a}, \lambda_A, \lambda)\bar{B}(\boldsymbol{c}, \lambda_B, \lambda)]\, \mathrm{d}\lambda_A\, \mathrm{d}\lambda_B\, \mathrm{d}\lambda \\ = & \int \rho(\lambda_A, \lambda_B, \lambda)\, [\bar{A}(\boldsymbol{a}, \lambda_A, \lambda)\bar{B}(\boldsymbol{b}, \lambda_B, \lambda)\, \{1 \pm \bar{A}(\boldsymbol{d}, \lambda_A, \lambda)\bar{B}(\boldsymbol{c}, \lambda_B, \lambda)\}] \\ & - \bar{A}(\boldsymbol{a}, \lambda_A, \lambda)\bar{B}(\boldsymbol{c}, \lambda_B, \lambda)\, \{1 \pm \bar{A}(\boldsymbol{d}, \lambda_A, \lambda)\bar{B}(\boldsymbol{b}, \lambda_B, \lambda)\}]\, \mathrm{d}\lambda_A\, \mathrm{d}\lambda_B\, \mathrm{d}\lambda. \end{aligned}$$

Taking the absolute value of both sides and noting that

$$\begin{aligned} |\bar{A}(\boldsymbol{a}, \lambda_A, \lambda)\bar{B}(\boldsymbol{b}, \lambda_B, \lambda)| &\leq 1 \\ |\bar{A}(\boldsymbol{a}, \lambda_A, \lambda)\bar{B}(\boldsymbol{c}, \lambda_B, \lambda)| &\leq 1 \end{aligned}$$

we obtain

$$|P(\boldsymbol{ab}) - P(\boldsymbol{ac})| \leq 2 \pm (P(\boldsymbol{dc}) + P(\boldsymbol{db})).$$

From this it follows that

$$|P(\boldsymbol{ab}) - P(\boldsymbol{ac})| + |P(\boldsymbol{dc}) + P(\boldsymbol{db})| \leq 2. \tag{7.17}$$

The above is Bell's inequality which must be satisfied for a local hidden variable theory to apply to our system of two particles with spin.* These inequalities make possible a test for locality on the basis of measurements of four sets of correlations.

---

*This symmetric form of the Bell inequality was first proposed by Clauser *et al.* [10].

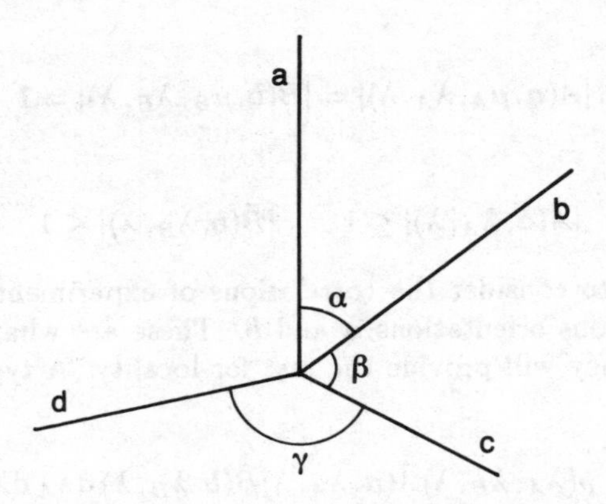

Figure 7.1: Angular relation between the axes of various spin measurements

The predictions of the quantum theory are that

$$P(\boldsymbol{ab}) = -\hat{\boldsymbol{a}} \cdot \boldsymbol{b}. \tag{7.18}$$

There is a range of angles for which (7.18) does not satisfy the inequality (7.17). For example consider the set of axes shown in figure 7.1. Here, $\alpha = \boldsymbol{b} - \boldsymbol{a}$ $\beta = \boldsymbol{c} - \boldsymbol{d}$ and $\gamma = \boldsymbol{d} - \boldsymbol{c}$. Putting these in (7.17), we find $|P(\alpha) - P(\alpha, \beta)| + |P(\gamma) + P(\beta + \gamma)| \leq 2$. Now if $\alpha = 60, \beta = 60$ and $\gamma = 0$ then the inequality on the left hand side is 5/2 which is certainly not less than 2 so that the inequality is violated.

Bell's inequality has been tested in a large number of experiments and generally speaking the inequality has been found to be violated. These experiments do not use particles with spin but instead use correlations of polarised photons produced in atomic cascades. It can be easily shown that this case is mathematically equivalent to that of spin. The first experiment of this kind was performed by Freedman and Clauser [11]. The latest and perhaps the most thorough set of experiments, has been performed by Aspect *et al.* [12]. In particular the latter were able to test whether the correlations were maintained even when the events of detection of the two photons were outside each other's light cones. Aspect *et al.* found the inequality was still violated. This implies that, independently of quantum mechanics, we have an experimental proof that if there are hidden variables they must be nonlocal.

Aspect's experiments have been criticised by some physicists [13]. The main criticism is that Bell's inequalities could have been absolutely proved to have been violated only if the photon detector had an efficiency close to unity. With the actual rather low efficiencies of these detectors, there seems to be room for assumptions concerning the hidden variables of the apparatus which could still preserve locality in spite of the experimental results. However, these assumptions seem rather arbitrary and artificial and, in fact, they give the impression of being contrived just to 'save the appearances'. The least we can say is that there is a strong ***prima facie*** case for nonlocality.

As we have seen earlier, the whole notion of taking Bell's inequality as a test for locality is physically relevant only in a context in which we are dealing with measurements on A and B that are outside each other's light cones. Aspect's third experiment [12] is the one that provides strong evidence that this inequality is still violated in this context. For included in this experiment is a switching device that could randomly change the orientations of the detectors in a time too short for a light-like signal to pass between them. Of course one could suppose along the lines we have discussed earlier, that there was an ambient field that could pass signals not only between the particles A and B, but also between these and the random switching device. The particles A and B could then 'know' what the random switching device was going to do and they could adjust their behaviour accordingly. But this does not seem to be very plausible especially since the switching device could, in turn, depend on other random devices going on in an indefinitely proliferating chain. This could even include the human being who chose to set these devices up in a certain way. If, for example, measurements were made when A and B were separated by millions of miles, giving the experimenters the chance to change the apparatus in an arbitrary way, we would then assume that the ambient field would have to 'inform' the particles A and B of these changes so that they could adapt accordingly to produce the proper correlated behaviour. Of course it is possible that somewhere along this chain quantum mechanics would fail, but then it is the whole purpose of Bell's inequality to show that any local hidden variable theory which allows for any kind of independent disturbance of a system from outside itself must imply the failure of quantum mechanics.

In our treatment we have only discussed purely deterministic local hidden variable theories. There has been a great deal of work done on obtaining inequalities of the Bell type for stochastic hidden variable theories [14]. Although the details of this work are more complex, the results are essentially the same. We will discuss a theory of stochastic hidden variables in chapter 9.

7.4 Extension of Bell's theorem to three particles

Recently there has been developed a new proof of the impossibility of local hidden variables based on the consideration of three or more particles. In this proof it is directly shown that even the states of perfect correlation lead to a contradiction with quantum mechanics so that no statistical criteria are required and no inequalities are involved.

We shall use here the very concise and simplified version of this proof presented by Mermin [16] who considers a system of three spatially separated particles which we denote by 1, 2, and 3 respectively. The argument of Bell was based on choosing a particular state, the singlet state, of a combined system of two particles and showing that under the assumption of locality the correlations of these particles are not compatible with those predicted by the quantum theory. However, to obtain this result, Bell had to work out certain features of the *statistical* correlations of the two particles. There is no way from the correlations observed in individual instances to show a contradiction with locality. But when there are three or more particles then one finds that even the perfect correlations cannot be assigned consistent values under the assumption of locality. And so the impossibility of local hidden variables is demonstrated without the need for any further statistical measurements.

Mermin begins by considering the three mutually commuting operators $\sigma_{1x}\sigma_{2y}\sigma_{3y}$, $\sigma_{1y}\sigma_{2x}\sigma_{3y}$ and $\sigma_{1y}\sigma_{2y}\sigma_{3x}$. He supposes that the wave function Φ is an eigenfunction of all three of these with eigenvalue 1. One can then immediately obtain a fourth operator, $\sigma_{1x}\sigma_{2x}\sigma_{3x}$, which is the negative of the product of all the above and therefore commutes with them. It clearly has the eigenvalue -1 for the state Φ. We can then write

$$m_{1x}m_{2y}m_{3y} = 1 \qquad m_{1y}m_{2x}m_{3y} = 1$$
$$m_{1y}m_{2y}m_{3x} = 1 \qquad m_{1x}m_{2x}m_{3x} = -1$$

where the ms are the eigenvalues of the corresponding individual spin operators. From these equations, we can obtain a set of perfect correlations. For example, if we measure σ_{1x} and σ_{2y} then the value of σ_{3y} is completely determined and so on.

If we multiply the four equations together, we obtain -1 on the right hand side, but $+1$ on the left hand side. Therefore there is a contradiction with our initial assumption that these values for any one particle are independent of what is measured for the other particles. From this we see that the assumption of locality is directly violated.

A weaker implicit assumption, namely, that of simple locality, is that the spin variables are at most locally context dependent (in the sense that the apparatus that measures a particular spin variable is indivisibly linked to that spin). However, what follows from the above proof is that each particle is also indivisibly linked to a context determined by the pieces of apparatus that measures the other particles. Each value such as m_{1x} appears twice in the above equations, combined with different values for the other particles (in this case $m_{2y}m_{3y}$ and $m_{2x}m_{3x}$). Therefore we are in essentially the same situation as that treated by Mermin for the case of two particles as discussed in chapter 6, section 6.4. Although the three operators σ_{1x}, σ_{2y} and σ_{3y} all commute as do σ_{1x}, σ_{2x} and σ_{3x}, nevertheless we cannot in general obtain the same value of m_{1x} when it is measured along with σ_{1x} and σ_{3y} as when it is measured along with σ_{2x} and σ_{3x}. As we have already pointed out the reason is that the operators other than σ_{1x} in one set do not commute with those in the other.

In the ontological interpretation this kind of context dependence would be evident because the measurements of non-commuting operators require different and mutually incompatible pieces of apparatus. Therefore there is no reason why σ_{1x} should have the same value no matter what other operators could be measured simultaneously by the chosen apparatus. Therefore there is no way in which the contradiction obtained in the proof could arise.

7.5 The EPR experiment according to the causal interpretation

It will be instructive to show in detail how an EPR-type experiment is to be understood in our approach for the case of two particles (no essentially new principles are involved in the case of three or more particles). Strictly speaking, to compare our results with the discussion that has been given here, we would need to use the causal interpretation for spin. Although we do develop this in chapters 10 and 12, we shall, for the sake of simplicity, consider here the case of a molecule of total angular momentum $\hbar$. This will give an EPR type correlation very similar to that obtained with spin. Although the Bell inequality cannot be directly applied to this case, an extension of Bell's idea to systems of higher angular momentum has actually been carried out by Mermin [17] and yields similar but more complex results.

Let us consider atom A whose centre of mass is represented by $\boldsymbol{R}_A$ and whose internal electron coordinate is represented by $\boldsymbol{r}_A$, while corresponding qualities for atom B are $\boldsymbol{R}_B$ and $\boldsymbol{r}_B$. After the molecule has been disintegrated, the wave function of the whole system can be built out of wave functions of the separated atoms. We assume the centre of mass of these atoms have wave functions given by $\phi_A(\boldsymbol{R}_A)$ and $\phi_B(\boldsymbol{R}_B)$. We choose internal states with angular momentum $\hbar$, which can be represented by

$g_A(\boldsymbol{r_A})\psi_i(\boldsymbol{r_A})$ and $g_B(\boldsymbol{r_B})\psi_i(\boldsymbol{r_B})$ where

$$\begin{aligned}\psi_1(\boldsymbol{r_A}) &= \frac{x_A + iy_A}{\sqrt{2}\boldsymbol{r_A}} \\ \psi_0(\boldsymbol{r_A}) &= \frac{z_A}{\boldsymbol{r_A}} \\ \psi_{-1}(\boldsymbol{r_A}) &= \frac{x_A - iy_A}{\sqrt{2}\boldsymbol{r_A}}.\end{aligned} \tag{7.19}$$

The above evidently represent the three possible eigenvalues of the angular momentum operator along the z-axis. The combined wave function is then

$$\begin{aligned}\Psi &= N\phi_A(\boldsymbol{R_A})\phi_B(\boldsymbol{R_B})g_A(\boldsymbol{r_A})g_B(\boldsymbol{r_B})\times \\ &\quad \times\frac{1}{\sqrt{3}}\left[\psi_1(\boldsymbol{r_A})\psi_{-1}(\boldsymbol{r_B}) - \psi_0(\boldsymbol{r_A})\psi_0(\boldsymbol{r_B}) + \psi_{-1}(\boldsymbol{r_A})\psi_1(\boldsymbol{r_B})\right] \quad \text{(a)} \\ &= -N\phi_A(\boldsymbol{R_A})\phi_B(\boldsymbol{R_B})g_A(\boldsymbol{r_A})g_B(\boldsymbol{r_B})\frac{1}{\sqrt{3}}\frac{\boldsymbol{r_A}\cdot\boldsymbol{r_B}}{r_A r_B}, \quad \text{(b)}\end{aligned} \tag{7.20}$$

where N is a normalisation constant. The above, being a scalar, corresponds to total angular momentum zero.

We assume for simplicity that all the ϕ and g are real so that both particles are at rest. As explained in chapter 4 this means that the quantum potential is balancing the total classical potentials that are responsible for the stability of the atoms.

Suppose now we made a measurement of the angular momentum of particle A in any direction. Because of the isotropy of the wave function, we can always choose this as the z-direction. Using the theory of measurement described in chapters 2 and 6, we may assume an interaction Hamiltonian

$$H_I = \lambda L_{z_A}\partial/\partial y, \tag{7.21}$$

where y is the relevant coordinate of the measuring apparatus and where λ is a suitable constant chosen so large that the measurement will be impulsive. The z-component of the angular momentum operator for particle A is

$$L_{z_A} = -i\left(x_A\frac{\partial}{\partial y_A} - y_A\frac{\partial}{\partial x_A}\right) = -i\frac{\partial}{\partial\phi_A}, \tag{7.22}$$

where ϕ_A is the azimuthal angle of particle A.

If $\Phi_A(y)$ represents the initial wave packet of the apparatus, then the total initial wave function will be

$$\chi_0 = \Phi_A(y)\Psi.$$

During the interaction, the wave function, χ, will be proportional to

$$\xi = \sum_j \psi_j(\boldsymbol{r}_A)\psi_{-j}(\boldsymbol{r}_B)\Phi_A(\boldsymbol{y} - j\alpha\Delta t) \tag{7.23}$$

where α is a constant proportional to λ. To obtain a measurement, $\alpha\Delta t$ must be large enough so that the $\Phi_A(\boldsymbol{y} - j\alpha\Delta t)$ do not overlap for different j. In this case there will be no interference between terms of different j, and the apparatus particle will enter one of the packets corresponding to a certain value of the angular momentum $j'\hbar$. Thereafter it will remain in that packet and the quantum potential will be determined by the corresponding wave function $\psi_{j'}(\boldsymbol{r}_A)\psi_{-j'}(\boldsymbol{r}_B)$. As explained in chapter 6, this means that everything has proceeded as if the wave function of the system had 'collapsed' to $\psi_{j'}(\boldsymbol{r}_A)\psi_{-j'}(\boldsymbol{r}_B)$. Not only will particle A have the angular momentum $j'\hbar$ but particle B will have the opposite angular momentum $-j'\hbar$, so that the two angular momenta will be correlated and this will happen for an arbitrary orientation of the apparatus. Thereafter particle B will behave in every way as if its angular momentum had been measured and found equal to $-j'\hbar$.

As we shall show this correlation is evidently both context dependent and nonlocal. Indeed we will find that B is dependent on the apparatus that measures A which provides the nonlocal context for B. Of course, both A and B are context dependent in the sense that the angular momentum of each is a potentiality that is realised according to the orientation of the apparatus that measures A and the state of the 'hidden' particle variables $\boldsymbol{y}_A$ belonging to this apparatus.

The cause of the correlation between the particles is evidently the quantum potential. It is interesting to follow in detail how this comes about during the period of interaction. However, to do this we shall use the guidance conditions which, as we recall, imply the quantum potential. This potential can indeed be calculated from $\partial S/\partial t$, but it is very complicated and it is more perspicuous to use the guidance conditions.

We begin by expressing the relevant part of the wave function (7.23) as the following

$$\begin{aligned}\xi \quad = \quad & \sin\theta_A \sin\theta_B \Big[\Phi_A(\boldsymbol{y} - \alpha\Delta t)\exp[i(\phi_A - \phi_B)] + \\ & + \Phi_A(\boldsymbol{y} + \alpha\Delta t)\exp[-i(\phi_A - \phi_B)]\Big] + 2\cos\theta_A \cos\theta_B \Phi_A(\boldsymbol{y}).\end{aligned} \tag{7.24}$$

According to equation (7.22), the z-component angular momentum of particle A is $p_{\phi_A} = \partial S/\partial\phi_A$ and of particle B is $p_{\phi_B} = \partial S/\partial\phi_B$ where S

is the phase of the wave function. For $\Delta t = 0$ the wave function is real, so that $S = 0$ and both particles have no angular momentum. More generally, the z-component angular momentum of particle A will be given by

$$p_{\phi_A} = \frac{1}{2mi|\xi|^2}\left(\xi^* \frac{\partial \xi}{\partial \phi_A} - \xi \frac{\partial \xi^*}{\partial \phi_A}\right)$$

and a similar expression holds for the z-component of the angular momentum of particle B. Assuming $\Phi_A(y)$ is real, we then obtain

$$|\xi|^2 p_{\phi_A} = Re\left\{\frac{\sin\theta_A \sin\theta_B}{m}\left(\xi^*[\Phi_A(y - \alpha\Delta t)\exp[i(\phi_A - \phi_B)] - \Phi_A(y + \alpha\Delta t)\exp[-i(\phi_A - \phi_B)]]\right)\right\}. \quad (7.25)$$

Because the wave function depends only on the combination $\phi_A - \phi_B$ it follows that

$$p_{\phi_A} = \frac{\partial S}{\partial \phi_A} = -\frac{\partial S}{\partial \phi_B} = -p_{\phi_B}.$$

Initially when $\Delta t = 0$, it is clear that $p_{\phi_A} = p_{\phi_B} = 0$. However, after a short time the contribution of $\Phi_A(y - \alpha\Delta t)$ and $\Phi_A(y + \alpha\Delta t)$ no longer cancel and the values of p_{ϕ_A} and p_{ϕ_B} will clearly then depend on y, which, as we have been assuming, has a random distribution with probability $|\Phi_A(y)|^2$. All other quantities being equal, the angular momentum variables for A and B will then separate depending on the value of y. But the net result also depends on θ_A and θ_B, and on $\phi_A - \phi_B$. It is clear that the behaviour of particle B for example will depend on the initial values of θ_A and ϕ_A, as well as on those of the apparatus. So we clearly have a nonlocal interaction. But it should be added that the net result for particle A will also depend on the initial properties of particle B. In other words the result we get for an individual measurement of A already anticipates to some extent the way in which particle B starts out. That is to say the nonlocality is reciprocal.

Because $p_{\phi_A} = -p_{\phi_B}$ we can be sure that eventually when the apparatus packets have separated one of the particles will end up with a unit of angular moment and other with the opposite unit. The probability of each result can be obtained by integrating the square of the net wave function over all the relevant coordinates including those of the apparatus, and as is easily shown, this comes out the same as in the conventional interpretation.

It is clear then that the assumption (7.14) underlying Bell's inequality is violated. For if we measure B, the result will depend, not only on the variables of particle A, but also on those of the apparatus that measures

A. From this it follows, as we have anticipated earlier, that this experiment involves not simple nonlocality but rather a dependence on a nolocal context.

Since the statistical results of our interpretation are the same as in the usual interpretation, we can see from the discussion in section 7.2 that the nonlocal quantum potential cannot be used to carry a signal, if this latter has to be detected statistically according to the rules of quantum theory. This means that there is no way to control the behaviour of atom *B* systematically by anything we might do to atom *A*.

7.6 Loss of nonlocality in the classical limit

We have thus far gone extensively into quantum nonlocality and have seen that it can arise in a great many contexts. At first sight one might be concerned as to whether we will then be able to understand why nonlocality is not encountered in our common experience of the world. Basically the answer to this question is quite simple. For our ordinary experience, both in the domain of common sense and in that of classical physics, is restricted to situations in which the quantum potential is very small, so that, in this context at least, it does not produce significant EPR correlations. For, as we have seen, quantum nonlocality is entirely the product of the quantum potential.

Nevertheless it may be instructive to see in more detail what are the conditions which favour nonlocality and what are those which do not. First of all it should be clear that EPR-type wave functions, which involve linear combinations of products of single particle wave functions, will be stabilised by suitable classical interaction potentials $V(\boldsymbol{r}' - \boldsymbol{r})$ between the particles. For example, in helium, the electrostatic interaction between electrons brings about a big energy difference, ΔE, between wave functions that are symmetric linear combinations of the one particle electron wave functions and those that are antisymmetric linear combinations of these wave functions.

In the interaction of the system with others which involve energies less than ΔE, these linear combinations of wave functions will remain stable. At the atomic level, ΔE is quite large and this explains why the chemical combination into molecules holds together until they are disturbed by, for example, thermal fluctuations with temperatures such that kT is not too much less than ΔE. With superconductors we also have the possibility of EPR-type wave functions. But since ΔE is small, these can only be maintained at low temperatures.

The EPR experiment is concerned with situations in which there are correlations between free particles for which $\Delta E = 0$. It may therefore be expected that such EPR correlations may be very difficult to maintain, for example, in the face of collisions and various kinds of external disturbances that could change the spins of atoms. Indeed if we regard measurement as a disturbance, we can see that it does indeed break the correlations and leaves us with product wave functions.

More generally we might consider what happens in a collision in which the angular momentum can be altered. We shall discuss the case of a molecule of total angular momentum zero which disintegrates into two atoms, each of which has angular momentum $\hbar$. While the atoms are flying apart the wave function is given by (7.20a). Suppose a collision takes place between particle A and another atom, C, whose centre of mass coordinates are $\boldsymbol{R}_C$ and whose internal coordinates are $\boldsymbol{r}_C$. To represent the full details of this process would require a complex notation that would not be very perspicuous. We shall here simply outline only the essential results. Let us suppose that atom C has the wave function $\phi_C(\boldsymbol{R}_C)g_C(\boldsymbol{r}_C)\psi_i(\boldsymbol{r}_C)$ so that the combined initial wave function for all three particles is $\phi_C(\boldsymbol{R}_C)g_C(\boldsymbol{r}_C)\psi_i(\boldsymbol{r}_C)\Psi$, where Ψ is given by equation (7.20a) and where $\phi_C(\boldsymbol{R}_C)$ represents a wave packet. In the collision process, the z-component of the angular momentum will be conserved, thus if $i \to i'$ and $j \to j'$, then $i + j = i' + j'$.

In the final wave function there will appear products such as

$$\phi'_{C_{i'}}(\boldsymbol{R}_C)\phi'_{A_{j'}}(\boldsymbol{R}_A)\phi_B(\boldsymbol{R}_B)g_C(\boldsymbol{r}_C)g_A(\boldsymbol{r}_A)g_B(\boldsymbol{r}_B)\psi_{i'}(\boldsymbol{r}_C)\psi_{j'}(rA)\psi_{-j'}(\boldsymbol{r}_B) \tag{7.26}$$

where $\phi'_{C_{i'}}(\boldsymbol{R}_C)$ represents a wave packet in which the centre of mass of particle C will be moving differently from what it does in its initial state, and similarly for $\phi'_{A_{j'}}(\boldsymbol{R}_A)$. Particles A and C will end up with energies that depend on their final and initial angular momenta. Therefore, if, for example, particle C interacts with a state of angular momentum $\hbar$ of particle A, its wave packet will move with a different speed from that of a particle that had interacted with an angular momentum of $-\hbar$, or 0. Sooner or later these wave packets will therefore cease to overlap and the system will behave as if particle A had 'collapsed' into a definite angular momentum, while particle B had collapsed into an opposite state. The effective wave function would now be a product of the type (7.26) and the EPR correlations would have vanished. This shows that even if EPR correlations should happen to arise under natural conditions, they would eventually be destroyed.

7.7 Symmetry and antisymmetry as an EPR correlation

A very important type of EPR correlation occurs with symmetric or antisymmetric wave functions. However, this type of correlation has an important feature that is different from those examples that we have thus far discussed.

To see what this difference is we first note that for equivalent particles, the Hamiltonian will be a symmetric function of all the particle variables. It follows then that if we have a solution of the wave equation, the interchange of two particles will generate another solution. We can then form linear combinations of solutions involving various permutations among the particles. Among these the symmetric and antisymmetric combinations are particularly interesting because all known particles are either bosons or fermions which have respectively symmetric and antisymmetric wave functions.

What is particularly significant in the context of our discussion here is that there is what is called a superselection rule implying that transitions between symmetric and antisymmetric wave functions can never take place (which follows from the symmetry of the Hamiltonian).

Therefore if one of the particles of such a system interacts with something else, for example, a measuring apparatus, it no longer follows that the unoccupied packet can have no further effect on the particles. Rather, there can be such an effect and this is indeed what is ultimately responsible for the Pauli exclusion principle.

To illustrate this, let us consider a measurement of a single particle operator O_A with eigenvalues a and eigenfunctions $\psi_a(\boldsymbol{x}_A)$. Because of the equivalence of particles, another particle B will have a corresponding operator O_B with eigenvalues b and eigenfunctions $\psi_b(\boldsymbol{x}_B)$. The wave function for the combined system can be expressed in terms of the products $\psi_a(\boldsymbol{x}_A)\psi_b(\boldsymbol{x}_B)$. If, for example, the wave function is antisymmetric, we can write

$$\Psi = \sum C_{ab}\psi_a(\boldsymbol{x}_A)\psi_b(\boldsymbol{x}_B),$$

with $C_{ab} = -C_{ba}$.

Let us suppose that we want to measure the operator O_A. According to the theory of measurement that we have given in chapter 6, to do this we would need an interaction with the measuring apparatus

$$H_I = \lambda O_A \frac{\partial}{\partial y},$$

where y is the relevant coordinate of the measuring apparatus and λ is a suitable constant. But for equivalent particles it is impossible to have such an interaction as this would destroy the symmetry of the Hamiltonian. To obtain a symmetric Hamiltonian we write

$$H_I^S = \lambda(O_A + O_B)\frac{\partial}{\partial y}.$$

If $\phi_0(y)$ is the initial wave packet of the apparatus we will obtain, after interaction,

$$\Psi = \sum C_{ab}\psi_a(x_A)\psi_b(x_B)[\phi_0(y-a) - \phi_0(y-b)].$$

Let us choose a certain pair of eigenvalues a_1 and b_1. Their contribution to ψ will be

$$[\phi_0(y-a_1) + \phi_0(y-b_1)]\,[\psi_{a_1}(x_A)\psi_{b_1}(x_B) - \psi_{b_1}(x_A)\psi_{a_1}(x_B)].$$

It is clear that the measurement has left the wave function antisymmetric.

Suppose we find that y is close to a_1. But from this information we cannot tell whether it is particle A or particle B that is left with the corresponding eigenfunction. It follows then that there is no way in such a measurement to distinguish between particle A and particle B.

This implies that quantum mechanically equivalent particles are indistinguishable whereas classically they might, in principle, be distinguishable by putting 'marks' on them. From this it has often been concluded that particles are not merely indistinguishable, but that they are identical. To do this is to go from an epistemological statement referring to our ability to distinguish things, to an ontological statement referring to what the particles are, i.e. one and the same. We recall from chapter 2 that in a positivist philosophy, epistemology and ontology are thus equated. But as we also stressed before, there is no real necessity to do this. Indeed in our ontological interpretation of the quantum theory, we must say that although the particles are indistinguishable, they are still different, and that each one exists continuously and therefore has its own identity.

In order to obtain the predicted results of the quantum theory, there is no need to assume such an identity, for these results all follow from the antisymmetry of the wave function alone. Indeed our interpretation gives the same results as those of the conventional ones, whether the wave function is symmetric, antisymmetric or has no particular symmetry. These results follow when we average over the positions of all the particles with the probability density in configuration space

$$P = |\phi(x_A, x_B)|^2.$$

To illustrate what is meant by the identity of a particle when the wave function is antisymmetric, let us again consider two particles A and B, with two possible wave functions

$$\phi_A(x_A) = \exp\left[-\frac{\alpha x_A^2}{2} + ik_1 x_A\right],$$
$$\phi_B(x_B) = \exp\left[-\frac{\alpha x_B^2}{2} + ik_2 x_B\right].$$

The antisymmetric wave function for the combined system is proportional to

$$\Psi_0 = \exp\left[-\frac{\alpha(x_A^2 + x_B^2)}{2}\right]\exp\left[\frac{i(k_1+k_2)(x_A+x_B)}{2}\right] \times \sin\left[\frac{(k_1-k_2)(x_A-x_B)}{2}\right].$$

The momenta of the particles are $p_A = \partial S/\partial x_A$ and $p_B = \partial S/\partial x_B$. It is clear that $p_A = p_B$ and that both particles are moving with the mean momentum $(k_1 + k_2)/2$. The probability that the particles are at the same place is zero and this exemplifies the exclusion principle. The quantum potential is

$$Q = -\frac{1}{2m}\frac{(k_1-k_2)^2}{2}.$$

This is a constant and therefore it is consistent with the fact that the particles have no relative motion. (Recall that what is described as relative kinetic energy in the usual approach is here attributed to the quantum potential.)

When the time-dependent Schrödinger equation is applied, one will find that these packets move and spread in a way described in chapter 4. We shall however suppose that this spread is negligible, so that the centres of the packets will move with velocities k_1/m and k_2/m respectively. Eventually the wave packets will separate and then the wave function will be proportional to

$$\Psi_t = \exp\left[\frac{i(k_1^2+k_2^2)t}{2m}\right] \times \left\{exp[i(k_1 x_A + k_2 x_B)]\exp\left[-\tfrac{\alpha}{2}\left(x_A - \tfrac{k_1 t}{m}\right)\right]\exp\left[-\tfrac{\alpha}{2}\left(x_B - \tfrac{k_2 t}{m}\right)\right] - exp[i(k_2 x_A + k_1 x_B)]\exp\left[-\tfrac{\alpha}{2}\left(x_A - \tfrac{k_2 t}{m}\right)\right]\exp\left[-\tfrac{\alpha}{2}\left(x_B - \tfrac{k_1 t}{m}\right)\right]\right\}.$$

At this point the two terms in the above wave function will no longer overlap, therefore they constitute separate channels in configuration space, into one of which the system as a whole must enter and remain thereafter. The unoccupied channel then has no effect on the quantum potential acting on the particles. Therefore the system is behaving as if it is in one of these channels alone, for example, the one in which the particle A has momentum $\boldsymbol{k}_1$ and particle B has momentum $\boldsymbol{k}_2$.

In this state, the quantum potential between the two particles is zero as it was in the original state. But evidently while the packets were separating, but still overlapping significantly, there had to be a quantum potential which accelerated particle A to momentum $\hbar \boldsymbol{k}_1$ and decelerated particle B to momentum $\hbar \boldsymbol{k}_2$. The antisymmetric wave function is therefore a particular case of the EPR-type wave function with its nonlocal interactions.

It is clear that once the particles have entered a particular channel in configuration space in which they are distinctly separated in ordinary space, each particle continues to exist with its own identity and continues this identity in its movement. We could in principle measure the position of one of these particles again and again, thus verifying that it continued to exist just as we could if the wave function were not antisymmetric. Although we could not know whether the particle we observed corresponded to the original label A or B, we would nevertheless be sure that we had detected the same particle in successive measurements and that the other, far away particle, could never become confused with the one that we had measured.

Up until now it would appear that the information in the unoccupied channel in configuration space had lost all potential for activity as it does for measurements in general in the way that we have explained in chapter 6. Suppose however that we brought the two particles back together by means of a suitable potential. This potential would also act on the unoccupied channel because of the symmetry of the Hamiltonian. So eventually the two channels would meet, and we would then lose the ability to keep track of the identity of each particle by further measurements, even though each particle would still retain its identity. The quantum potential would then be affected by the channel that had been unoccupied as well as by the one that had been occupied. And the particles would display the effects of the Pauli exclusion principle so that they would never be in the same place.

As pointed out earlier, the reason why the information in the unoccupied channel does not totally lose its potential for activity is because the antisymmetry of the wave function constitutes a superselection rule that is maintained in all interactions and therefore in all measurement processes as well. When there is no superselection rule, the unoccupied channel is treated sufficiently differently from the occupied one so that it would be extremely improbable for them ever to interfere significantly again. But with a superselection rule this will, as we have seen, take place.

It follows that the Pauli exclusion principle and all other effects of the antisymmetry of the wave function are really manifestations of nonlocality. In fact, this nonlocality is so extreme that even particles that are very far apart must maintain unoccupied packets that allow the wave function to be antisymmetric and it is just this whole set of packets that may later come together in such a way as to guarantee that the particles will obey the Pauli exclusion principle whenever these packets meet again.

7.8 On objections to the concept of nonlocality

We have seen that some kind of nonlocality is common to all interpretations of the quantum theory which we have discussed thus far. Nevertheless there have been persistent and strong objections to the consideration of nonlocal theories of any kind, usually with the hope that a local interpretation will sooner or later be possible.

In ancient times it does not seem that strong objections to nonlocality were very common. However, by the sixteenth century nonlocality did not seem to be an acceptable concept. For example, Newton referred to it as a 'philosophical absurdity' [18]. This sort of view continued to modern times. Thus, as we have already pointed out earlier, when Einstein noticed the possible nonlocal implications of the quantum theory, he referred to it as 'spooky' action at a distance [5]. The general mode of explanation that is currently acceptable in science is either action through contact or else action propagated continuously by fields. Indeed anything more than this is often regarded as incompatible with the very possibility of doing science.

We cannot see any well-founded reason for such objections to the concept of nonlocality. Rather they seem to be more or less of the nature of a prejudice which developed together with the growth of modern science. First of all, the concept of nonlocality is perfectly rational in the way that we have used it in our interpretation of the quantum theory. For it has led neither to internal logical contradiction nor to disagreement with any facts. Moreover there is nothing in what we have done that disagrees fundamentally with the current scientific approach. Some people might object that if everything is strongly and nonlocally connected, there is no way to do science, because we will not be able to isolate any system sufficiently to study it. But we have shown that in the context of the large scale level in which physical investigations are carried out, nonlocal effects are not significant. Therefore our interpretation allows for exactly the same degree of separability of systems that is required for the kind of scientific work that is actually being done. We then made nonlocal inferences about the finer and more subtle aspects of material processes, but these are based on ordinary rational scientific inferences. Indeed it is just by using the well-established methods of science that we have been led to make such inferences of nonlocality.

Another objection to nonlocality is that it would violate special relativity. We shall discuss this question in more detail in chapter 12, where we shall show that our interpretation is, in fact, compatible with special relativity. For the present we merely note that, as has been shown in section 7.2, quantum nonlocality will not allow a signal to be transmitted faster than light by means of the statistical measurements that are basic to the current quantum theory in its conventional interpretation. Since our interpretation leads to the same results as do the conventional ones, it follows that through quantum nonlocality there will be no way to exert instant control over what happens at far away places, nor to transmit signals to such places.

We have not yet found what we could regard as a valid logical or scientific reason for dismissing nonlocality. We are therefore led to ask whether there could not be some other kind of reason. It may well be that one of the main reasons that people dislike the concept of nonlocality can be found in the history of science. For in the early period of the development of science there was a long struggle to get free from what may perhaps have been regarded as primitive superstitions and magical notions in which nonlocality clearly played a key part. Perhaps there has remained a deep fear that the mere consideration of nonlocality might reopen the flood gates for what are felt to be irrational thoughts that lurk barely beneath the surface of modern culture. But as we have argued, there is no inherent reason why this should happen. However, even if there were such a danger, this would not by itself constitute a valid argument against nonlocality. Indeed if one were to accept such arguments one might equally well say, for example, that nuclear energy should not even be considered for further investigation because it might lead not only to nuclear war, but also to radioactive contamination and to destructive genetic mutations. While it is necessary to be aware of all these dangers, we still have to be able to enquire freely into all these avenues of research.

7.9 References

1. A. Einstein, B. Podolsky and N. Rosen, *Phys. Rev.* **47**, 777–780 (1935).

2. J. S. Bell, *Speakable and Unspeakable in Quantum Mechanics*, Cambridge University Press, Cambridge, 1987.

3. D. Bohm, *Quantum Theory*, chapter 22, section 15, Prentice-Hall,

Englewood Cliffs, New Jersey, 1951.

4. A. Einstein, in *The Born-Einstein Letters*, ed. I. Born, Macmillan, London, 1971, 158.

5. N. Bohr, *Phys. Rev.* **48**, 696–702 (1935).

6. J. von Neumann, *The Mathematical Principles of Quantum Mechanics*, Princeton University Press, Princeton, 1955.

7. P. H. Eberhard, *Nuovo Cimento* **46B**, 392 (1978).

8. J. S. Bell, *Speakable and Unspeakable in Quantum Mechanics*, chapter 1, Cambridge University Press, Cambridge, 1987.

9. D. Bohm and B. J. Hiley, *Found. Phys.* **11**, 529–546 (1981).

10. J. Clauser, M. Horn, A. Shimony and R. Holt, *Phys. Rev. Lett.* **26**, 880–884 (1969).

11. S. Freedman and J. Clauser, *Phys. Rev. Lett.* **28**, 934–941 (1972).

12. A. Aspect, P. Grangier and G. Roger, *Phys. Rev. Lett.* **47**, 460–466 (1981); *Phys. Rev. Lett.* **49**, 91–94 (1982); A. Aspect, J. Dalibard and G. Roger, *Phys. Rev. Lett.* **44**, 1804–1807 (1982).

13. T. W. Marshal, E. Santos and F. Selleri, *Phys. Lett.* **98A**, 5 (1983).

14. F. Selleri and G. Tarozzi, *La Rivista Nuo. Cim.* **4**, 1–53 (1981).

15. D. M. Greenberger, M. A. Horne, A. Shimony and A. Zeilinger, *Am. J. Phys.* **58**, 1131–1143 (1990).

16. N. D. Mermin, *Phys. Rev. Lett.* **65**, 3373–3376 (1990).

17. N. D. Mermin, *Phys. Rev.* **D22**, 356–361 (1980).

18. I. Newton, *Principia Mathematica*, ed. Mothe-Cajori, London, 1713.

Chapter 8
The large scale world and the classical limit of the quantum theory

8.1 Introduction

We have been discussing what may be called quantum reality as existing on its own without needing any observers, but capable nevertheless of containing observing instruments. We have suggested in several places that there is a continuous transition between this total quantum reality and the large scale world of ordinary experience. The question before us now is whether we can explain how this large scale world as a whole comes out of the basic quantum laws, or at any rate, those features of this large scale world that can be described in terms of classical physics. In other words, is classical physics contained in our interpretation of the quantum theory as a limiting case?

In order to answer this question it is essential to show that quantum mechanics contains a classical level which is not presupposed as in the usual approach, but which follows as a possibility within the quantum theory itself. Here we must point out that this is by no means self-evident. It is of course clear that we will obtain classical behaviour whenever the quantum potential is negligible. But as we shall see, quantum theory generally implies the possibility that even for very high quantum numbers, the quantum potential may still be large. The behaviour at the large scale level is therefore in general not necessarily classical. However, we shall show that the general conditions in the universe are actually such as to imply an approximately classical behaviour for objects containing a sufficiently large number of particles (which means that there is a classical level that is essentially independent of measuring instruments). This is a distinct advantage of the ontological interpretation, for as we shall bring out later, all other interpretations presuppose a classical level (chapter 13) or else imply the assumption of some kind of change in the basic theory, which makes possible the classical level (Stapp [1], Ghirardi, Rimini and Weber [GRW] [2] and Gell-Mann and Hartle [3]).

The classical level is necessary not only because it is observed to exist, but also because within it, quantum mechanical processes can manifest themselves in phenomena which are essentially independent of further observation. As we shall see, it is this which makes it possible for people to establish and share knowledge of an objective nature in spite of the nonlocality of the fundamental laws.

8.2 In what way is classical physics contained as a limit of the quantum theory?

It has commonly been assumed that the classical level arises where $\hbar \to 0$. But evidently, since $\hbar$ is fixed, it cannot go to zero. What we have to do is to derive the classical laws as approximations to the theory in which $\hbar$ is given its current value. But as we have already pointed out, it has to hold under certain conditions (which will be discussed throughout this chapter).

With regard to this point, we repeat that it is clear that wherever the quantum potential can be neglected, the classical limit will hold as an approximation. For in this case the wave function plays no significant part in the determination of the motion of the particles, which then obey the simple Newtonian law

$$\frac{dp}{dt} = -\nabla V. \tag{8.1}$$

In so far as this law holds, the ontology effectively reduces to a set of particles (and of course ultimately, fields, as we shall see in chapter 11) obeying the laws of classical physics.

Moreover, as stated in chapter 3, there is a well-known approximation, the WKB approximation, which gives rise to solutions describing classical trajectories. As we have already pointed out, this approximation becomes valid when the quantum potential is small compared with the kinetic energy. One can show [4] that this requirement will be satisfied whenever the potential varies smoothly enough so that the change in de Broglie wavelength $\Delta\lambda$ within a wavelength is small. One can also show that satisfaction of this condition will guarantee the smallness of the quantum potential.

We emphasise again that this criterion was developed for the special case of a wave going in a definite direction. However, a linear combination of such solutions corresponding to waves going in opposite directions will still be a solution of Schrödinger's equation with the same energy. But such a linear combination will not in general have a negligible quantum potential even though each of the functions that makes it up does have a negligible quantum potential.

Suppose for example that we have a system that varies only in the x-direction and whose properties are uniform in the y- and z-directions. Suppose further that we have an electron moving in the x-direction which is bounded in a space whose edges are defined by a potential $V(x)$ (which we assume to vary slowly enough so that we can apply the standard WKB techniques). One solution for the wave function is then

$$\psi_+(x) = \frac{A}{\sqrt{p(x)}} \exp\left[\frac{i}{\hbar}\int_0^x p(x)\,dx\right] \tag{8.2}$$

where $p(x) = \sqrt{2m(E - V(x))}$.

Another solution is:

$$\psi_-(x) = \frac{B}{\sqrt{p(x)}} \exp\left[-\frac{i}{\hbar}\int_0^x p(x)\,dx\right]. \tag{8.3}$$

Each of these, of course, has a negligible quantum potential except at $x = a$ and at $x = 0$ (where these are the turning points of the particle at which $p(x) = 0$). To find the stationary states of the particle, we take a linear combination of the above two solutions (8.2) and (8.3) in which $|A| = |B|$, while the phases of A and B are chosen so that the wave function fits a decaying exponential beyond $x = a$ and $x = 0$. One finds that solutions are possible only for certain quantum numbers, n, representing the number of nodes of a wave function satisfying the Bohr-Sommerfield condition (see Bohm [4]).

The solution inside the box is then

$$\psi(x) = \frac{|A|}{\sqrt{p(x)}} \sin\left[\frac{1}{\hbar}\int_0^x p\,dx + \phi\right] \tag{8.4}$$

where ϕ is the appropriate phase factor which brings about the fit.

With this linear combination we obtain a quantum potential which may be quite large. A simple calculation shows that this is

$$Q = \frac{p^2(x)}{2m}. \tag{8.5}$$

Since the wave function is real, the particle is at rest inside the box and its kinetic energy has been transformed to the energy of the quantum potential (as explained in more detail in chapter 4).

To simplify the discussion we may consider the case in which the potentials become infinitely high at $x = 0$ and at $x = a$. This is essentially a box of size a and the solutions are

$$\psi(x) = A \sin \frac{n\pi x}{a}. \tag{8.6}$$

It is clear for these solutions that the velocity of the particle is zero and that there are $(n - 1)$ nodes in the wave function at which the probability is zero. The picture according to our interpretation is that the particle is sitting at a particular position so that it is consistent to say that it never reaches a node. For small values of n this perhaps does not violate our intuition. But when n gets large we customarily expect that classical behaviour will result. Yet we can see that, no matter how large n is, we do not go to a classical limit.

Einstein [5] objected to this result because he felt that when n becomes sufficiently large, we should always approach a classical behaviour in which the particle is moving back and forth with equal probability in both directions. Nevertheless independently of our interpretation we can see that this is impossible with the wave function (8.6) because it has nodes at which the probability density is zero. If the particles were moving back and forth at uniform velocity these nodes would not be possible. They might however be possible if the particles speeded up to infinite velocity where the wave function is zero, but this would violate classical intuition even more than would our assumption that the particles are at rest. Einstein's picture could only hold, therefore, if the quantum theory itself began to fail at sufficiently high quantum numbers (for example in the way that could be implied in later suggestions of Rosen [6], as well as by those of Leggett [7] and Penrose [8]).

As we have already indicated, however, we are exploring the hypothesis that quantum mechanics is universally valid, and it is therefore necessary to see whether this is coherent with our general experience at the classical level. Of course it is not part of the quantum theory to say that classical physics *always* holds at the large scale level (or at high quantum numbers). Indeed there are strong reasons to suppose that it should not hold always in this case.

To bring this out, one could assume, for example, that the walls of the box did not correspond to an infinite potential, but rather to a finite, though large, barrier of limited thickness which could, therefore, be penetrated. In this case the system would be analogous to a Fabry-Perot interferometer. Therefore if electrons were incident on one side of the barrier, they could not only penetrate, but could also show transmission resonances for certain incident momenta (see Bohm [4]). These are characteristic quantum

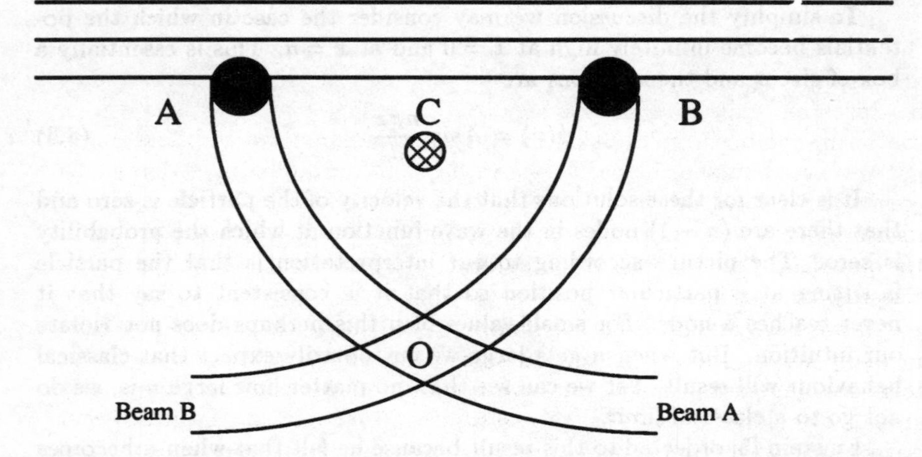

Figure 8.1: Deflection of two different portions of a plane wave by an inhomogeneous field centred at C

mechanical effects. Indeed if these effects are not found even at arbitrarily large distances and for arbitrarily high quantum numbers, we will have to conclude that, as Einstein has implicitly assumed, quantum theory itself must fail here. However, it should be noted that Tonomura *et al.* [9] have demonstrated experimentally with electrons that the quantum theory holds for quite high quantum numbers.

It is necessary nevertheless to explain why such non-classical behaviour at the large scale level is not more commonly encountered. At first sight one might suppose that this could be understood by considering the fact that our 'Fabry-Perot' interferometer would require extremely accurately machined parts and extremely accurately held positions. Clearly these will not be found in nature and will be very difficult to produce in the laboratory. However, this argument is not adequate. This is basically because the criticism raised by Einstein implies more generally that non-classical behaviour will be obtained whenever waves of appreciably different wave numbers cross. In the case of the particle in a box there are only two such waves going in opposite directions. But actually there is a wide variety of solutions in which many waves could cross.

For example we may have a wave function spreading over a large area with a fairly well-defined momentum satisfying the WKB approximation so that the particle will be behaving classically. Then by means of inhomogeneous forces occurring naturally, contributions from two different parts of the wave front may come to overlap, as shown in figure 8.1.

In the region, O, of overlap, the orbits will become highly irregular and the particle will jiggle rapidly because the quantum potential is both large and rapidly varying. Indeed a particle originally passing through A may be deflected while it is in the region O in such a way that it leaves in the beam coming from B. This means that the non-classical behaviour is not restricted to the region O where the quantum potential is large. Of course as we have shown in the previous chapter the statistical results of measurements will still come out consistently without violating the laws of the quantum theory. But the individual processes will have highly non-classical implications.

As we have remarked earlier however, such situations can readily arise in many ways. Indeed as we shall show, even if we consider systems as large as planets and stars, the wave function of the centre of mass may spread over large areas. As long as contributions coming from different parts of the wave front do not cross the behaviour of the objects concerned will be essentially classical. But an inhomogeneous gravitational field, for example, can bring about just such a crossing. And when this happens, the stellar or planetary trajectories would jiggle violently and be deflected in non-classical ways.

If the positions of these objects were measured (e.g. by telescope), such interference would be destroyed. However, it does not seem to be satisfactory in an ontological theory to obtain classical behaviour of large scale and distant objects solely with the aid of measurements.

8.3 Large scale objects in the classical limit

It is clear then that we need to go further to develop a theory of a classical level that arises naturally in the universe. This should give an ontological explanation of not only planetary and stellar behaviour, but also that of large scale objects in general (including, of course, measuring instruments).

Large scale objects in general are described by collective coordinates. Moreover if the classical behaviour of stars or planets would depend on measuring instruments it would still be necessary to show independently that the ontological interpretation of the quantum theory implies the possibility of having measuring instruments which will be stable and behave classically without the need of further instruments to establish these properties.

This question evidently has to be treated in a cosmological context. It is not necessary for our purposes to go all the way back to the big bang. Rather it will be convenient to begin with that phase of the universe that followed the inflationary period. By this time protons and electrons have been formed in states of high energy. But as the universe cooled down it became possible for atoms and molecules to form (the excess energy being given off as radiation which was disposed of by its spreading through the empty space created by the expansion of the universe). From then on as the universe cooled further, larger aggregates of matter were formed leading to galaxies containing stars, planets and other objects of this nature. Later still the temperature dropped so as to form solid matter that had a stable atomic structure (e.g. crystals). Whichever interpretation of quantum theory we use, this result will follow. For as the energy of aggregation is disposed of in the form of field quanta, we will finally be left with the wave functions involving stable configurations of fairly low energy (a detailed account of such processes will be found in Chew [10]).

Once we have stable solid structures then we can form arrangments of matter which constitute measuring apparatuses and we have the possibility of leaving an irreversible macroscopic residue which can constitute a record held in this stable structure. In chapter 6 we did this in terms of the discharge of an ion chamber. Another example would be a photographic plate which would respond to a single particle by irreversibly producing a set of grains of silver. It is thus clear that the existence of the observing apparatus can be understood in terms of our ontological interpretation.

Moreover, wherever stable solid matter is formed, each atom is localised relative to the others within a fairly well-defined range of the order of the interatomic spacing. Because the spread of the wave function is so limited, there will not be significant interference of contributions from different parts of the wave function (which was described earlier in connection with the galaxies, stars and planets). Thus there will be no problem in showing the classical behaviour of various parts of the earth relative to each other, and of all the various macroscopic objects that are attached to the earth (of course, including measuring apparatus, laboratories etc.).

Evidently this sort of localisation of atoms relative to each other depends on the strong forces of mutual interaction between them. However, when it comes to the centre of mass of the whole object there are in general no such strong forces. There is then no reason why the wave function of the centre of mass should not spread out over indefinitely large areas so that its various parts may come to cross in the manner described earlier with resulting non-classical behaviour.

One of the basic questions to be addressed is therefore that of finding out what determines the centre of mass wave functions of the various objects in the universe which need not be strongly localised relative to each other. Consider, for example, the case of galaxies, stars and planets discussed earlier— what determines the spreads of the wave functions of their centres of mass?

To answer this question we have to go back to the process in which these objects were formed from elementary particles. Ideally we would need to know the forms that these wave functions took when the particles were first created. At present there is no well-defined answer to this question, but we may explore a range of possible answers.

Case A If for example the particles are created with localised wave functions, the wave packets would spread so that we would be left with very nearly plane waves in each region of space. The quantum potential would be negligible except when parts of the waves were made to cross as discussed earlier.

Case B The other extreme possibility is that particles are created as standing waves covering the whole of the closed universe (which would, for example, be spherical in the case of a universe described by a Robinson-Walker metric). In this situation the quantum potential would be large everywhere and would remain large as the universe expanded. Even if there were inhomogeneities which broke the spherical symmetry, there would still be a great deal of criss-crossing of waves having different wave numbers and directions, so that the quantum potential would remain generally large. Therefore we would not have an approximately classical level at this stage of evolution of the universe.

To see how a classical level may develop in both cases let us now consider what happens as the particles aggregate to form large scale objects. The mean momentum of the centre of mass will be the sum of the mean momenta of all the particles that have come together. In the case of approximately plane waves resulting from the spread of the original individual wave packets (Case A), we will obtain a well-defined momentum for the centre of mass while the possible positions will spread, in principle, over the whole universe.

In the case of standing waves (Case B), however, the momentum of the centre of mass will not be well defined, but will have a spread due to the spread of the momenta in the wave numbers of the original particles. If we make the reasonable assumption that the momenta of the different particles combine in random ways, the mean spread $|\delta \boldsymbol{k}|$ in the wave number $\boldsymbol{k}$ of the centre of mass will be $|\delta \boldsymbol{k}| = \sqrt{N}|\delta \boldsymbol{k}_0|$ where $|\delta \boldsymbol{k}_0|$ is the average spread in the wave numbers of the individual particles. The quantum potential will be of the order of

$$Q \approx \frac{\hbar^2}{2M}|\delta k|^2 = \frac{\hbar^2|\delta k_0|^2}{2m_0}$$

where $M = Nm_0$ and m_0 is the mass of the original particles. No matter how large N becomes, Q will thus remain more or less the same value as it would be for one atom. In a large system this is evidently negligible. The wave function will then be close to a plane wave (or more generally to a WKB approximate solution of the form $\psi \propto \exp[iS/\hbar]$ in which there is no significant crossing of contributions from different parts of the wave front).

To make a quantitative estimate of the spread of momenta, let us consider the extreme assumption that immediately after inflation the protons have a spread of wave vectors corresponding to a wave number $m_0 c/\hbar$. With this assumption even a small object weighing one gram would have a spread of velocities of a few centimetres per second while an object as large as the earth would have a spread of velocities of 10^{-9} times smaller. Clearly then (both for Case A and Case B) the overall wave function of the whole system will take the form

$$\Psi = f(X)\phi(x_1 \ldots x_{N-1})$$

where $f(X)$ represents the wave function of the centre of mass and $\phi(x_1 \ldots x_{N-1})$ represents the wave function of the internal state of the system. Therefore in general large scale objects in free space will have wave functions spreading over large areas. However, these objects will nevertheless behave classically, except when contributions from different parts of the wave front are made to overlap by the action of inhomogeneous gravitational fields. In this case the wave function becomes

$$\Psi = \left[\psi_{\mathrm{A}}(X)\Lambda_{\mathrm{A}_f}(\Phi) + \psi_{\mathrm{B}}(X)\Lambda_{\mathrm{B}_f}(\Phi)\right] \phi(x_1 \ldots x_{N-1})$$

where $\psi_{\mathrm{A}}(X)$ represents the beam coming from A and $\psi_{\mathrm{B}}(X)$ that coming from B. This wave function gives a quantum potential that is not negligible and we have the possibility of a non-classical behaviour of objects as large as stars and planets. We must evidently find further arguments to eliminate this possibility.

To prevent interference of contributions from different parts of the wave front, we must have some natural process analogous to the making of an irreversible record in the apparatus as a result of a measurement. Essentially what is required is that the general environment be such as to allow the possibility of changes in it to take place which remain stable and which are different for different possible positions of the object in question. There are a number of ways that this might come about, but we will discuss just one of these. What we shall show here is that if there is a suitable radiation falling on these objects we can achieve our objective.

To simplify the discussion we shall begin by considering a planet that is illuminated by light from a single star. If the planet is far from the star in question the light will be parallel. The planet will therefore cast a shadow which will be essentially a cylinder extending in the direction of the light.

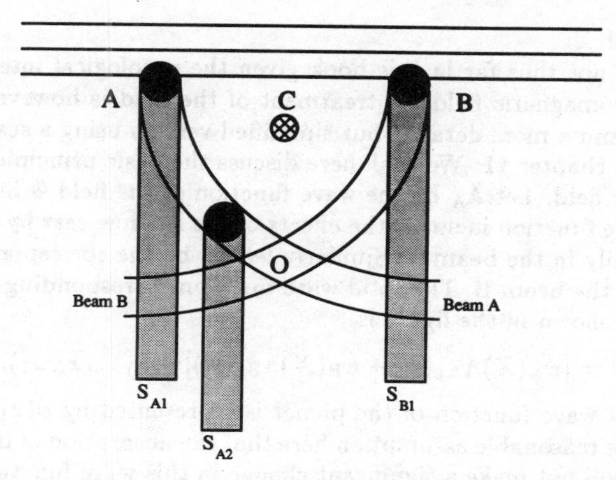

Figure 8.2: Shadows S_{A_1}, S_{B_1} etc. arising from possible positions of planet A, B etc. when illuminated by light from a distant star

Because of diffraction, some light will enter the shadow. The diffraction angle will be of the order λ/a, where λ is the wavelength of the light and a is the size of the object. Then as we move along the cylinder it will begin to fill with light and at a distance of a^2/λ, this shadow will more or less disappear. Therefore the volume of the shadow is of the order a^4/λ. With a typical planet, $a \sim 10^9$ cms and $\lambda \sim 10^{-5}$ cms, the volume is $\sim 10^{32}$ cm^3.

Let us now consider two possible positions of the planet, A and B. Suppose there is a centre, C, of gravitation attraction which causes the two parts of the wave function respectively passing through A and B to overlap at O. Without the starlight we would have interference at O with resulting non-classical behaviour of planets centre of mass (see figure 8.2).

To see what happens in the starlight we must take the shadow into account. If the planet is at A it will have a shadow S_{A_1} but there will be no shadow in the path B. Conversely if the particle is at B, it will have a shadow S_{B_1}, but now there will be no shadow in the path A. Let us consider the case when the planet is at A, then as it moves, its shadow moves with it from S_{A_1} to, say, S_{A_2} as shown in figure 8.2. There will be no shadow associated with the path B. In order to see the effect of these shadows on the quantum potential, we note that the total wave function must now include not only that of the wave function of the planet, $\Psi_P = \psi_P(x)\phi_P(x_1 \ldots x_{N-1})$, but also that of the electromagnetic field as modified by the shadows. We will now show that it is the presence of the wave function of the electromagnetic field that is responsible for the quantum potential acting on the planet to remain small.

We have not thus far in this book given the ontological interpretation of the electromagnetic field. A treatment of the field is however given in Bohm [11], and a more detailed but simplified version using a scalar field is presented in chapter 11. We shall here discuss the basic principles in terms of the scalar field. Let Λ_A be the wave function of the field Φ in the beam A. This wave function includes the effects of the shadow cast by the planet if it is actually in the beam A. Similarly let Λ_B be the corresponding wave function for the beam B. The total wave function corresponding to the two possibilities shown in the figure is

$$\Psi = \left[\psi_A(X)\Lambda_{A_f}(\Phi) + \psi_B(X)\Lambda_{B_f}(\Phi)\right]\phi(x_1 \dots x_{N-1}).$$

The internal wave function of the planet is represented by $\phi(x_1 \dots x_{N-1})$. We make the reasonable assumption here that the absorption of this light by the planet will not make a significant change in this wave function because the planet is so large and the intensity of starlight is so small. Evidently this wavefunction does not depend on whether the planet is at A or at B.

With the arguments that are given in the appendix to chapter 11, we can show that there can be negligible overlap between the overall wave functions corresponding to the two beams. The condition for this will be seen to be essentially that each shadow contains a large number of quanta. When the beams cross there will then be no interference and we will have a classical behaviour.

This condition is in fact satisfied with the light from a single star. To see this we note that the number of quanta in a cylinder of volume V is equal to V/c. With our value of $V \sim a^4/\lambda$ and with the estimated intensity from a typical star as 10^4 quanta per cm^2 per second, we obtain a total number of quanta of the order 10^{25}. Clearly the condition for no interference is satisfied for planets and indeed for objects that are quite a bit smaller, even when the effects of the light from a single star are taken into account.

We have thus far assumed that the star which is the source of the light is at a well-defined position and that there will be no possibility of interference with light coming from the star at another possible position. In effect we have been supposing that the wave function of the star is a small packet which spreads negligibly because its mass is so large. But according to our earlier argument, we cannot guarantee that such a wave function will have arisen by evolution from the early phases of the development of the universe. Therefore we must consider the possibility that even the star has an indefinitely extended wave function, which, however, according to our previous arguments must have a fairly well-defined wave number.

But we can see this will not change our conclusion. Thus according to figure 8.3, if the distance $AA' << D$, the change in angle of the light will be

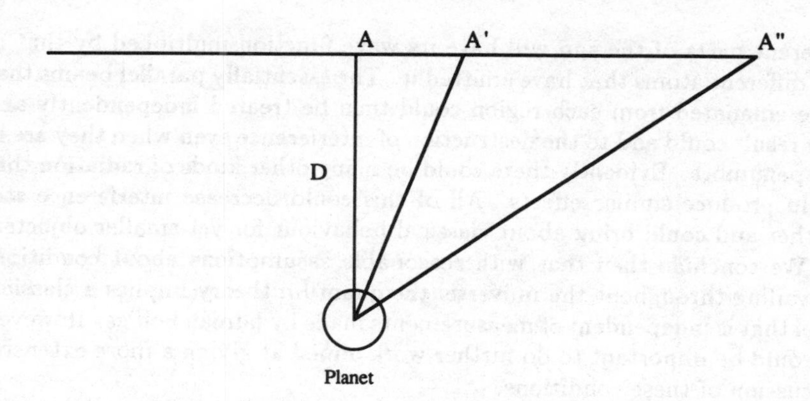

Figure 8.3: Effect of a light source at a finite distance from planet

negligible and everything will happen as before. If, however, AA$''$ $\sim$ D, then the direction of the light will be appreciably different. However, interference between contributions from different planetary positions is only possible if there is some inhomogeneous force which could bring contributions of *the wave function of the star* from A$'$ and from A$''$ together. But these distances are so large that we cannot see any reasons to suppose that this could happen to any significant extent.

At this point it is interesting to ask what will happen with light sources that are closer than the stars and therefore correspondingly stronger, e.g. the sun. Here the light will not be parallel, but will have a certain range of angles α. The range of the shadow is a/α but there is of course the penumbra that extends indefinitely with decreasing intensity. If we consider the shadow alone, the number of quanta in it will be the order of $(a^3/\alpha c)I$, where I is the intensity of the light. In the case of sun light, $\alpha \sim 0.01$ radians and the intensity of light will be $\sim 10^{10}$ quanta per cm^2 per second. For an object of the order of 1 cm in size, this will give rise to something like 10^{10} quanta in the shadow. Clearly for objects of this size and down to about 10^{-3} cms, interference will be destroyed and the behaviour will be essentially classical. Moreover once interference has been destroyed, then the wave packet will not subsequently spread appreciably because the objects are so heavy. (These considerations could thus also be applied to small objects on the surface of the earth that have become free of the solid structure of the earth.)

We have thus far ignored the possible contribution of the penumbra to the destruction of interference and the resultant establishment of classical behaviour. Indeed further investigation shows that light, for example, from different parts of the sun will have its wave function multiplied by that of the different atoms that have emitted it. The essentially parallel beams that have emanated from each region could then be treated independently and as a result could add to the destruction of interference even when they are in the penumbra. Evidently there could be many other kinds of radiation that could produce similar effects. All of this could decrease interference still further and could bring about classical behaviour for yet smaller objects.

We conclude then that with reasonable assumptions about conditions prevailing throughout the universe, the quantum theory implies a classical level that is independent of measurements made by human beings. However, it would be important to do further work aimed at giving a more extensive discussion of these conditions.

8.4 Illustration of the destruction of interference in terms of streams of particles

In order to give a simple explanation of how interference may be destroyed by a beam of light quanta, we shall here discuss a similar process in which an object is exposed to a stream of particles, e.g. neutrons or protons, from a distant source. (In the appendix to Chapter 11 we shall see that essentially the same kind of reasoning will apply to light quanta.)

In this example we will evidently have then a very nearly parallel beam of particles incident on the object in question. We will assume that if a particle strikes the object it will be absorbed and that otherwise there is no interaction with the object. The incident wave function of the nth particle will then be a plane wave $\lambda_n \equiv \exp[ik_n x_n]$. The total wave function of the incident particles will be

$$\Lambda_{\text{incident}} \equiv \prod_n \exp[ik_n x_n].$$

But actually the wave function of each particle will be modified because the object will leave a fairly well-defined shadow, limited only by the diffraction angle $\alpha \sim \lambda/a$. The location of this shadow will depend on the location $\boldsymbol{y}$ of the object. Specifically, the wave function of the nth particle will be $\lambda_n \sim \exp[ik_n x_n]$ except for the region of the shadow where it is zero. The wave function of the nth particle can then be written as

$$\lambda_n(x_n) \sim S(\boldsymbol{y}) \exp[ik_n x_n]$$

where $S(\boldsymbol{y})$ is a function that is unity everywhere except in a cylinder of base of the order of a, which latter is centred at $\boldsymbol{y}$.

Let us now consider the wave function of the object (which corresponds to the planet in our previous example as illustrated in figure 8.2). Our question is whether contributions $\psi_A(y)$ and $\psi_B(y)$ to this wave function from two different regions A and B can be made to overlap in some region O. The net wave function representing these two contributions together with that of the nth incident particle will be

$$\Psi = \psi_A(y)\lambda_{nA}(x_n) + \psi_B(y)\lambda_{nB}(x_n) \quad (8.7)$$

where λ_{nA} is the wave function of the nth particle as affected by the object at A and λ_{nB} is that as affected by the object at B.

In order to make a calculation possible, let us suppose that the wave functions λ_{nA} and λ_{nB} are limited to some box of side L, where L is very large. Interference between the two contributions to Ψ in equation 8.7 will be possible only for those values of x_n for which $\lambda_{nA}(x_n)$ and $\lambda_{nB}(x_n)$ overlap, but will not be possible for those values of x_n for which one of these functions is zero. Since the wave functions $\lambda_{nA}(x_n)$ and $\lambda_{nB}(x_n)$ spread over very large areas, most of the values of x_n will produce overlap, but a fraction of the order of $F = a^3/\alpha L^3$ of the possible values of x_n will not give rise to overlap. This function is essentially the ratio of the volume of the cylinder whose base is the object in question and whose height is the distance $a/\alpha = a^2/\lambda_n$ within which diffraction will not produce an appreciable widening of the shadow.

It is clear then that if we consider just one single particle there will only be a slight probability $P_0 = F$ that interference will be destroyed, and a corresponding probability $P = 1 - F$ that it is not destroyed. But if we consider N particles together, the probability that interference is not destroyed will be the product

$$P_N = (1 - F)^N \sim e^{-NF}$$

(since F is small). Evidently NF is just the number of particles in the shadow, so if this number is large enough interference will be destroyed.

If the incident number of particles per cm^2 per second is I and their mean velocity is V, then the total number of particles in the box will be L^3I/V. The probability P_N will then be

$$P_N \simeq e^{-a^3I/\alpha V} \simeq e^{-a^4I/\lambda_n V}.$$

We can see that interference will be destroyed if I is large enough so that there is a large number of particles in the shadow. This criterion for negligible interference is basically the same as the one we give earlier for light quanta

8.5 The extent of non-classical behaviour

We have already seen that non-classical behaviour could occur for individual particles in many situations in which the quantum numbers are large. Even for larger objects containing many particles there may still be non-classical behaviour except under conditions in which quantum interference is destroyed, for example, by suitable radiation. In this section we shall go in more detail into the nature of this non-classical behaviour.

Let us begin by considering the case of a microparticle moving in a box. This particle is constantly reflecting off the walls of the box. Even if these walls are smooth and regular, the corresponding wave function will reflect back and forth many times so that a great many waves of large wave number going in different directions will criss-cross. If the walls are irregular, this breakup into smaller waves will be even stronger. Eventually we may expect a nearly isotropic distribution of waves whose velocity fluctuates around a mean of zero. At this point there is little kinetic energy in the particle, and most of the energy will be in the quantum potential. Such a behaviour, which is evidently non-classical, can occur with quite large boxes. This behaviour in which the microparticle has a small fluctuating velocity is evidently an extension of Einstein's example of a particle in a one-dimensional box. It could perhaps give an approximate description, for example, of an electron in a metal.

In order to illustrate this behaviour in a simple case, let us consider a particle in a two-dimensional box of side L. We shall simplify further by neglecting the molecular constitution of the walls. The wave function can then be built out of a set of eigenfunctions of the energy of the free particle. That is

$$\psi_{mn} = A \sin \frac{m\pi x}{L} \sin \frac{n\pi y}{L} \exp \left[-\frac{i}{2M} \hbar \frac{\pi^2}{L^2} \left(m^2 + n^2 \right) t \right].$$

Although we are neglecting the molecular constitution of the walls in determining the stationary states, we can still say that the walls produce a net wave function that is a complex linear combination of eigenstates in which all the ψ_{mn} are involved. Let us write this as

$$\psi = \sum_{mn} C_{mn} \psi_{mn}.$$

We have already seen that for a single eigenfunction ψ_{mn}, the particle is at rest. But this need not be so for a general linear combination. Indeed the velocity of the particle is

$$\boldsymbol{v} = \frac{\hbar}{2Mi} \frac{\psi^* \nabla \psi - \psi \nabla \psi^*}{\psi^* \psi}.$$

With $\psi^*\psi = \rho$ we can write for the component of the current

$$\rho v_x = \frac{\hbar\pi}{2MLi}\left[\sum_{\substack{mn\\m'n'}} \exp\left[-\frac{i}{2M}\hbar\frac{\pi^2}{L^2}(m^2+n^2-m'^2-n'^2)t\right]\right.$$
$$\times\left\{m\cos\frac{m\pi x}{L}\sin\frac{n\pi y}{L}C_{mn}C^*_{m'n'}\sin\frac{m'\pi x}{L}\sin\frac{n'\pi y}{L}\right.$$
$$\left.\left. - m'C^*_{mn}C_{m'n'}\cos\frac{m'\pi x}{L}\sin\frac{n'\pi y}{L}\sin\frac{m\pi x}{L}\sin\frac{n\pi y}{L}\right\}\right]$$

and similarly for v_y.

It is clear that for the general C_{mn} this does not vanish. To understand why this is so we may first note that if we take different eigenvalues of the energy we will obtain a probability density that is a function of time. This evidently implies that there is a net current. But to obtain such a current, different eigenvalues of the energy will not be needed as long as these are degenerate. For example we can consider a system with spherical symmetry in which there are three eigenfunctions of the energy corresponding to a p-state. If these are taken as real, they are proportional to x, y and z. For each of these wave functions, the velocity is zero, but for the linear combination $x + iy$, the z-component of the angular momentum is $\hbar$ corresponding to a circulation of the particle around the centre.

To return to the case of a particle in the box, we see from the above equation that the current (and therefore the velocity) is constituted of a large number of terms, each of which is small. As the particle moves, the sin and cosine terms will change rapidly, especially for large m and n. The situation is in someway similar to that of the Lissajous figures because v_x and v_y will contain many terms of very high frequency which are in complex phase relations with each other. Although the wave numbers are in integral relationship here, these are so high that we are approaching the case in which the frequencies of the x- and y-motions are not commensurable. The velocity is thus built of many terms whose phases will have extremely complex relations with each other.

Moreover for a real box, there would no longer be a simple integral relationship of the wave numbers of the eigenfunctions of the Hamiltonian. Any irregularity would destroy this, as would also a change to a more complex shape of the box. This would imply that the frequencies and wave numbers were irrationally related. As is well known, under these conditions in which a motion is constituted of many submotions with very complex phase relationships, the particle will follow an irregular trajectory which can eventually come arbitrarily close to any point in the box. Indeed even if we restricted ourselves to a linear combination of eigenfunctions of the same energy, this sort of result would still follow. In this way we see that going only to the two-dimensional case, we obtain for eigenfunctions of the energy a quasi-chaotic movement. As we shall show in the next chapter, this conclusion is strengthened when we go to more complex systems.

Even if such objects are suspended in liquids or gases, the possibilities of interference will be limited because their mobilities are so low. It is clear that there will be an interesting area of study in the mesoscopic range, between the classical and quantum domains. It is perhaps significant that in this range the simplest forms of life are to be found.

Of course if we try to use precise measurement to observe the behaviour of such objects, they will be put into well-defined positions and will behave essentially classically, provided that they are heavy enough so that their wave packets do not spread significantly during the period of the experiment. This means that the study of the mesoscopic region will require more subtle means than can be provided by exact measurements of the usual kind.

8.6 The quantum world and its classical sub-world

We are now in a position to give at least the general outlines of an overall world view that is implied by our interpretation of the quantum theory.

Our first experience is, of course, of a world that is revealed to us fairly directly by our senses in relation to our outward actions and inward reflections in thought. The immediate experience in this world is that which is described by what is called common sense, but later, as Bohr has pointed out, this is refined, where necessary, to the more exact description of classical physics. Within the domain of such experience it may be said that this world is *manifest*. According to its Latin root, the word manifest would signify what can be held in the hand. More generally it is what can be held in the hand, the eye, and, of course, scientific instruments. Its basic characteristic is that it contains certain relatively stable structures that make the holding possible. These structures must not only be relatively stable, but also essentially local. Everything in this world can be ultimately constituted of such structures which are outside of each other in every sense and which interact only locally.

Without such a world we would not be able to make sense of our observations of matter, nor to assign causes in any orderly way. Indeed it is for this reason that there is a natural reluctance to consider ideas such as nonlocality and indivisible wholeness, i.e. non-separability, into independently existent component parts.

When we come to the underlying quantum world, we find that it has a radically different nature. To be sure we still assume a particle, which at first sight would appear to be what is also done in classical physics. But we now say further that this particle is profoundly affected by the wave function, i.e. through the quantum potential and the guidance condition. The action of the quantum potential depends only on its form and not on its magnitude, so that its effect may be dominant even when the wave intensity is small. This implies the possibility of a strong nonlocal connection of distant particles and a strong dependence of the particle on its general environmental context. The forces between particles depend on the wave function of the whole system, so that we have what we may call 'indivisible wholeness'. This means that for different wave functions we can have radically different connections between particles (not expressible, for example, in terms of a predetermined interaction potential). Thus there is a kind of objective wholeness, reminiscent of the organic wholeness of a living being in which the very nature of each part depends on the whole.

All this behaviour is very different from what is to be expected classically. Indeed as we have already explained in chapter 3, classical physics depends not only on Newton's laws of motion for the particles, but also on the nature of the potentials acting on them. The quantum potential implies so fundamental a difference in the behaviour of particles that we cannot regard the latter in general as remotely classical under most circumstances.

Because of nonlocality, quantum jiggling under quantum interference conditions and other quantum properties to which we have alluded, we may say that the quantum world is *subtle*. According to the dictionary this means "rarified, highly refined, delicate, elusive, indefinable". Its root meaning is based on the Latin *subtexlis* which signifies "finely woven". Clearly the quantum world as we have described it cannot be held in the hand or in any other way. The very effort to hold it, e.g. in measurement, produces thoroughgoing unpredictable and uncontrollable changes in it. Each element participates irreducibly in all the others. The absence of mutual externality and separability of all the elements makes this world very elusive to the grasp of our instruments. It slips through the ordinary 'nets' that we have devised to hold it. Nevertheless we are proposing that it is real and indeed that it constitutes a more basic reality than does the classical 'world'. Indeed as we have shown, this classical 'world' comes out of the theory as relatively autonomous. This autonomy arises wherever the quantum potential can be neglected so that the classical world can be treated on its own as if it were independently existent. But according to our interpretation it is actually an abstraction from the subtle quantum world which is being taken as the ultimate ground of existence.

In the quantum 'world' there is no way to obtain an objective public display of results because everything in it is so irregularly mobile, subtle and mutually interdependent. But in the classical sub-world, events can happen that are negligibly affected by our measurements and observations. This sub-world includes not only particles, but also fields. We have not yet discussed the fields (though, as we have already pointed out, this will be done in some detail in chapter 11). Nevertheless we can say that the movement of classically describable distribution of charge will produce classically describable movements of fields.

This world of fields and particles is what conveys information to our senses in a well-defined way. Indeed we know of no other conditions in which this could happen. For example it is well known that the eye can be sensitive to a few quanta at a time, but the reception of a small number of quanta gives only the vaguest sense of optical stimulation. Meaningful perception requires a large number of quanta and therefore, along the lines we have already explained this will imply an essentially classical behaviour. Meaningful communication between people also requires classically describable processes involving a large number of quanta. Thus it is not that we are *assuming* that the brain responds only to the states of particles and not to their wave functions. Rather we are simply calling attention to the observed fact that meaningful sense perception and communication has to go through the classical level in which the effects of this wave function can be consistently left out of account.

But, of course, this does not mean that we cannot learn about the quantum world. Instead, as we have shown, for example, through a process of amplification and recording in the stable structure of a measuring apparatus, the overall quantum 'world' can manifiest itself in the more limited classical 'sub-world'.

In demonstrating this possibility we have not had to assume any kind of 'cut' between these two 'worlds' such as that supposed by von Neumann. Rather there is only one overall quantum world which contains an approximately classical 'sub-world' that gradually emerges under conditions that have been described throughout this chapter. As we have said earlier we do not believe that this kind of result has been achieved by other interpretations without making assumptions that go beyond the scope of the current quantum theory.

Moreover it is significant to note that it is just the most characteristic quantum properties such as nonlocality and undivided wholeness that bring about the classical world with its locality and separability into distinct components. Thus for example, in a measurement it is the nonlocal quantum potential produced in interaction with what may be a distant piece of apparatus that separates the wave function of the 'observed system' into distinct channels that can be treated locally. Similarly, with planets, it is the wave function of the incident radiation with its distant shadow that destroys interference between contributions from different parts of the wave front and thus brings about a simple local behaviour.

In a certain sense we could say that the overall quantum world measures and observes *itself*. For the classical 'sub-world' that contains the apparatus is inseparably contained within the subtle quantum world, especially through those nonlocal interactions that bring about the classical behaviour. In no sense is the 'observing instrument' really separate from what is observed. The relative autonomy of the classical level that we have already discussed is then what makes it possible for the total quantum world to manifest and reveal itself within itself in a measurement. Thus in contrast to the classical notion of measurement we should regard a quantum measurement as a manifesting process.

How do we understand what happens when the human being perceives the classical world with its implications, that may simultaneously manifest and reveal the quantum world? The first step is through the senses, which as far as physiological sciences now know, can be explained classically. Sensory impulses are then carried into the brain, and even here a great many processes have been studied which likewise can be understood classically. Most neuroscientists seem to believe however that the brain can be *completely* treated in terms of classical concepts. At present this is evidently a speculative assumption. However, some neuroscientists, notably Eccles [12], have suggested that quantum processes may be important in understanding the more subtle activities of the brain. For example as has already been pointed out, we know that retinal cells respond to a few quanta at a time and that this response leads to a multiplication of their effects to a classical level of intensity. But the retina is just an extension of the brain. There could evidently be other parts of the brain in which such a sensitivity may exist, e.g. in certain kinds of synapses. If this were the case, then the brain would be a system that could, like a measuring apparatus, manifest and reveal aspects of the quantum world in the overall processes. Such quantum sensitivity would imply that in more subtle possibilities of behaviour of the brain, a classical analysis would break down.

All this means that as the processes of perception unfolds into the brain, it may as it were connect to the subtle quantum domain which latter may in turn reconnect to the classical domain, as outgoing action is determined through amplification of quantum effects.

Moreover it does not necessarily follow that quantum effects will only be of importance in the domain of very small energies. For as we have seen it is just through certain kinds of nonlocality that locality can emerge, e.g. in a measuring apparatus. Similar nonlocality may be required for the brain to have a local and essentially classical sub-domain of function.

Finally there is no reason to believe that quantum theory is an ultimate truth. Rather, like other theories, it probably has a limited domain of validity. Indeed various authors including ourselves have already suggest new theories containing quantum theory as an approximation within it and some of these will be discussed in chapter 15. Once again it may well be that a fuller understanding of the brain will require such more extensive theories going beyond the quantum mechanics.

8.7 References

1. H. P. Stapp, 'Einstein Time and Process Time', in *Physics and the Ultimate Significance of Time*, ed. D.R. Griffin, State University Press, New York, 1986, 264–270.

2. G. C. Ghirardi, A. Rimini and T. Weber, *Phys. Rev.* **D34**, 470–491 (1986).

3. M. Gell-Mann and J. B. Hartle, 'Quantum Mechanics in the Light of Quantum Cosmology', in *Proc. 3rd Int. Symp. Found. of Quantum Mechanics*, ed. S. Kobyashi, Physical Society of Japan, Tokyo, 1989.

4. D. Bohm, *Quantum Theory*, Prentice-Hall, London, 1951; also available in Dover Publications, New York, 1989.

5. A. Einstein, in *Scientific Papers Presented to M. Born on his retirement*, Oliver and Boyd, London, 1953.

6. N. Rosen, *Found. Phys.* **14**, 579 (1984).

7. A. J. Leggett, *Prog. Theor. Phys., supp.* **69**, 80 (1980).

8. R. Penrose, *The Emperor's New Mind : Concerning Computers, Minds and the Laws of Physics*, Oxford University Press, Oxford, 1989.

9. A. Tonomura, T. Matsuda, J. Endo, H. Todokoro and T. Komoda, *J. Electron Micros.* **28**, 1–11 (1979).

10. G. F. Chew, *Phys. Rev.* **45A**, 4312–4318 (1992).

11. D. Bohm, *Phys. Rev.* **85**, 166–193 (1952).

12. J. C. Eccles, *Proc. Roy. Soc.* **227B**, 411–428 (1986).

Chapter 9
The role of statistics in the ontological interpretation of the quantum theory

9.1 Introduction

In our interpretation, the primary significance of the wave function is that it is a quantum field which determines information that is active on the particles in each individual system. However, we also say that the wave function determines the probability density in a statistical ensemble through the relationship $P = |\Psi|^2$. But as already pointed out in chapter 3, section 3.4, this is regarded as a secondary significance of the wave function. In principle there is no reason why the probability could not be different from $|\Psi|^2$, even though it is equal to $|\Psi|^2$ in all cases that we have encountered so far.

What we have to explain then is why P should tend to approach $|\Psi|^2$ in typical situations that are currently treated in physics (i.e. situations in which the quantum laws are valid). In this chapter we shall give such an explanation showing that one can understand how an arbitrary probability density, P, may approach $|\Psi|^2$ even on the basis of our deterministic theory because the latter leads to chaotic motions under a wide range of conditions. We shall then show how the overall statistical approach may be generalised to include, not only what are usually called pure states, but also what are usually called mixed states (which are at the basis of quantum statistical mechanics). Finally we shall extend this study and show how the approach of P to $|\Psi|^2$ could further be justified on the basis of an underlying stochastic process in the movement of the particles.

9.2 Chaotic behaviour of particles in many-body systems

We have already seen in chapter 8, section 8.5 that even for a single particle in a two-dimensional box, we obtain a fairly complex and chaotic motion. With N particles in a three-dimensional box, this will evidently become still more chaotic.

To illustrate what happens in such cases, let us consider a cubic box of side L. However, instead of supposing that the walls are impenetrable, we shall make the more realistic assumption that the particles can actually touch the walls where they could interact with its constituent atoms. We shall not, however, treat these interactions in detail, but we shall merely suppose that, as in chapter 8, section 8.4, they lead to a complex linear combination of stationary wave functions. To simplify calculations (which are intended to indicate what sort of thing may happen), we also assume that the wave functions are periodic with period L. For a single particle the basic eigenfunctions are $\exp[i\boldsymbol{k}\cdot\boldsymbol{x}]\exp[-i\omega_k t]$ where $\boldsymbol{k}_x = 2\pi l/L$, $\boldsymbol{k}_y = 2\pi m/L$ and $\boldsymbol{k}_z = 2\pi n/L$, and where $\omega_k = k^2/2m$.

For the N-particle system, the eigenfunctions of the energy are $\exp[i\sum_n \boldsymbol{k}_n\cdot\boldsymbol{x}_n]$ where $\boldsymbol{x}_n$ represents the position of the nth particle and $\boldsymbol{k}_n$ is proportional to its momentum. The general wave function can then be written as

$$\Psi = \sum_{k_1,k_2\ldots k_N} C_{k_1,k_2\ldots k_N}\exp\left[i\sum_n \boldsymbol{k}_n\cdot\boldsymbol{x}_n\right]\exp\left[-i\sum\omega_k t\right]. \qquad (9.1)$$

For equivalent particles, the Cs will have to be restricted to anti-asymmetric functions. But here we need not specify what is the symmetry as this will be included in the values of C.

The coefficients C can be very complex, for example interaction with the molecules constituting the walls may produce interfering combinations of wave packets which will involve a very wide range of possible values of k in the manner discussed in chapter 8, section 8.3. We may expect that after the system has been left by itself for a long time, there will indeed be such a linear combination.

If we write $\rho = |\psi|^2$ and let $\boldsymbol{v}_s$ be the velocity of the sth particle, we get

$$\rho\boldsymbol{v}_s = \frac{1}{2mi}\left[\psi^*\boldsymbol{\nabla}_s\psi - \psi(\boldsymbol{\nabla}_s\psi^*)\right]. \qquad (9.2)$$

This then gives

$$\rho v_s = \frac{1}{2m}\sum_{\{k\}}\sum_{\{k'\}} C^*_{\{k\}} C_{\{k'\}}(k_s + k'_s)$$
$$\times \exp\left[-i\left(\sum_{n'} k_n x_n - k'_{n'} x'_{n'}\right)\right] \exp\left[-i\sum_{mm'}(\omega_m - \omega'_m)t\right] \tag{9.3}$$

where $C_{\{k\}} = C_{k_1, k_2 \ldots k_N}$.

It is clear that the above is built out of a very large number of terms each of which oscillates with a phase that is in effect random relative to those of the others. The point in configuration space corresponding to $\boldsymbol{x}_1 \ldots \boldsymbol{x}_n$ moves in a chaotic way through all the configuration space that is accessible to it.

Intuitively we could expect that for statistical averages, this motion would be equivalent to some kind of probability distribution in the configuration space. What would this probability distribution be?

To enquire into this, we first note that the function $\rho = \psi^* \psi$ satisfies a conservation equation in configuration space which is

$$\frac{\mathrm{d}\rho}{\mathrm{d}t} + \sum_n \boldsymbol{\nabla}_n \cdot \rho \boldsymbol{v}_n = 0. \tag{9.4}$$

Of course this does not mean that ρ itself is necessarily the probability. We may therefore write this probability density as $P = \rho F$. By definition, P satisfies the same conservation equation as ρ. We then obtain

$$\frac{\partial F}{\partial t} + \sum \boldsymbol{v}_n \cdot \boldsymbol{\nabla}_n F = 0 = \frac{\mathrm{d}F}{\mathrm{d}t}. \tag{9.5}$$

This means that F is a constant of the motion. Suppose for example that we started with a distribution which was localised in an element of configuration space $\Delta\Omega = (\Delta\boldsymbol{x}_1, \Delta\boldsymbol{x}_2 \ldots \Delta\boldsymbol{x}_N)$. This element is chosen to be small relative to macroscopic dimensions, but large in relation to the wavelengths implied by a typical $\boldsymbol{k}$ vector. Because F is a constant of the motion, we can see what happens to the element $\Delta\Omega$ by following the various particle trajectories on its boundaries. Consider for example two trajectories initially separated by $\delta\boldsymbol{x}$. These will correspond to large phase differences between most of the contributions to the velocity in equation (9.3). Therefore the two trajectories will correspond to very different quasi-random paths that will in general lead them very far from each other. It follows that a simple element $\Delta\Omega$ will turn into a very complex, long drawn out and enmeshed thread-like structure. Because any trajectory eventually comes near to any given point in configuration space, it also follows that in the sense of a coarse grained average, this thread-like structure will move in such a way as to more or less cover the whole of this space. Therefore in any process that does not depend on the complex and chaotic fine details of the motion, F may be taken as effectively a constant. In such a process we may say the probability is effectively $\rho = |\psi|^2$.

There are a very wide range of conditions in which the sort of process described above may take place. These include metals, gases, plasmas etc., so that not only in the laboratory, but also in the stars and interstellar space, we may reasonably expect that quantum theory itself contains processes that tend to produce distributions near $P = |\psi|^2$. Therefore it is not necessary to regard $P = |\psi|^2$ as an assumption as is done in conventional interpretation. Although our interpretation has the additional assumption of particles, this is balanced by the fact that it does not require the usual assumptions on probability. So at least on the basis of a formal count of the number of assumptions, we cannot say that either interpretation is favoured over the other by the principle of Occam's razor.

It is instructive at this point to consider once again our example of electrons emerging from a metal and going through a set of collimating slits and velocity selectors to enter a slit system in an interference experiment. To simplify the discussion, suppose we have cold emission of electrons brought about by a strong electric field near the surface of the metal. This combined with the wave function gives rise to a barrier of finite width which can be penetrated. The wave function (9.1) implies that electrons are constantly striking this barrier and a small part of the wave function penetrates as explained in chapter 5, section 5.1. Some electrons will go through if their initial conditions are right. If an electron does go through, its wave function can then be regarded as simply the part that has penetrated, multiplied by the wave function of the metal containing the remaining electrons. It is therefore factorised so that the metal and the electrons behave independently from this point on.

Let us now consider an ensemble of cases in which electrons have left the metal in the way described above. There will be a distribution of electrons, each of which has its own wave function and its own initial position. Because the electrons are independent of each other and of the metal, we can follow the propagation of each of their wave functions according to Schrödinger's equation for a one-body system. This implies that they will move onward and pass through the collimator and velocity selector. When they have done this all the electrons will have wave functions of practically the same form. Because the distribution $P = |\psi|^2$ is also propagated by Schrödinger's equation, it follows that our statistical ensemble of electrons will enter the slit system with the probability $P = |\psi|^2$ where ψ is the common wave function of all these electrons. In this way we understand how the ordinary probability distributions arise through the chaotic motions of the particles in their source and through the selection of a common wave function by the apparatus that prepares the experimental ensemble.

9.3 Statistics of wave functions

In the previous section we have discussed a case in which electrons were selected so that they all have the same wave function. This corresponds to what has commonly been called a pure state. But this state is still a statistical distribution, since it refers to an ensemble of particles with different positions even though they have wave functions of the same form. But is it actually appropriate to call this ensemble a 'state'? For this word generally connotes some property of an individual system rather than of a selected collection of systems. In the interests of clarity it would perhaps be better to call this collection of particles a pure ensemble rather than a pure state.

Suppose however that we have a situation in which electrons are selected by a procedure that does not determine the wave function very precisely, e.g. the collimating slits are wide and the velocity selectors permit a large range of velocities to go through. In this case the wave function itself will in general be significantly different for each particle. In terms of our example of electrons coming out of a metal, we can see that the wave function in the metal contains effectively a random distribution of eigenfunctions of the total energy. Electrons will therefore emerge with some statistical distribution of wave functions, as well as of particle positions. This will correspond to what is commonly called a mixed state, although we shall, in accordance with our proposed terminology, call it a mixed ensemble.

The distinction between pure and mixed ensembles clearly applies only in a statistical context. An individual system is described neither by a pure nor by a mixed ensemble. Rather it is characterised by specifying the wave function and the positions of all the particles in the system (as well as values of the relevant field variables). Thus, for example, the state of the universe has to be understood in this way because we are not assuming an ensemble of universes.

Any quantum experiment always involves a statistical selection of similar sub-systems within the universe. Such a selection will give rise, in the way just described, either to a pure or to a mixed ensemble. More generally under certain conditions a similar selection might even take place naturally so that, for example, a set of portions of an interstellar plasma might be treated in terms of suitable mixed ensembles.

Our basic concept is, however, that for each member of the ensemble the wave function has some definite form to which we can apply our interpretation of the quantum theory in the usual way. Up until now we have been discussing in this book only the case of a pure ensemble for which averages are calculated by integrating the probability density $P = |\psi|^2$ over all the particle positions. However, in a mixed ensemble, the wave function is distributed over a range of possible forms. The statistical properties of such an ensemble will have to be calculated by averaging over the distribution of wave functions, in addition to averaging over the particle positions for each wave function.

The general wave function may be represented by an expression in terms of an orthonormal basis which, for convenience, we shall take to be eigenfunctions of the energy. In the previous section we did this for free particles in a box and the wave function was given by equation (9.1). More generally we may write

$$\Psi = \sum_{E\alpha} c_{E\alpha} \psi_{E\alpha}(x)$$

where E is the energy and α runs over all the states with the same energy. Any individual system must have a wave function described by a particular set of values of $c_{E\alpha}$. It will be useful also to write

$$c_{E\alpha} = R_{E\alpha} \exp[i\phi_{E\alpha}], \tag{9.6}$$

so that

$$\Psi = \sum_{E\alpha} R_{E\alpha} \exp[i\phi_{E\alpha}] \psi_{E\alpha}(x). \tag{9.7}$$

If we have an ensemble of wave functions, there will have to be a corresponding probability distribution of the $R_{E\alpha}$ and the $\phi_{E\alpha}$. To describe this distribution we shall introduce the concept of wave function space. A point in wave function space is determined by all the $c_{E\alpha}$, or alternatively by all the $R_{E\alpha}$ and $\phi_{E\alpha}$. Each individual system then corresponds to a point in wave function space.

Let us now define the element of volume in wave function space. Writing $c_{E\alpha} = u_{E\alpha} + iv_{E\alpha}$, we obtain for this element

$$\mathrm{d}\Omega = \prod_{E\alpha} \mathrm{d}u_{E\alpha} \mathrm{d}v_{E\alpha}. \tag{9.8}$$

Since $P_{E\alpha} = R_{E\alpha}^2 = u_{E\alpha}^2 + v_{E\alpha}^2$ and since $\phi_{E\alpha} = \tan^{-1}(v_{E\alpha}/u_{E\alpha})$, we have

$$\mathrm{d}\Omega = \prod_{E\alpha} \mathrm{d}p_{E\alpha} \mathrm{d}\phi_{E\alpha}. \tag{9.9}$$

The mean number, $\mathrm{d}Z$, of systems in this element is then

$$\mathrm{d}Z = P(\ldots R_{E\alpha}\ldots;\ldots\phi_{E\alpha}\ldots)\,\mathrm{d}\Omega \tag{9.10}$$

where $P(\ldots R_{E\alpha}\ldots;\ldots\phi_{E\alpha}\ldots)$ is the probability density of systems in wave function space.

For a given wave function ψ, physical averages will evidently be determined by quantities such as

$$\overline{O} = \int \psi^* O\psi\,\mathrm{d}x = \int \sum_{E_1E_2}\sum_{\alpha_1\alpha_2} c^*_{E_1\alpha_1}c_{E_2\alpha_2}\psi^*_{E_1\alpha_1}O\psi_{E_2\alpha_2}\,\mathrm{d}x. \tag{9.11}$$

If we have a statistical distribution of wave functions, we evidently have to average over this ensemble too. We then obtain

$$\begin{aligned}\overline{\overline{O}} \;=\; & \int\!\!\int\sum\sum P(\ldots R_{E_1\alpha_1}\ldots R_{E_2\alpha_2}\ldots;\ldots\phi_{E_1\alpha_1}\ldots\phi_{E_2\alpha_2}\ldots) \\ & \times R_{E_1\alpha_1}R_{E_2\alpha_2}\mathrm{e}^{i(\phi_{E_2\alpha_2}-\phi_{E_1\alpha_1})}\psi^*_{E_1\alpha_1}O\psi_{E_2\alpha_2}\,\mathrm{d}x\,\mathrm{d}\Omega.\end{aligned} \tag{9.12}$$

Let us define

$$\int \psi^*_{E_1\alpha_1}O\psi_{E_2\alpha_2}\,\mathrm{d}x = O_{E_1\alpha_1,E_2\alpha_2} \tag{9.13}$$

and

$$\begin{aligned}\rho_{E_1\alpha_1E_2\alpha_2} \;=\; & \int P(\ldots R_{E_1\alpha_1}\ldots R_{E_2\alpha_2}\ldots;\ldots\phi_{E_1\alpha_1} \\ & \ldots\phi_{E_2\alpha_2}\ldots)R_{E_1\alpha_1}R_{E_2\alpha_2}\mathrm{e}^{i(\phi_{E_2\alpha_2}-\phi_{E_1\alpha_1})}\,\mathrm{d}\Omega.\end{aligned} \tag{9.14}$$

We then obtain

$$\overline{\overline{O}} = tr(\rho O). \tag{9.15}$$

From this it follows that ρ is the usual density matrix since all averages are obtained from it in the usual way.

All that we have said holds both in our interpretation and in the conventional interpretation. For thus far we have not used the assumption of a definite particle position. In both interpretations ρ has only an essentially statistical significance. We have to emphasise, however, that in our interpretation the particle velocity in each individual case is determined by the actual wave function for this case which is

$$\Psi = \sum c_{E\alpha}\psi_{E\alpha}(\boldsymbol{x}).$$

Writing

$$P(\boldsymbol{x}) = \sum\sum c^*_{E\alpha} c_{E'\alpha'} \psi^*_{E\alpha}(\boldsymbol{x}) \psi_{E'\alpha'}(\boldsymbol{x})$$

we obtain the following expression for the momentum $\boldsymbol{p}(\boldsymbol{x})$ at the point $\boldsymbol{x}$

$$P(\boldsymbol{x})\boldsymbol{p}(\boldsymbol{x}) = \tfrac{1}{2}\sum\sum c^*_{E\alpha} c_{E'\alpha'} \psi^*_{E\alpha}(\boldsymbol{x}) \psi_{E'\alpha'}(\boldsymbol{x}) \left[\nabla\lambda_{E\alpha}(\boldsymbol{x}) + \nabla\lambda_{E'\alpha'}(\boldsymbol{x})\right],$$

where

$$\psi_{E\alpha}(\boldsymbol{x}) = \chi_{E\alpha} \exp\left[i\lambda_{E\alpha}(\boldsymbol{x})\right].$$

Clearly the momentum for each case depends on all the coefficients $c_{E\alpha}$ which determine the wave function for that case. The probability density $P(\ldots R_{E\alpha}\ldots;\ldots\phi_{E\alpha}\ldots)$ then evidently implies some corresponding distribution over the momenta $\boldsymbol{p}(\boldsymbol{x})$. If we carry out this average we will obtain the average momentum density

$$P(\boldsymbol{x})\boldsymbol{p}(\boldsymbol{x}) = tr\left\{P_{E\alpha E'\alpha'} \psi^*_{E\alpha}(\boldsymbol{x}) \psi_{E'\alpha'}(\boldsymbol{x}) \tfrac{1}{2}\left[\nabla\lambda_{E\alpha}(\boldsymbol{x}) + \nabla\lambda_{E'\alpha'}(\boldsymbol{x})\right]\right\}.$$

$P_{E\alpha E'\alpha'}$ will therefore give only a statistical average of the momentum over the various wave function coefficients.

Although the density matrix is sufficient to determine, not only the average of the momentum, but all relevant physical averages over the ensemble, it evidently does not determine the probability distribution $P(\ldots R_{E\alpha}\ldots;\ldots\phi_{E\alpha}\ldots)$ itself. That is to say, physical averages are extremely insensitive to the detail of the distribution P. Indeed it is clear that to a given density matrix, a very wide range of such distributions is possible. Therefore within the framework of the statistical mechanics of quantum theory, the function P is highly ambiguous. Yet conceptually it is, in principle, determined for each ensemble. It is true that in doing any calculations, it is not necessary to say anything about P over and above what is said about the density matrix. Yet both in the conventional interpretation and in our interpretation, conceptual clarity requires the notion of a probability distribution in wave function space. In our interpretation this is clearly needed because we have to suppose the wave function has a given form $\psi = Re^{iS}$ before we can say what is the particle velocity, $v = \nabla S/m$. We then consider a statistical distribution of such forms. In the usual interpretation the basic postulate also begins with the wave function and the density matrix is derived from this. The density matrix is not a fundamental concept in any of the interpretations of the quantum theory (though some, such as Prigogine [1] and Hawking [2], have proposed new theories which are extensions of quantum mechanics in which the density matrix does play a fundamental role).

To illustrate the ambiguity of the statistical distributions underlying the density matrix, let us consider an atom of total angular momentum

$\hbar$. In the z-direction, for example, the three possible eigenfunctions of the angular momentum will be denoted by $\psi_j(\boldsymbol{x})$ with $j = -1, 0, +1$. The general wave function may be written as

$$\psi = \sum c_j \psi_j(\boldsymbol{x}) = \sum R_j e^{i\phi_j} \psi_j(\boldsymbol{x}). \tag{9.16}$$

The probability distribution in wave function space can then be written as $P(\ldots R_j \ldots; \ldots \phi_j \ldots)$. The density matrix is then

$$\rho(\boldsymbol{x}, \boldsymbol{x}') = \int \sum_{jk} P(\ldots R_i \ldots; \ldots \phi_i \ldots) R_j R_k e^{i(\phi_j - \phi_k)} \psi_j(\boldsymbol{x}) \psi_k(\boldsymbol{x}') \, d\Omega. \tag{9.17}$$

If P does not depend on ϕ_i, this signifies random phases. In this case the density matrix reduces to

$$\begin{aligned} \rho(\boldsymbol{x}, \boldsymbol{x}') &= \int \sum_j P(\ldots R_i \ldots) R_j^2 \psi_j^*(\boldsymbol{x}) \psi_j(\boldsymbol{x}') \, d\Omega \\ &= \sum_j \overline{R_j^2} \psi_j^*(\boldsymbol{x}) \psi_j(\boldsymbol{x}'). \end{aligned} \tag{9.18}$$

This is evidently diagonal in the angular momentum quantum numbers j. If we further assume that the distribution is a symmetric function of $\ldots R_j \ldots$, all the $\overline{R_j^2}$ will be the same and the density matrix will be

$$\rho(\boldsymbol{x}, \boldsymbol{x}') = \sum_j \overline{R^2} \psi_j^*(\boldsymbol{x}) \psi_j(\boldsymbol{x}'). \tag{9.19}$$

A very wide range of $P(\ldots R_j \ldots)$ can evidently lead to the above density matrix. This matrix has an isotropic significance because it takes the same form even if we choose our z-axis in some other direction. This means that the distribution of angular momenta obtained in a measurement will be independent of the orientation of the apparatus. But from this distribution we cannot even say that the original $P(\ldots R_j \ldots)$ is necessarily isotropic, since, as we have seen, these results follow merely from the symmetry under interchange of any pair of variables R_j and R_k. More generally it is evident that from the observable distribution of the results of measurements, we can say very little about the details of the distribution over the $\ldots R_j \ldots$.

To further bring out the ambiguity of the $P(\ldots R_j \ldots)$, suppose that we had three ensembles in each of which the z-component of the angular momentum was well defined to be $-1, 0, +1$, respectively. If these are mixed with equiprobability, this will give rise to the diagonal density matrix (9.19). But suppose the mixed ensemble was formed by having a sequence in which

+1 is always followed by 0 and this by −1, which in turn is followed by +1, and so on. The mixed ensemble would still imply isotropic averages, so that an experimenter would obtain the same statistical distribution no matter in which direction he measured the spin. Yet if he chose to measure only every third member of the ensemble, he would obtain a non-isotropic result corresponding to a spin in the z-direction of $+1, 0$ or -1. But the density matrix itself for the mixed ensemble does not show any sign of such a possibility. It is thus clear that many different physical situations may give rise to the same density matrix for the same overall ensemble.

Even if the various sub-ensembles are chosen by a random process this may still hold. For example, one could use a random number generator to determine which of the above three sub-ensembles are chosen to be fed into the mixed ensemble. An experimenter who knew the order determined by the random number generator could then obtain a non-isotropic result for the measurements of the angular momentum. Of course in natural situations, we cannot know how such a random selection occurs and in such cases we will always obtain an isotropic result. But this does not change the fact that a given density matrix leaves open a vast range of physical states that could give rise to it.

There may be yet further ambiguities in the physical state underlying the density matrix. Thus in many cases, our ensembles have come from larger systems, such as, for example, the electrons emerging from a metal. In principle there may still be correlations of one kind or another between the electrons of interest and those left in the metal. To be accurate, we should therefore base our considerations on the larger system containing both of these. Hence we should start with a probability distribution over the wave function coefficients of the combined system to form a corresponding density matrix for this combined system. We may write this as $\rho(\boldsymbol{x} \ldots, \boldsymbol{y} \ldots, \boldsymbol{x}' \ldots, \boldsymbol{y}' \ldots)$ where $\boldsymbol{y}$ represents the coordinates of electrons left in the metal.

As long as we are concerned only with those electrons that have emerged from the metal, we can simplify the treatment by integrating over the coordinates of the electrons that are left in the metal, while equating $\boldsymbol{y}$ with $\boldsymbol{y}'$. This will give us a reduced density matrix $\rho(\boldsymbol{x} \ldots, \boldsymbol{x}' \ldots)$ from which the statistical properties of the electrons that have left the metal can be calculated. (This sort of density matrix has, for example, been used by Feynman [3].) As we have already indicated, the original density matrix may contain correlations of all kinds, including, for example, EPR correlations, but these no longer appear explicitly in the reduced density matrix, although their effects are indirectly taken into account.

9.4 The density matrix as determining all the physically significant results

We emphasise then that the density matrix, in general, contains a condensation of information in which most of the details of the individual system do not appear, but which contains just that information which is relevant statistically. In particular the density matrix determines, not only the statistics of all processes of measurement and of all processes of transition in general, but also that of large scale manifestations of the microlevel, and ultimately all collective large scale averages, the latter being the basis of the description of the manifest world discussed in chapter 8.

Firstly we see from equation (9.15) that the density matrix determines the statistical results of all measurements. To show how it determines the statistics of transition processes, let us, for example, once again consider the case of cold emission from a metal through barrier penetration. This will, in general, give rise to a mixed ensemble described by a suitable density matrix in the manner that we have described earlier. (With the aid of further selection processes through collimating slits etc., we can, of course, approach as close as we please to a pure ensemble.) Each individual member of the mixed ensemble corresponds to a well-defined wave function, ψ, with a further ensemble of particle positions $P = |\psi|^2$. For such a wave function, the probability of transition can be obtained by integrating the part of $|\psi|^2$ corresponding to final state over the particle positions. We then have to integrate this over the distribution of wave functions. The final result will then be the same as that obtained by starting with the initial density matrix and integrating the trace of the final matrix over the ensemble corresponding to the final set of possible results of the transition process.

Another important physical significance of the quantum theory is that it determines probabilities of large scale manifestations of microsystems. We have already given an example of this in chapter 6, where we discussed an ion chamber in which a single atom could initiate a cascade leading to a current of macroscopic magnitude. Such a process depends on an unstable situation created by the electric field which allowed the cascade to develop and eventually produce a large scale result. But this sort of process does not require an observing apparatus or any other set of conditions arranged by human beings. For example, in an interstellar plasma there could be a strong electric field in which individual electrons could initiate such cascades. In all these cases, the probability of any given fluid distribution of the macroscopic currents would depend ultimately on the ensemble of particle positions in the atoms and the ensemble of wave functions of these atoms. However, this dependence is such that the net statistical result could always be obtained from the density matrix in the usual way.

Let us now come to the large scale averages of the collective variables themselves. As we have already pointed out, the variables will include the energy, momentum and angular momentum of the whole system and of its relevant sub-systems, as well as certain collective functions of the position (such as centre of mass and Fourier coefficients of the particle density which describe sound waves). Typically at the large scale level, systems will be close to stationary states. It will therefore be convenient to express wave functions in terms of eigenfunctions of the energy, $\psi_{E\alpha}(\boldsymbol{x})$.

We begin with a pure ensemble. The general wave function for such an ensemble is then

$$\Psi(\boldsymbol{x},t) = \sum c_{E\alpha}\psi_{E\alpha}(\boldsymbol{x})\mathrm{e}^{-iEt/\hbar}.$$

The mean value of the energy is

$$\overline{E} = \sum c^*_{E\alpha}c_{E\alpha}E. \tag{9.20}$$

But a typical large scale system has to be described by a mixed ensemble (e.g. because it is generally abstracted from a larger context in the way that we have already discussed). Let the probability distribution of wave functions in this mixed ensemble be $P(\ldots R_{E\alpha}\ldots;\ldots\phi_{E\alpha}\ldots)$. The density matrix will then be

$$\begin{aligned}\rho_{E\alpha E'\alpha'} &= \int P(\ldots R_{E\alpha}\ldots;\ldots\phi_{E\alpha}\ldots)\\ &\quad\times R_{E\alpha}R_{E'\alpha'}\mathrm{e}^{i(\phi_{E\alpha}-\phi_{E'\alpha'})}\mathrm{e}^{-i(\omega_{E\alpha}-\omega_{E'\alpha'})t}\,\mathrm{d}\Omega.\end{aligned} \tag{9.21}$$

In general this density matrix is diagonal as a function of E, because the phases will be random. We may always choose a representation in which it is also diagonal in terms of α and α'. We therefore write

$$\rho_{E\alpha E'\alpha'} = P_{E\alpha}\delta(E-E')\delta(\alpha-\alpha'). \tag{9.22}$$

We then obtain

$$\begin{aligned}\overline{E} &= tr(\rho E)\\ &= \sum_{E,\alpha} E\rho_{E\alpha}.\end{aligned} \tag{9.23}$$

To illustrate how such averages determine physical properties, let us consider a collection of particles in a box constituting a perfect gas, assuming periodic boundary conditions for convenience. The general wave function for such a system is given by equation (9.1).

A typical eigenfunction of the energy is $\exp\left[i \sum_n \boldsymbol{k}_n \cdot \boldsymbol{x}_n\right]$. The energy corresponding to this eigenfunction is

$$E = \sum_s \frac{k_s^2}{2m}.$$

For a given particle, $\boldsymbol{k}_x = 2\pi l/L$, $\boldsymbol{k}_y = 2\pi m/L$ and $\boldsymbol{k}_z = 2\pi n/L$.

We can apply the above to calculate the pressure in a box, which is

$$p = -\frac{\partial E}{\partial V}.$$

With $V = L^3$, we have

$$p = -\frac{\partial E}{\partial L}\frac{\partial L}{\partial V} = -\frac{1}{3}\left(\frac{\partial E}{\partial L}\right) V^{-2/3}. \tag{9.24}$$

As the box expands, the energy decreases, and a simple calculation shows that

$$p = \tfrac{2}{3}\overline{E} \tag{9.25}$$

where $\overline{E}$ is determined from the density matrix as $tr(\rho E)$.

The above shows how typical macroscopic properties will follow from the density matrix. For the case of $\overline{E}$ we do not need to know the precise form of the density matrix to obtain the relationship between ρ and $\overline{E}$ that we have just calculated. But more generally to treat thermal properties, we need a more detailed knowledge of ρ. It is well known that for thermodynamic equilibrium, ρ is proportional to $\mathrm{e}^{-E/kT}$. If we write

$$\rho = \exp\left[\frac{\psi - E}{kT}\right] \tag{9.26}$$

where ψ is the normalisation factor for ρ, we get

$$\psi = -kT \ln Z \tag{9.27}$$

where

$$Z = tr\left[\mathrm{e}^{-E/kT}\right]. \tag{9.28}$$

In the well-known way we can obtain all the thermodynamic averages from the free energy ψ.

The problem of justifying that in a state of thermodynamical equilibrium, $\rho \propto \mathrm{e}^{-E/kT}$, is essentially the same in our interpretation as it is in the conventional interpretation. Various treatments of this problem have already been given. (See, for example, Tolman [4].) However, in this book it is not our main purpose to discuss quantum statistical mechanics, but rather to focus on the fact that an ontological interpretation of the basic quantum theory is possible. What we have done in this section is to show the general lines along which one can include quantum statistical mechanics within the framework of our approach.

9.5 The stochastic explanation of quantum probabilities

We have thus far been explaining quantum probabilities in terms of chaotic motions that are implied by the quantum laws themselves, with pure ensembles representing chaotic motions of the particles and mixed ensembles bringing in also chaotic variations in the quantum field. Whenever we have statistical distributions of this kind, however, it is always possible that these chaotic motions do not originate in the level under investigation, but rather that they arise from some deeper level. For example, in Brownian motion, small bodies which may contain many molecules undergo chaotic velocity fluctuations as a result of impacts originating at a finer molecular level. If we abstract these chaotic motions and consider them apart from their possible causes we have what is called a stochastic process which is treated in terms of a well-defined mathematical theory [5]. We can have two attitudes to such a stochastic process. The first is that it is a result of deeper causes that do not appear at the level under discussion. The second is that there is some intrinsic randomness in the basic motions themselves. In so far as we apply the ordinary mathematical treatment, we need not commit ourselves to either attitude. But of course, if we are thinking of possible models for the process then our attitude may make a difference, because the assumption of deeper causes implies that the stochastic treatment will break down at the finer level at which these causes are operating. (We shall discuss this possibility in chapter 14.)

A stochastic explanation of the quantum theory along these lines was first suggested by Bohm and Vigier [6], who proposed a sub-quantum mechanical level which would bring about chaotic behaviour of the particles. Later a similar theory was proposed by Nelson [7] which, however, had certain important differences which we have discussed in detail in a subsequent paper (Bohm and Hiley [8]).

The basic assumption of Bohm and Vigier was that the velocity of an individual particle i is given by

$$v_i = \frac{\nabla_i S(x_i)}{m} + \xi_i(t), \qquad (9.29)$$

where $\xi_i(t)$ represents a chaotic contribution to the velocity of that particle which fluctuates in a way that may be represented as random but with zero average. This means that in the causal form of ontological interpretation that we have given thus far, we should regard $(\nabla_i S(x_i))/m$ as the average velocity of the particle i rather than as its actual velocity. It is through this average velocity that the quantum potential comes in. For as we have shown in chapter 3, if the average velocity

$$\overline{v_i} = \frac{\nabla_i S(x_i)}{m} \tag{9.30}$$

and if $\psi = R\exp[iS]$ then Schrödinger's equation implies that

$$\frac{\mathrm{d}\overline{v_i}}{\mathrm{d}t} = -\nabla_i(V+Q). \tag{9.31}$$

The stochastic process simply adds the random motion $\xi_i(t)$ so that

$$\frac{\mathrm{d}\overline{v_i}}{\mathrm{d}t} = -\nabla_i(V+Q) + \frac{\mathrm{d}\xi_i(t)}{\mathrm{d}t}. \tag{9.32}$$

However, as we shall see, in that branch of mathematics known as stochastic theory, it is not necessary to use explicitly the above expression for the acceleration and it will be sufficient to start with equation (9.29) from which equation (9.32) follows in the way described above.

What stochastic theory typically does is to begin with a statistical ensemble of particles which are assumed to have some probability distribution given by $P(x,t)$. In such a theory, the notion of probability that is used is entirely classical and has nothing whatsoever to do with the quantum mechanical subtleties involved in the uncertainty principle and in the 'collapse' of the wave function in the conventional interpretations of quantum theory.

What we want to do here is to begin by assuming that our probabilities likewise have no intrinsic relation to the wave function and that they are simple 'classical-like' probabilities in the kind of ensemble that is treated so widely by standard stochastic theory. However, our aim in doing this is to determine the properties of the random motions described by $\xi_i(t)$ so that an arbitrary initial ensemble with a probability distribution $P(x,t)$ will approach an ensemble with a probability distribution $P(x,t) = \rho(x,t) = |\psi(x,t)|^2$ after a suitable interval of time. We then suppose that under typical conditions obtaining in quantum mechanical measurements, the interval of time will have been long enough so that the probability distribution will be given by $|\psi(x,t)|^2$. In this way we understand the fact that while $P(x,t)$ and $|\psi(x,t)|^2$ are basically independent concepts their numerical values will experimentally turn out to be equal in quantum mechanical experiments. In addition of course the chaotic processes occurring at the quantum level independently of the stochastic process which we have already discussed in this chapter will then contribute further to bringing this result about.

9.6 Detailed mathematical treatment of the stochastic model

To make a detailed treatment of the stochastic process, we shall assume that, whatever its origin may be, it can be represented as simple diffusion. To illustrate what this means we shall first consider a classical context in which we treat the diffusion of Brownian motion particles in a gravitational field using the simple theory of Einstein. If P is the probability density of particles, then there is a diffusion current

$$J^{(d)} = -D\nabla P, \tag{9.33}$$

where D is the diffusion coefficient. If this were all there was, the conservation equation would be

$$\frac{\partial P}{\partial t} = -D\nabla^2 P \tag{9.34}$$

and clearly this would lead to a uniform equilibrium distribution. However, as Einstein showed, in a gravitational field in the z-direction there is what he called an osmotic velocity

$$u_0 = D\frac{mg}{kT}z. \tag{9.35}$$

The conservation equation then becomes

$$\frac{\partial P}{\partial t} = -D\nabla\left[\frac{mg}{kT}zP + \nabla P\right]. \tag{9.36}$$

For the equilibrium distribution, $\partial P/\partial t = 0$, so that

$$\frac{\nabla P}{P} = \frac{mg}{kT}z + \text{const.} \qquad \text{or} \qquad P = Ae^{-mgz/kT} \tag{9.37}$$

which is the well-known Boltzmann factor.

The picture implied by the Einstein model of diffusion is that the particle is drifting downward in the gravitational field and that the net upward diffusive movement balances this to produce equilibrium.

It is worthwhile to provide a still more detailed picture of this process. To do this let us consider a simple one-dimensional model in which there is a unique free path λ and a unique speed v. We assume further that after collisions the velocities are randomly distributed in positive and negative directions while the speed is still v.

Let us now consider two layers which are separated by λ. The net diffusion current between these layers will be

$$J^{(d)} = -\frac{v}{2}\left[P_z - P_{z-\lambda}\right] \cong -\frac{v}{2}\frac{\partial P}{\partial z}\lambda. \tag{9.38}$$

Between collisions the average velocity gained from the gravitational field is

$$u_0 = -\frac{g}{2}\frac{\lambda}{v}. \tag{9.39}$$

The above is evidently the osmotic velocity. The total current is

$$J = -\frac{g}{2}\frac{\lambda}{v}P - \frac{v}{2}\frac{\partial P}{\partial z}\lambda = -\frac{\lambda v}{2}\left[\frac{g}{v^2}P + \frac{\partial P}{\partial z}\right].$$

Writing $v^2 = kT/m$ and $D = v\lambda/2$ we have

$$J = -D\left[\frac{mg}{kT}P + \frac{\partial P}{\partial z}\right] \tag{9.40}$$

and this is just the equation that Einstein assumed.

It is important to emphasise that, at least in this case, the osmotic velocity is produced by a field of force. Without such a field there would be no reason for an osmotic velocity.

For the stochastic interpretation of the quantum theory we would like to have a random diffusion process whose equilibrium state corresponds to a probability density $P = |\psi|^2 = \rho$ and to a mean current $j = \rho v = \rho(\nabla S/m)$. Such a state is a consistent possibility if $\psi = \sqrt{\rho}e^{iS/\hbar}$ satisfies Schrödinger's equation because this implies the conservation equation

$$\frac{\partial \rho}{\partial t} + \nabla \cdot j = 0.$$

In order to have $P = |\psi|^2$ as an equilibrium density under such a random process, we will have to assume a suitable osmotic velocity. We do not have to suppose, however, that this osmotic velocity is necessarily produced by a force field, similar to that of the gravitational field in our example given above, but rather as we have already indicated, it may have quite different causes. (Thus as suggested by Nelson [9], it could be some kind of background field that would produce a systematic drift as well as a random component of the motion.) At this stage it will be sufficient simply to postulate a field of osmotic velocities, $u_0(x, t)$, without committing ourselves as to what is its origin. We therefore assume an osmotic velocity

$$\boldsymbol{u}_0 = \frac{D\nabla\rho}{\rho} \tag{9.41}$$

and a diffusion current

$$\boldsymbol{j}^{(d)} = -D\nabla P. \tag{9.42}$$

The total current will be

$$\boldsymbol{j} = \frac{P}{m}\nabla S + \frac{D\nabla\rho}{\rho} - D\nabla P. \tag{9.43}$$

The conservation equation is then

$$\frac{\partial P}{\partial t} + \nabla\cdot\left(P\frac{\nabla S}{m} + DP\frac{\nabla\rho}{\rho} - D\nabla P\right) = 0. \tag{9.44}$$

In the above equations there is a systematic velocity

$$\boldsymbol{v}_s = \frac{\nabla S}{m} + D\frac{\nabla\rho}{\rho}. \tag{9.45}$$

This is made up of two parts, the mean velocity $\overline{\boldsymbol{v}} = \nabla S/m$ and the osmotic velocity $\boldsymbol{u}_0 = D\nabla\rho/\rho$. The mean velocity $\overline{\boldsymbol{v}}$ may be thought to arise from de Broglie's guidance condition. As we have just explained, the osmotic velocity will arise from some other source such as a field, but the main point is that it is derivable from a potential $D\ln\rho$ where ρ is a solution of the conservation equation

$$\frac{\partial P}{\partial t} + \nabla\cdot\left(\rho\frac{\nabla S}{m}\right) = 0. \tag{9.46}$$

It follows from equation (9.44) that there is an equilibrium state with $P = \rho$ in which the osmotic velocity is balanced by the diffusion current so that the mean velocity is $\overline{\boldsymbol{v}} = \nabla S/m$.

But now we must raise the crucial question as to whether this equilibrium is stable. In other words, will an arbitrary distribution P always approach ρ? Writing $P = F\rho$ we obtain the following equation

$$\rho\frac{\partial F}{\partial t} = D\nabla\cdot(\rho\nabla F) = D\nabla\rho\cdot\nabla F + D\rho\nabla^2 F. \tag{9.47}$$

We shall show that F must eventually approach 1 everywhere by proving that the maxima of F must always decrease and the minima must always increase. Maxima and minima are characterised by $\nabla F = 0$. At such points we have

$$\rho \frac{\partial F}{\partial t} = D\rho\nabla^2 F. \tag{9.48}$$

At a maximum, $\nabla^2 F$ is negative and clearly F must be decreasing at this point. At a minimum, $\nabla^2 F$ is positive and therefore F must be increasing. This can only cease when $F = 1$ everywhere.

For times much longer than the relaxation time, τ, of this process we will therefore quite generally encounter the usual quantum mechanical probability distribution. It is implied, of course, that at shorter times this need not be so and therefore the possibility of a test to distinguish this theory from the quantum theory is in principle opened up. We shall discuss the conditions under which such a test may be possible in chapter 14.

The mean velocity is, of course, $\overline{v} = \nabla S/m$. As we have seen in connection with equation (9.31), this by itself does not determine the acceleration. In addition we need a differential equation for S. Then as we have seen in chapter 3 and also earlier in the present section, provided ψ satisfies the Schrödinger equation, S will satisfy the extended Hamilton-Jacobi equation (3.7) which contains the quantum potential. It is thus clear that in the present approach, the quantum potential is actually playing a secondary role and this is indeed why it does not have to be included explicitly in the stochastic approach (though it provides a very useful insight into the meaning of the process especially in the non-stochastic approach). But in all cases the fundamental dynamics depend first of all on the guidance condition, and, in a stochastic theory, we bring in also the osmotic velocity and the effects of random diffusion. All these work together to keep a particle in a region in which $P = |\psi|^2$ is large and where its average velocity fluctuates around $\nabla S/m$. As explained earlier the quantum potential merely represents the mean self-acceleration of the particle under the influence of the de Broglie guidance conditions, and this will be valid only if ψ satisfies Schrödinger's equation. If ψ had satisfied another wave equation, the mean accelerations would have been different. In fact we shall extend this theory in chapter 12 so that the wave function satisfies the Dirac equation, for which a very different mean acceleration is implied.

To bring out the physical significance of the stochastic model, let us consider the two-slit interference experiment. A particle undergoing random motion will go through one slit or the other, but it is affected by the Schrödinger field coming from both slits. In the interpretation that we gave earlier, this effect was expressed primarily through the quantum potential. In the stochastic interpretation it is primarily expressed through the osmotic velocity which reflects the contributions to the wave function

ψ coming from both slits. Near the zeros of the wave function the osmotic velocity approaches infinity and is directed away from the zeros. Thus a particle diffusing randomly and approaching a zero is certain to be turned around before it can reach this zero. This explains why no particle ever reaches the points where the wave function is zero. And, as we have indeed already pointed out, the osmotic velocity is constantly pushing the particle to the regions of highest $|\psi|^2$ and this explains why most particles are found near the maxima of the wave function.

Without assuming an osmotic velocity field of this kind, there would be no way of explaining such phenomena. As a result of random motions, for example, a particle just undergoing a random process on its own would have no way of 'knowing' that it should avoid the zeros of the wave function.

To obtain a consistent overall picture we must consider the random background field. This is assumed, as we have already pointed out, to be the source of the random motions of the particle, but in addition it must determine a condition in space which gives rise to the osmotic velocity. Indeed, a single particle in random motion cannot contain any information capable of determining, for example, that every time it approaches the zero of a wave function it must turn around. This information could be contained only in the background field itself. Therefore, the stochastic model does not fulfil the expectations that would at first be raised by its name, that is to provide an explanation solely in terms of the random movements of a particle without reference to a quantum mechanical field (which may be taken as ρ and S or as $\psi = \sqrt{\rho}e^{iS/\hbar}$).

In the interpretation that we have been giving throughout this book, this field has the property that its effect does not depend on its amplitude. As we have suggested in earlier chapters, this behaviour can be understood in terms of the concept of active information, i.e. that the movement comes from the particle itself, which is however 'informed' or 'guided' by the field. In the stochastic interpretation there is a further effect of the quantum field through the osmotic velocity, which is also independent of its amplitude. We can therefore say that along with the mean velocity field, $\nabla S/m$, the osmotic velocity field constitutes active information which determines the average movement of the particle. This latter is however modified by a completely random component due to the fluctuations of the background field.

Clearly then, there are basic similarities between our original essentially causal interpretation and the stochastic interpretation. Nevertheless, there are also evidently important differences. One of the key differences can be seen by considering a stationary state with $S =$ const., e.g. an s-state. In the stochastic model the particle is executing a random motion which would bring about diffusion in space, but the osmotic velocity is constantly drawing it back so that we obtain the usual spherical distribution as an average. But now the basic process is one of dynamic equilibrium. The average velocity, which is zero, is the same as the actual velocity in the causal interpretation. Such a view of the s-state as one of dynamic equilibrium seems to fit in with our physical intuition better than one in which the particle is at rest.

9.7 Stochastic treatment of the many-body system

The extension of this model to the many-particle system is straightforward. The wave function, $\psi(x_{ir})$, which is defined in a 3N-dimensional configuration space, satisfies the many-body Schrödinger equation. We assume that the mean velocity of the nth particle is

$$\overline{v}_{in} = \frac{1}{m}\frac{\partial}{\partial x_{in}} S(x_{ir}). \tag{9.49}$$

In addition we assume an arbitrary probability density $P(x_{ir})$ and a random diffusion current of the nth particle

$$j_{in}^{(d)} = -D\frac{\partial P}{\partial x_{in}}. \tag{9.50}$$

We then make the further key assumption that the osmotic velocity component of the nth particle is

$$u_{in}^{(o)} = \left(\frac{D}{\rho}\right)\frac{\partial \rho}{\partial x_{in}} \tag{9.51}$$

where $\rho = |\psi|^2$.

From here on the theory will go through as in the one-particle case and it will follow that the limiting distribution will be $P = |\psi|^2$.

Equation (9.50) describes a stochastic process in which the different particles undergo statistically independent random fluctuations. However, in equation (9.51) we have introduced an important connection between the osmotic velocities of different particles. For the general wave function that does not split into independent factors, the osmotic velocities of different particles will be related and this relationship may be quite strong even though the particles are distant from each other. For this model requires an effectively instantaneous nonlocal connection which brings about the related osmotic velocities of distant particles (according to $m\dot{v}_n = -\nabla_n(V+Q)$).

It is clear then that the stochastic approach brings in yet another kind of nonlocality over and above that which is present in the causal interpretation (in which, as we have seen in chapters 3 and 4, the acceleration is given by $m\dot{v}_n = -\nabla_n(V + Q)$, and Q is the generally nonlocal quantum potential). However, by its very name, the stochastic approach seems to imply a theory in which quantum mechanics could be explained in terms of particles undergoing random processes and thus one could easily be led to expect that at least in this theory there would be no need for nonlocality. For example, Nelson [9], who has worked intensively on this sort of interpretation, has written "But the whole point (of the stochastic theory) was to construct a physically realistic picture of microprocesses, and a theory that violates locality is untenable" [10]. From this it is clear that Nelson expected the stochastic theory to be local even though he finally came to the conclusion, in agreement with us, that it has to be nonlocal.

To bring out the full meaning of this nonlocality it is useful to discuss the Einstein-Podolsky-Rosen experiment in terms of the stochastic interpretation. To do this we first recall that, as shown in chapter 7, the essential feature of a measurement process is that the particles constituting the measuring apparatus can be shown to enter a definite channel, and that after this they cannot leave the channel in question because the wave function is zero between the channels. From this point on the wave function of the whole system effectively reduces to a product of the wave function of the apparatus and of the observed system. It was demonstrated that thereafter, the remaining empty channels will not be effective. It follows that the net result is the same as if there had been a collapse of the wave function to a state corresponding to the result of the actual measurement.

It is clear that a similar result will follow for the stochastic interpretation because here too there is no probability that a particle can enter the region between the channels in which the wave function is zero.

Let us now consider the EPR experiment in the stochastic interpretation. We recall that in this experiment we have to deal with an initial state of a pair of particles in which the wave function is not factorisable. A measurement is then made determining the state of one of these particles, and it is inferred from the quantum mechanics that the other will go into a corresponding state even though the Hamiltonian contains no interaction terms that could account for this. In the causal interpretation the behaviour of the second particle was explained by the nonlocal features of the quantum potential which could provide for a direct interaction between the two different particles that does not necessarily fall off with their separation and that can be effective even when there are no interaction terms in the Hamiltonian.

As we have seen in connection with equation (9.31), in the stochastic interpretation all the above nonlocal effects of the quantum potential are still implied, but now we have seen that there is also a further nonlocal connection through the osmotic velocities. And when the properties of the first particle are measured the osmotic velocities of both will be instantaneously affected in such a way as to help bring about the appropriate correlations of the results.

It is clear that the stochastic interpretation and the causal interpretation treat the nonlocal EPR correlations in a basically similar way. The essential point is that in an independent disturbance of one of the particles, the fields acting on the other particle (osmotic velocities and quantum potential) respond instantaneously even when the particles are far apart. It is as if the two particles were in instantaneous two-way communication exchanging active information that enables each particle to 'know' what has happened to the other and to respond accordingly.

Of course, in a non-relativistic theory it is consistent to assume such instantaneous connections. We shall, however, show later how these considerations can be extended to take into account the fact that the theory of relativity has been found to be valid in a very broad context.

9.8 References

1. I. Prigogine, *From Being to Becoming: Time and Complexity in the Physical Sciences*, Freeman, San Francisco, 1980.
2. S. W. Hawking, *Commun. Math. Phys.* **87**, 395–415 (1982).
3. R. P. Feynman, *Statistical Mechanics: A Set of Lectures*, chapter 2, Benjamin, London, 1972.
4. R. C. Tolman, *The Principles of Statistical Mechanics*, Dover, New York, 1979.
5. E. Nelson, *Dynamical Theories of Brownian Motion*, Princeton University Press, Princeton, 1972.
6. D. Bohm and J-P. Vigier, *Phys. Rev.* **96**, 208 (1954).
7. E. Nelson, *Phys. Rev.* **150B**, 1079 (1966).
8. D. Bohm and B. J. Hiley, *Phys. Reports* **172**, 93–122 (1989).
9. E. Nelson, *Quantum Fluctuations*, Princeton University Press, Princeton, 1985.
10. *Ibid.*, p.127.

Chapter 10
The ontological interpretation of the Pauli equation

10.1 Introduction

We have thus far developed a systematic ontological interpretation of non-relativistic Schrödinger theory for particles. In this and the next two chapters of the book, we shall be concerned principally with extensions of the theory to include

- the non-relativistic Pauli equation
- field theories considered non-relativistically
- relativistic treatment of both particle and field theories.

We begin this chapter with a treatment of the non-relativistic Pauli equation. Firstly, we shall discuss some previous suggestions for an ontological interpretation of this equation based on the idea that the particle is an extended rigid object which makes a 'spin' contribution to the total angular momentum [1,2]. We then give reasons showing why this assumption does not give rise to a satisfactory theory. We are led instead to regard the Pauli equation as a non-relativistic limit of the Dirac equation. When we do this we can see that, in addition to the straightforward contribution to the current implied by the previous models (cf. equation 5.4), there is a further circulating current in which the orbit precesses in a way that depends on the wave function. This makes a contribution to the total angular momentum which accounts for that which has usually been attributed to the particle's spin. This means that the spin angular momentum does not arise from an extended particle structure at all, but that rather it is a further and more subtle feature of the orbital motion of the particle.

It is thus implied that the spin is a context-dependent property of the particle, rather than an intrinsic property. As such it is, as we shall see, readily related to a model originally proposed by Bell [3].

We develop our model and exhibit its consequences for a variety of typical situations. We then extend it to a many-body system to provide an explanation of the EPR experiment with two atoms of spin one-half constituting a molecule of total spin zero (thus extending the results of chapter 7 for particles of spin one).

10.2 The Bohm, Schiller and Tiomno model

We now give a sketch of the original Bohm, Schiller and Tiomno model, [BST], for spin [1]. We shall give some reasons why we regard this model as ultimately inadequate, but it is nevertheless useful to consider it here as it will help make clear the model that we are actually going to adopt.

We begin with the Pauli spinor

$$\psi_j(x) = \begin{pmatrix} \psi_1 \\ \psi_2 \end{pmatrix}$$

which satisfies the Pauli equation

$$i\frac{\partial \psi_j}{\partial t} = -\frac{(\nabla - ie\boldsymbol{A})^2 \psi_j}{2m} + V\psi_j + \mu(\boldsymbol{\sigma} \cdot \boldsymbol{B})\psi_j, \tag{10.1}$$

where $\boldsymbol{\sigma}$ are the Pauli spin matrices, $\boldsymbol{A}$ the vector potential and $\boldsymbol{B}$ the applied magnetic field. (We shall put $\hbar = c = 1$ throughout this chapter.)

To develop this interpretation, we start with the special case where $\boldsymbol{A} = 0$ and $\boldsymbol{B} = 0$. In this case the Pauli equation reduces to two uncoupled Schrödinger equations. (It is readily seen that when $\boldsymbol{B} \neq 0$, the equations will generally be coupled.) Then setting $\psi_j = R_j \exp[iS_j]$ we obtain

$$\frac{\partial S_j}{\partial t} + \frac{(\nabla S_j)^2}{2m} + V - \frac{1}{2m}\frac{\nabla^2 R_j}{R_j} = 0 \tag{10.2}$$

and

$$\frac{\partial R_j^2}{\partial t} + \nabla \cdot \left(R_j^2 \frac{\nabla S_j}{m} \right) = 0. \tag{10.3}$$

We can picture the meaning of the above therefore as two independent probability distributions, each of which is conserved and in which the velocities of the particle are, in general, different. When $\boldsymbol{B} \neq 0$, neither of these probability distributions is conserved by itself because the two wave functions are coupled. However, it is easily shown that their sum

$$P = \sum R_j^2 = |\psi_1|^2 + |\psi_2|^2 = \rho \tag{10.4}$$

is conserved.

We may then define a mean velocity

$$\boldsymbol{v} = \frac{(P_1(\nabla S_1/m) + P_2(\nabla S_2/m))}{P_1 + P_2} \tag{10.5}$$

where $P_1 = |\psi_1|^2$ and $P_2 = |\psi_2|^2$. The conservation equation is then

$$\frac{\partial P}{\partial t} + \nabla \cdot (P\boldsymbol{v}) = 0. \tag{10.6}$$

On this basis we can say that the first step in obtaining an ontological interpretation is to assume a particle following a definite trajectory with the guidance condition (10.5) (which is clearly a generalisation of the guidance condition for a particle without spin). The probability distribution of such particles will be assumed to be $P = \rho$. (By a generalisation of the argument used with Schrödinger's equation, it is clear that one can show that if there are chaotic processes in the motion of the particle, then, if P is not initially equal to ρ, it will approach ρ in the long run (see chapter 9).)

Thus far we have not provided enough physical concepts to give a meaning to the R_i and S_i, which constitute four quantities. The simplest physical idea would be to assume (as was indeed done in the original paper of BST [1]) that the particle is a spinning object whose orientation is specified by the three Euler angles (θ, ϕ, χ). It is well known that a two component spinor can be put into correspondence with these angles. Indeed we can write

$$\psi = \rho^{\frac{1}{2}} \begin{pmatrix} \cos\frac{\theta}{2} \exp\left[i(\chi + \phi)/2\right] \\ i \sin\frac{\theta}{2} \exp\left[i(\chi - \phi)/2\right] \end{pmatrix}. \tag{10.7}$$

Assuming that the object is spherical, we may propose that θ and ϕ represent the orientation of the angular momentum vector, while χ represents the angle of rotation of the object about this axis. If the angular momentum is one half, we can write the spin-vector as

$$\boldsymbol{S} = \frac{1}{2}\frac{\psi^\dagger \sigma \psi}{\rho}. \tag{10.8}$$

This spin-vector is then easily shown to be in the direction determined by θ and ϕ. It is clear that

$$\chi = \frac{S_1 + S_2}{2}. \tag{10.9}$$

The spinor is now fully interpreted. We are evidently regarding the Hamilton-Jacobi phase, S, as being equal to the angle of rotation of the rigid body around an axis in the direction of the spin.

We can obtain an interesting physical picture here. If we add to the energy of the particle the rest energy m_0c^2, Schrödinger's equation (in this non-relativistic treatment) can be written as

$$i\frac{\partial \psi_j}{\partial t} = -\frac{1}{2m}(\nabla - e\boldsymbol{A})^2\psi_j + V\psi_j + m_0c^2\psi_j + \mu(\boldsymbol{\sigma}\cdot\boldsymbol{B})\psi_j. \quad (10.10)$$

With $\boldsymbol{B} = 0$ and $V = 0$, a particle with zero momentum will have the wave function

$$\psi \propto \exp[-imt]\begin{pmatrix} \cos\frac{\theta}{2}\exp[i\phi/2] \\ i\sin\frac{\theta}{2}\exp[-i\phi/2] \end{pmatrix}. \quad (10.11)$$

Thus the angular frequency is

$$\dot{\chi} = m$$

which is the frequency corresponding to the Compton wavelength. When $\boldsymbol{k}$ is not zero, we get

$$\psi \propto \exp\left[-i\left(mc^2 + \frac{\boldsymbol{k}^2}{2m}\right)t\right]\exp[i\boldsymbol{k}\cdot\boldsymbol{x}]\begin{pmatrix} \cos\frac{\theta}{2}\exp[i\phi/2] \\ i\sin\frac{\theta}{2}\exp[-i\phi/2] \end{pmatrix}. \quad (10.12)$$

This implies a motion in which, while $\boldsymbol{x}$ is increasing with time, the particle is simultaneously rotating around the axis determined by the spin direction. For example, in the case in which the spin is directed along the line of velocity, an element in the spherical structure of the particle executes a helical motion with the screw sense determined by the spin direction. (In general the motion is, of course, more complicated.)

There will be a generalisation of the Hamilton-Jacobi equation for $S = \chi$. To obtain this we first transform the velocity (10.5) with the aid of (10.7), and we find that

$$\boldsymbol{v} = \frac{1}{2m}(\nabla S + \cos\theta\nabla\phi) - e\boldsymbol{A}. \quad (10.13)$$

It can then be shown (see BST p.63) that S satisfies the equation

$$\frac{\partial S}{\partial t} - \cos\theta(\boldsymbol{v}\cdot\nabla)\phi + \frac{(\nabla S + \cos\theta\nabla\phi - e\boldsymbol{A})^2}{2m} - \frac{1}{2m}\frac{\nabla^2\rho^{\frac{1}{2}}}{\rho^{\frac{1}{2}}}$$
$$+ \; V + \frac{1}{2m}\left[(\nabla\theta)^2 + \sin^2\theta(\nabla\phi)^2\right] + \mu\widehat{\boldsymbol{S}}\cdot\boldsymbol{B} = 0. \quad (10.14)$$

In the above we have a generalisation of the quantum potential which is given by

$$Q = -\frac{1}{2m}\frac{\nabla^2 \rho^{\frac{1}{2}}}{\rho^{\frac{1}{2}}}. \tag{10.15}$$

In addition, the orbital motion depends on the angles θ and ϕ which determine the spin-vector $\boldsymbol{S}$. This is a quantum mechanical spin-orbit coupling. Evidently it is also necessary that θ and ϕ depend on the orbital motion. What happens is that the spin-orbit coupling brings about a quantum torque acting on the spin-vector (we shall not go into details here, but the relevant equations are given in BST p.64).

We can illustrate the meaning of these equations in terms of a measurement process in which we begin with a spin pointing in an arbitrary direction and measure it in the z-direction. As the wave packets corresponding to positive and negative spins respectively begin to separate in the z-direction, it is found, first of all, that the quantum torque eventually lines the particle spin up with the corresponding wave packet into which the particle ultimately goes. Where the particle will go depends not only on its initial position, but also on the initial values of θ and ϕ. However, with the probability distribution that has been assumed, it is easily shown that the probability that the particle will end up in a given final packet, $\psi_f(x)$, is $\int |\psi_f(x)|^2 \, dx$. And this, of course, will be the same as one obtains in the usual interpretations.

The relevant orbits have been worked out for two cases (see Dewdney *et al.* [4]). In both, one assumes the mass, m, of the particle is so great that we can neglect the spread of the wave packets. In Case 1, the spin points initially in the x-direction. The trajectories are shown in figure 10.1, while the turning of the spin directions are shown in figure 10.2. For Case 2, similar results are given for a certain initial spin direction that yields a 75% probability for spin-up in the z-direction and a 25% probability for spin-down. These trajectories are shown in figure 10.3, while the spin directions are shown in figure 10.4.

10.3 The many-body Pauli equation

As shown in BST [1,2], the model described here works consistently and provides an adequate ontological interpretation of the Pauli equation for a one-body system. However, this model gets into very serious difficulties when we try to extend it to a many-body system. To show what these difficulties are, let us begin by considering the general wave function for such a system, $\Psi = \Psi_{i_1 \ldots i_N}(\boldsymbol{x}_1 \ldots \boldsymbol{x}_N)$. This has 2^N complex components (or 2^{N+1} real components).

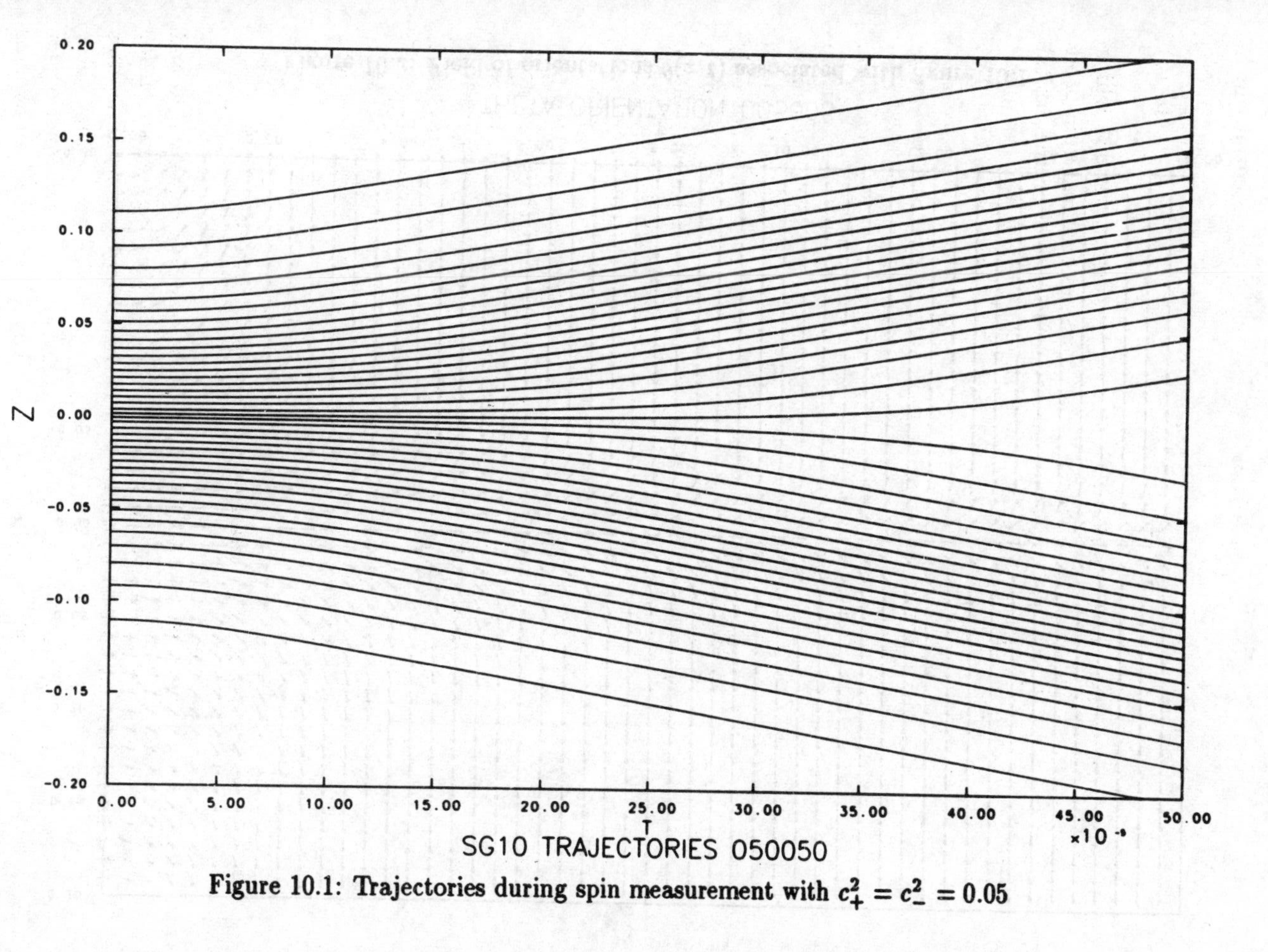

Figure 10.1: Trajectories during spin measurement with $c_+^2 = c_-^2 = 0.05$

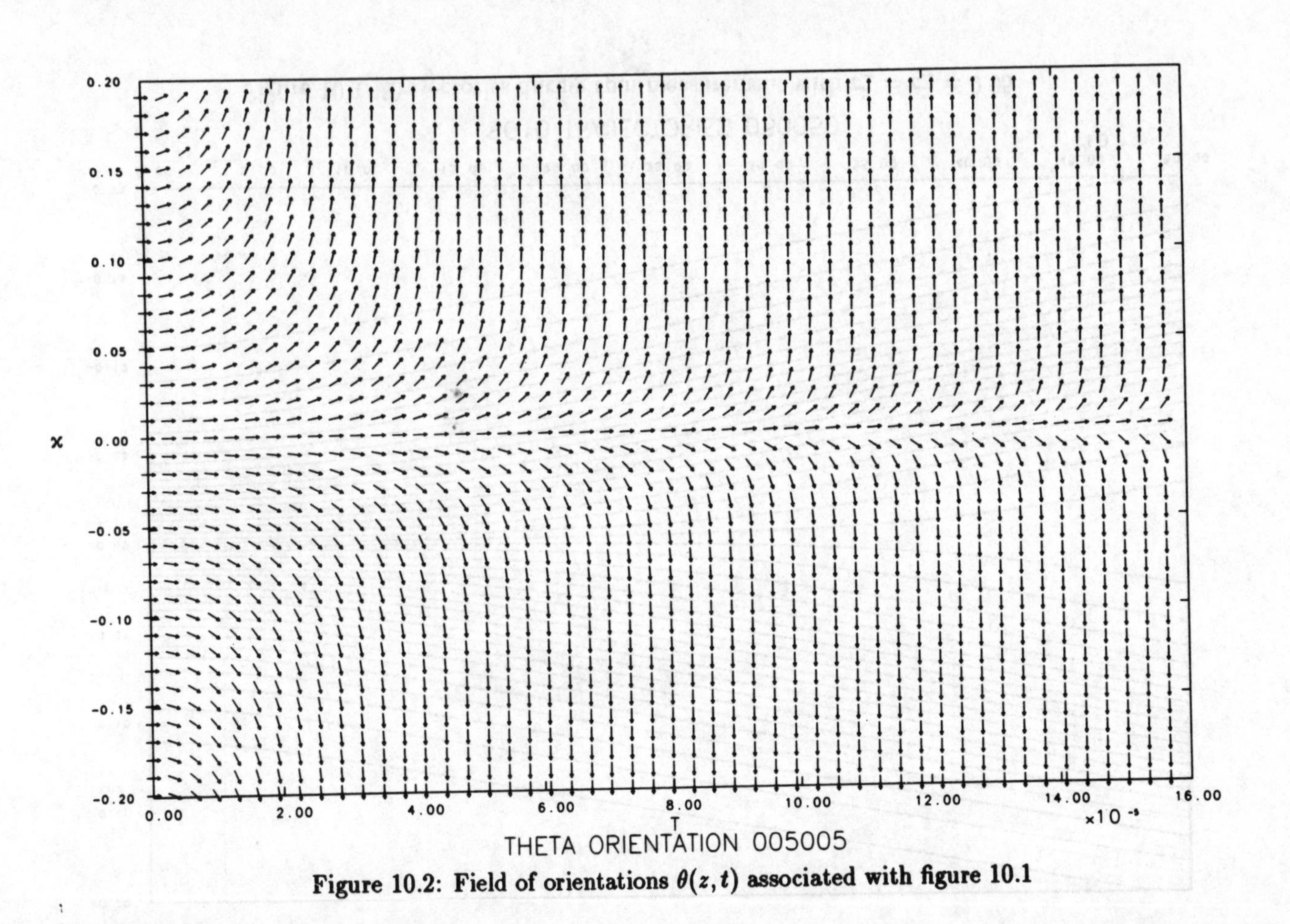

Figure 10.2: Field of orientations $\theta(z,t)$ associated with figure 10.1

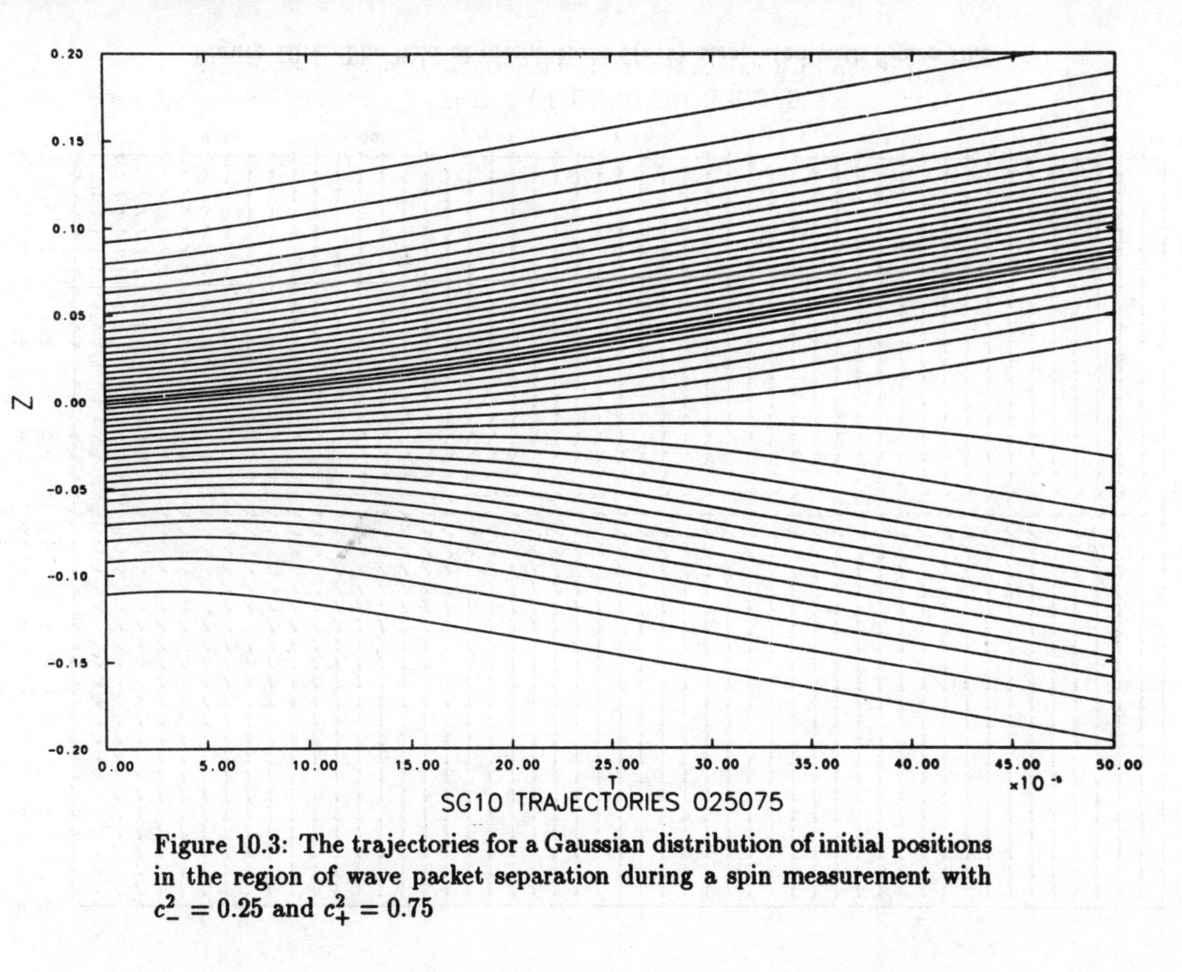

Figure 10.3: The trajectories for a Gaussian distribution of initial positions in the region of wave packet separation during a spin measurement with $c_-^2 = 0.25$ and $c_+^2 = 0.75$

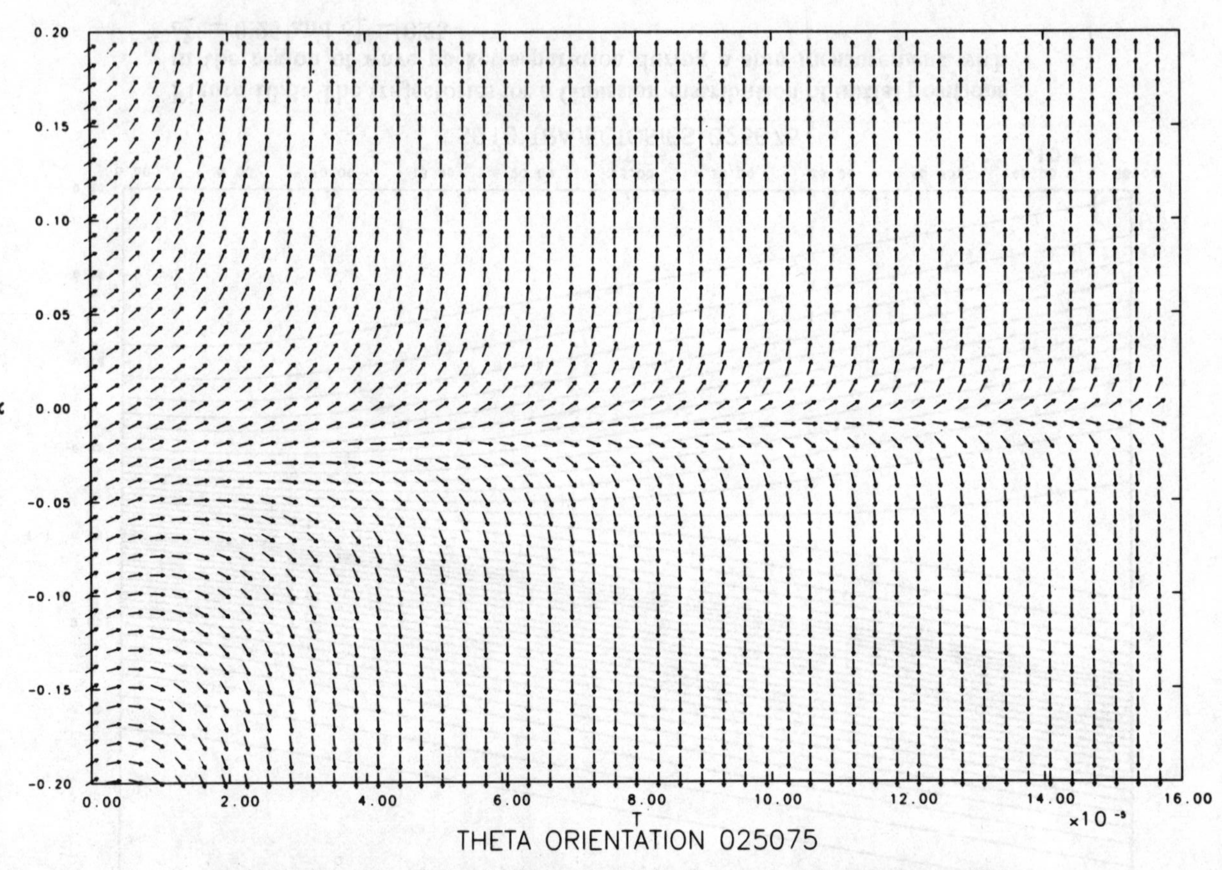

Figure 10.4: The field of orientations $\theta(z,t)$ associated with figure 10.3

The many-body Pauli equation is then (assuming $\boldsymbol{B} = 0$)

$$i\frac{\partial \Psi}{\partial t} = \left[-\frac{1}{2m} \sum_{i=1,\dots,N} \nabla_i^2 + V(\boldsymbol{x}_1 \dots \boldsymbol{x}_N) \right] \Psi. \tag{10.16}$$

From this we obtain the net probability density in the configuration space of all the particles

$$P = \sum_{i_1 \dots i_N} |\Psi_{i_1 \dots i_N}|^2 = \rho. \tag{10.17}$$

(See section 10.4 for a more detailed discussion of the many-body wave function.)

Let us then define the current density in configuration space as

$$\boldsymbol{j}_n = Im \sum_{i_1 \dots i_N} \frac{1}{2m_i} \Psi^\dagger_{i_1 \dots i_N} \nabla_n \Psi_{i_1 \dots i_N}. \tag{10.18}$$

It then follows from the Pauli equation that

$$\frac{\partial P}{\partial t} + \sum_n (\nabla_n \cdot \boldsymbol{j}_n) = 0. \tag{10.19}$$

The above expresses the conservation of probability in configuration space. In analogy with what was done in the many-body Schrödinger equation, this suggests that we define the particle velocity as

$$\boldsymbol{v}_n = \frac{\boldsymbol{j}_n}{\rho}. \tag{10.20}$$

The above can be regarded as a generalisation of the guidance condition, while the conservation equation (10.19) implies that particles guided in this way will have a conserved probability density. But to complete the interpretation, we need to define the spin-vectors. In agreement with what was done with the velocities, the spin of the nth particle, $\boldsymbol{S}_n$, has to be defined as

$$\boldsymbol{S}_n = \frac{\psi_n^\dagger \sigma_n \mathbf{1}_{n'} \psi_n}{\rho} \tag{10.21}$$

where σ_n represents the spin matrix of the nth particle, and $\mathbf{1}_{n'}$ represents the unit matrix on the spin indices of all the other particles. This will define at most $3N$ functions. We have also the definition of the velocities $\boldsymbol{v}_n$, which will add up to give at most $6N$ functions (though not all of these may be independent). These are all the properties belonging to the individual particles alone that our spin model can allow for. However, as we have already pointed out, the wave function contains, in general, 2^{N+1} independent real parameters. Thus for the 2-body system there are 8 independent parameters. But there are 6 spin vectors and 6 velocity components. Therefore it should be possible to give a physical meaning to all the parts of the wave function for this case (in which, of course, not all the physical quantities need be independent). Dewdney *et al.* [5] have, in fact, considered this case and have given some detailed results.

It is clear however that as we increase N indefinitely, the number of physical properties of the individual particles will only increase as N, while the number of independent parameters of the wave function will increase as 2^{N+1}. Therefore there is no way to give a physical meaning to all the parameters along these lines. One might consider properties belonging to sets of 2, 3 or 4 particles etc., but it is difficult to see what this could mean physically. Such properties would be nonlocal and might be significant even when the particles are far apart. In any case, the simple model of a set of spinning particles would no longer apply.

The above difficulty makes it doubtful that a consistent interpretation of the many-body Pauli equation can be obtained in terms of a model of spinning particles. But there is a further difficulty even with the one-body Pauli equation, when we consider the fact that this must be the non-relativistic limit of a relativistic theory, i.e. the theory of the Dirac equation. The fact is that we cannot account properly for the value of the angular momentum for a model of this sort. Using the relation for the frequency of rotation of a particle that we obtained earlier, we can easily see that to have a spin one-half, we would need an electron of the size of the Compton wavelength. But we know that this cannot be allowed because of the experimental data on scattering. In fact this data implies that the electron cannot be larger than about 10^{-16} cm. To obtain the angular momentum, $\hbar/2$, a spinning particle of this size would have to have a peripheral velocity very much greater than that of light. Since we have to regard the Pauli equation as the non-relativistic limit of the Dirac equation, which is fully relativistic, we cannot allow such velocities.

For these two reasons we conclude that this model of a spinning particle will not be an adequate one.

10.4 The Pauli equation as the non-relativistic limit of the Dirac equation

We can obtain some further insight into what would be a proper model for the Pauli equation by considering in some detail how it comes out as the non-relativistic limit of the Dirac equation. We begin with the Dirac spinor

$\Psi = \Psi_a$ where $a = 1, 2, 3, 4$. The Dirac equation is

$$\frac{\partial \Psi}{\partial t} = -\boldsymbol{\alpha} \cdot (\nabla - ie\boldsymbol{A})\Psi - iV\Psi - im\beta\Psi \tag{10.22}$$

where α, β are the Dirac matrices.

We choose a representation in which β is diagonal and write

$$\Psi_a = \begin{pmatrix} \psi_1 \\ \psi_2 \end{pmatrix}$$

where ψ_1 and ψ_2 are two-component spinors corresponding respectively to $\beta = +1$ and $\beta = -1$. The Dirac equation becomes

$$\begin{aligned} \frac{\partial \psi_1}{\partial t} &= -\sigma \cdot (\nabla - ie\boldsymbol{A})\psi_2 - i(V + m)\psi_1 \qquad \text{(a)} \\ \frac{\partial \psi_2}{\partial t} &= -\sigma \cdot (\nabla - ie\boldsymbol{A})\psi_1 - i(V + m)\psi_2 \qquad \text{(b)} \end{aligned} \tag{10.23}$$

where σ are the Pauli matrices.

For low velocities and positive energies, $\psi_2 \ll \psi_1$. With $\psi_2 \propto exp[-iEt]$ we obtain

$$(E - V + m)\psi_2 = -i\sigma \cdot (\nabla - ie\boldsymbol{A})\psi_1. \tag{10.24}$$

Putting equation (10.24) into (10.23a) we find

$$i\frac{\partial \psi_1}{\partial t} = \frac{-\sigma \cdot (\nabla - ie\boldsymbol{A})\sigma \cdot (\nabla - ie\boldsymbol{A})\psi_1}{(E - V + m)} + (V + m)\psi_1. \tag{10.25}$$

This becomes

$$i\frac{\partial \psi_1}{\partial t} = \frac{-(\nabla - ie\boldsymbol{A})^2\psi_1 + e(\nabla \times \boldsymbol{A}) \cdot \sigma\psi_1}{(E - V + m)} + (V + m)\psi_1. \tag{10.26}$$

Writing $E = m + \Delta E$ and using

$$\frac{1}{2m + \Delta E - V} \cong \frac{1}{2m}\left(1 - \frac{\Delta E - V}{2m}\right), \tag{10.27}$$

we obtain

$$i\frac{\partial \psi_1}{\partial t} = -\frac{1}{2m}(\nabla - ie\boldsymbol{A})^2\psi_1 + \frac{e}{2m}\sigma \cdot \boldsymbol{B}\psi_1 + (V + m)\psi_1. \tag{10.28}$$

This is clearly the Pauli equation with a constant rest mass term added.

To give an ontological interpretation of the Pauli equation considered as a non-relativistic limit of the Dirac equation, we begin with the Dirac current

$$\boldsymbol{j} = \psi^{\dagger}\alpha\psi \tag{10.29}$$

which satisfies the conservation equation

$$\frac{\partial\rho}{\partial t} + \nabla\cdot\boldsymbol{j} = 0. \tag{10.30}$$

As we have been doing in previous cases, we take advantage of the fact that when a conservation equation is satisfied, we can always assume a particle with a velocity satisfying our now generalised guidance condition

$$\boldsymbol{v} = \frac{\boldsymbol{j}}{\rho} \tag{10.31}$$

which has a probability distribution given by $P = \rho$. To do this it is, of course, necessary that ρ be positive definite. But $P = |\psi|^2$ is clearly positive definite. (Alternative definitions of the current density which do not, in general, lead to positive definite probabilities have been considered by de Broglie [6]. But evidently we cannot obtain a consistent ontological interpretation if we admit negative probability densities for the particles.)

To simplify the subsequent discussion, we shall from now on restrict ourselves to a free particle. From equations (10.24) and (10.27) we then obtain

$$\psi_2 \cong -i\frac{\sigma\cdot\nabla}{2m}\psi_1 \tag{10.32}$$

(also assuming energies low enough so the $\Delta E/m$ can be neglected). We then obtain

$$\boldsymbol{j} = -\frac{i}{2m}\left[\psi_1^{\dagger}\sigma(\sigma\cdot\nabla)\psi_1 - (\sigma\cdot\nabla)\psi_1^{\dagger}\sigma\psi_1\right], \tag{10.33}$$

which reduces to the sum of two terms, $\boldsymbol{j}_A + \boldsymbol{j}_B$, where

$$\boldsymbol{j}_A = \frac{1}{2mi}\left[\psi_1^{\dagger}(\nabla\psi_1) - (\nabla\psi_1^{\dagger})\psi_1\right], \tag{10.34}$$

and

$$\boldsymbol{j}_B = \frac{1}{2m}\nabla\times(\psi_1^{\dagger}\sigma\psi_1). \tag{10.35}$$

The velocity correspondingly reduces to the sum of two parts $\boldsymbol{v} = \boldsymbol{v}_A + \boldsymbol{v}_B$ where

$$\boldsymbol{v}_A = \frac{\boldsymbol{j}_A}{\rho} \tag{10.36}$$

$$\boldsymbol{v}_B = \frac{\boldsymbol{j}_B}{\rho}. \tag{10.37}$$

The first part of the current, $\boldsymbol{j}_A$, is the same as it was in the BST interpretation of the Pauli equation (and $\boldsymbol{v}_A$ gives the same velocity as in this interpretation). The second part, $\boldsymbol{j}_B$, satisfies the identity $\nabla \cdot \boldsymbol{j}_B = 0$ and this evidently corresponds to some kind of additional circulation. To see what this means consider the case in which ψ is a spherical wave packet of radius a, centred at the origin and falling off rapidly at distances of the order $r = a$. The term

$$\boldsymbol{v}_B = \frac{1}{2m} \nabla \times \frac{(\psi^\dagger \sigma \psi)}{\rho} \tag{10.38}$$

represents a velocity in the direction $\hat{\boldsymbol{r}} \times \boldsymbol{s}$ (where $\boldsymbol{s} = \psi^\dagger \sigma \psi$). This implies a circular movement around the direction of 'spin' which has a maximum velocity near $r = a$.

This additional velocity, $\boldsymbol{v}_B$, gives rise to a further contribution to the mean magnetic moment over and above that coming from the orbital motion. Indeed the average of the total magnetic moment, $\overline{M}$, is

$$\overline{M} = \frac{e}{2} \int \boldsymbol{r} \times \boldsymbol{j} \, \mathrm{d}\tau = \frac{e}{2} \int \boldsymbol{r} \times \boldsymbol{j}_A \, \mathrm{d}\tau + \frac{e}{2} \int \boldsymbol{r} \times \boldsymbol{j}_B \, \mathrm{d}\tau. \tag{10.39}$$

The first term in the above is just

$$\overline{M}_1 = \frac{e}{2im} \int \boldsymbol{r} \times [\psi^\dagger \nabla \psi - \psi \nabla \psi^\dagger] \, \mathrm{d}\tau. \tag{10.40}$$

We can write this as

$$\overline{M}_1 = \frac{e}{2} \int (\boldsymbol{r} \times \boldsymbol{v}_A) \rho \, \mathrm{d}\tau. \tag{10.41}$$

The above is just the orbital momentum of the current. The second term is

$$\begin{aligned} \overline{M}_2 &= \frac{e}{4m} \int \boldsymbol{r} \times \nabla \times (\psi^\dagger \sigma \psi) \, \mathrm{d}\tau \\ &= \frac{e}{4m} \int (\boldsymbol{r} \times \boldsymbol{v}_B) \rho \, \mathrm{d}\tau. \end{aligned} \tag{10.42}$$

We can readily transform this into

$$\overline{M}_2 = \frac{e}{2m} \int (\psi^\dagger \sigma \psi) \, \mathrm{d}\tau. \tag{10.43}$$

This is the average magnetic moment due to the circulating motion, which evidently comes out equal to the average obtained in the usual interpretation.

We see then that, in the Dirac theory, the magnetic moment usually attributed to the 'spin' can actually be attributed to a circulating movement of a point particle, and not that of an extended spinning object. The total value is, however, independent of the size of the wave packet because, although the velocity of circulation goes down by a factor $1/a$ for large wave packets, the moment of momentum contains a further factor r which compensates for this.

This implies that the magnetic moment is not an intrinsic localised property of an extended particle structure at all. Rather it comes about as a result of the general motion of the particle through space as determined by the Dirac guidance condition (10.31). This means that the magnetic moment, like so many quantum properties that we have discussed in earlier chapters, is a context dependent property of a point particle. This property depends on the quantum field, i.e. the Dirac spinor, Ψ_a, and therefore ultimately it has to be regarded as a potentiality that depends on the context of the overall environment in a way that we have discussed in chapter 4.

In the usual approach, the magnetic moment is regarded as the product of spin and therefore it is implied that whatever movement produces the magnetic moment should also be responsible for what is generally called the spin angular momentum.

It seems natural to suggest, therefore, that there is some relationship between the circulatory motion that we have described above and what has commonly been called the spin angular momentum. In seeking this relationship, it is crucial to note that, for the Dirac particle, the velocity and the momentum are not in the same direction. Indeed the momentum of the particle follows from equation (10.34)

$$m\boldsymbol{v}_A = \frac{1}{2i}\frac{\left[\psi_1^\dagger(\nabla\psi_1) - (\nabla\psi_1^\dagger)\psi_1\right]}{\rho}$$

from which we can express the velocity as

$$\boldsymbol{v} = \frac{\boldsymbol{p}}{m} + \boldsymbol{v}_B. \tag{10.44}$$

The orbital angular momentum is

$$\boldsymbol{L}_0 = \boldsymbol{r} \times \boldsymbol{p}. \tag{10.45}$$

In an isolated system one expects the total angular momentum to be a constant. But if we compute $\dot{\boldsymbol{L}}_0$ we obtain

$$\begin{aligned}\dot{\boldsymbol{L}}_0 &= \dot{\boldsymbol{r}}\times\boldsymbol{p}+\boldsymbol{r}\times\dot{\boldsymbol{p}}\\ &= \dot{\boldsymbol{r}}\times\boldsymbol{p}+\boldsymbol{r}\times\boldsymbol{F}\\ &= \dot{\boldsymbol{r}}\times\boldsymbol{p}+\boldsymbol{T},\end{aligned} \tag{10.46}$$

where $\boldsymbol{F}$ is the applied force and $\boldsymbol{T}$ is the applied torque. Moreover even if $\boldsymbol{T}=0$ we will still not have conservation of the orbital angular momentum for the case of a Dirac particle.

In the usual interpretation a similar question arises. Thus the orbital angular momentum operator $\boldsymbol{L}=\boldsymbol{r}\times\boldsymbol{p}$ is not conserved even for a free particle because $\dot{\boldsymbol{L}}=\dot{\boldsymbol{r}}\times\boldsymbol{p}=\boldsymbol{\alpha}\times\boldsymbol{p}\neq 0$. This fact implies that there is an additional 'intrinsic' angular momentum operator $\boldsymbol{s}$ and, as is well known, if we choose $\boldsymbol{s}=\boldsymbol{\sigma}/2$ we obtain $\dot{\boldsymbol{s}}=-\boldsymbol{\alpha}\times\boldsymbol{p}$ so that $\boldsymbol{J}=\boldsymbol{L}+\boldsymbol{\sigma}$ is conserved.

Returning to our model we want to show that what has been usually called spin angular momentum can be attributed consistently to the circulatory motion that comes out of the non-relativistic limit of the Dirac equation. What is usually called the mean spin angular momentum density is just

$$\boldsymbol{s}=\psi^{\dagger}\frac{\boldsymbol{\sigma}}{2}\psi. \tag{10.47}$$

Let us now consider the quantity

$$\boldsymbol{\Sigma}=\boldsymbol{r}\times(\nabla\times\boldsymbol{s}), \tag{10.48}$$

which has the dimensions of an angular momentum density. Using three index tensors with indices i, j etc., we can readily show that

$$\boldsymbol{\Sigma}_j=\frac{\partial}{\partial x_i}(\boldsymbol{r}_i\boldsymbol{s}_j)-\frac{\partial}{\partial x_i}\,\delta^i_j(\boldsymbol{r}_k\boldsymbol{s}_k)+2\boldsymbol{s}_j, \tag{10.49}$$

so that we obtain for what is usually called intrinsic angular momentum density

$$s_j=\frac{1}{2}\left[\boldsymbol{\Sigma}_j-\frac{\partial}{\partial x_i}(\boldsymbol{r}_i\boldsymbol{s}_j)+\frac{\partial}{\partial x_i}\,\delta^i_j(\boldsymbol{r}_k\boldsymbol{s}_k)\right]. \tag{10.50}$$

We see then that s_j differs from $\frac{1}{2}\boldsymbol{\Sigma}_j$ only by a perfect divergence. Therefore it is clear that their averages will be the same (because the integral of a divergence is zero), but in addition, $\frac{1}{2}\boldsymbol{\Sigma}_j$ will also be conserved. To see that this is so we note that replacing $\boldsymbol{s}_j$ by $\frac{1}{2}\boldsymbol{\Sigma}_j$ merely adds a term to J_j which is equal to

$$+\frac{1}{2}\frac{\partial}{\partial x_i}(r_is_j)-\frac{1}{2}\frac{\partial}{\partial x_i}\,\delta^i_j(r_ks_k). \qquad \text{(a)} \tag{10.50}$$

If T_j^i is the tensor current density for the total angular momentum density J_j, the conservation equation for the latter can be written as

$$\frac{\partial J_j}{\partial t} + \frac{\partial T_j^i}{\partial x_i} = 0.$$

Adding (10.50a) to T_j^i leads to

$$\frac{\partial J_j}{\partial t} + \frac{\partial T_j^i}{\partial x_i} + \frac{1}{2}\frac{\partial}{\partial x_i}\left[r_i \frac{\partial s_j}{\partial t} - \delta_j^i r_k \frac{\partial s_k}{\partial t}\right] = 0.$$

It is then clear that the total angular momentum will be conserved if the angular momentum density of circulatory motion as defined through $\frac{1}{2}\boldsymbol{\Sigma}_j$ undergoes detailed conservation. Therefore $\frac{1}{2}\boldsymbol{\Sigma}_j$ is just as good a candidate for the 'intrinsic' angular momentum density as $\boldsymbol{s}_j$.

If we accept $\frac{1}{2}\boldsymbol{\Sigma}_j$ as the intrinsic angular momentum density, we obtain for the total angular momentum

$$\boldsymbol{J}_{\text{Total}} = \tfrac{1}{2i}\left[\psi^\dagger(\boldsymbol{r} \times \nabla)\psi - \psi(\boldsymbol{r} \times \nabla)\psi^\dagger\right] + \tfrac{1}{2}(\boldsymbol{r} \times \nabla \times \boldsymbol{s}). \qquad (10.51)$$

In this way we are able to account for the intrinsic angular momentum as produced by the circulatory motion. Note however that this angular momentum is only half that which we might have expected at first sight from our expression for the magnetic moment (10.43). The reason is basically that when $\boldsymbol{v}_B$ is not in the direction of the momentum, there are intrinsic torques. These reduce the contribution of this motion to the total angular momentum below that which one might have expected from our general intuition (this latter, of course, does not include such cases in which the velocity and the momentum are not parallel).

It follows then, that like the magnetic moment, the spin is also a context dependent property of the particle rather than an intrinsic property. It might be better to call it a result of the context dependent circulatory motion of the trajectories which is over and above that which determines the usual orbital angular momentum.

To sum up then, the interpretation that flows out of our discussion is that the electron must still be regarded as a simple point particle whose only intrinsic property is its position. The particle has a velocity which is determined by the guidance condition (10.29). In the non-relativistic approximation, this includes that assumed in the BST model, along with an additional circulatory contribution $\nabla \times (\psi_1^\dagger \sigma \psi_1)/2m\rho$. In this approximation the probability density is $P = \rho = |\Psi_1|^2$.

This model is very close to the one that was first proposed by Bell [3] who likewise assumed a point particle, but who assumed that the mean velocity is the same as that of the BST model, where

$$\boldsymbol{v}_A = \frac{\psi^\dagger \nabla \psi - (\nabla \psi^\dagger)\psi}{2mi\rho}.$$

The Bell model differs from the one that we have just developed, because it does not include the circulatory motion corresponding to the magnetic moment and the spin angular momentum. But since $\nabla \cdot \boldsymbol{j}_B = 0$, both models will give the same probability density. Therefore as long as we are only interested in probability densities, the circulating spin-current can be neglected, and we can use Bell's model instead.

In the models that have just been described above, we restrict ourselves to determining the motion of a particle through suitable guidance conditions, and we do not attempt to account in detail for the accelerations through something like a quantum potential or some form of spin-orbit interaction. To do this latter will give insight only if the acceleration can be given as some simple function of the relevant physical variables. For example, in the original causal interpretation, we find that the acceleration is determined by $m\ddot{\boldsymbol{x}} = -\nabla[V + Q]$. The first of these terms represents the part of the acceleration that can be accounted for in terms of classical forces that are generally local and that produce an effect that is proportional to the intensity of the field. The term $-\nabla Q$ represents the quantum mechanical contribution that depends on the form of the field and not on the intensity, and that acts in a way that is generally nonlocal.

In the BST model for the one-particle system there were additional quantum mechanical terms corresponding to explicit expressions for the spin-orbit coupling. But in the many-particle system, we could not find any acceptable physical interpretation for the vast number of terms that entered into the determination of the acceleration. In addition we found that we could not consistently attribute the spin angular momentum to an extended particle. We have therefore reached a point where the attempt to account for the acceleration in terms of various forces no longer provides any useful insight. Nevertheless the theory still works consistently as long as we have a suitable guidance condition.

Given that there is no extended particle that is spinning and that what we call spin represents only some average property of the circulating orbital motion, how do we understand what we usually call a measurement of the spin, for example, in a Stern-Gerlach experiment? To discuss this, we may conveniently consider a neutral hydrogen atom with one orbital electron in an s-state. From equation (10.28) we can see that the operator for the interaction energy of the electron 'spin' with the magnetic field is $(e/2m)\boldsymbol{\sigma}\cdot\boldsymbol{B}$. (We neglect nuclear spin here.) If $\boldsymbol{B}$ is inhomogeneous, that is $\boldsymbol{B} = \boldsymbol{B}(z)$, it is readily shown [7] that this will cause the initial wave packet to break into two parts that move in opposite directions along the axis determined by $\boldsymbol{B}(0)$. For example, suppose the initial 'spin' was in the x-direction, so that the spin-current was circulating around the x-axis. After interaction with the inhomogeneous magnetic field in the z-direction, two packets begin to separate in this direction and eventually cease to overlap. Clearly the particle must end up in one of these packets. The results in each individual case depend, of course, on the precise initial conditions. The probability for each possible result will come out as one-half, given our assumptions about the probability distribution of the initial conditions. If the result corresponds to spin $\hbar/2$, the current will circulate around the positive direction, while if the result is $-\hbar/2$, the circulation is opposite. In this way we understand what happens in a so-called spin measurement as an essentially participatory process. In this process the net properties of the orbital motion are transformed through this participation in a way that depends on the precise initial conditions as well as on the 'environment' supplied by the measuring apparatus. The value that is attributed to the 'spin' as a result of a measurement process is thus a potentiality whose realisation depends on these conditions and on the general environment, including the measuring apparatus.

10.5 Extension to the many-body system

We shall now extend this model to a system of many particles. In doing so we shall prove that the non-relativistic limit of the many-body Dirac equation implies that the current for each particle is just the one that we obtain for the many-body Pauli equation given in section 10.3, together with a circulating spin current for each particle. This latter is basically just an extension of the spin current obtained in the one-particle system.

We begin by defining the Dirac wave function for a system of many particles. The treatment is very similar to that of the many-body Pauli equation. The total wave function can be written as

$$\Psi = \Psi_{a_1 \ldots a_N}(\boldsymbol{x}_1 \ldots \boldsymbol{x}_N, t), \tag{10.52}$$

where each index a_i runs from 1 to 4. (We are not for the moment interested in the question of Lorentz invariance, which we shall discuss in a later chapter.) The Dirac equation for this system is then

$$i\frac{\partial \Psi}{\partial t} = \sum_n [\alpha_n \cdot \nabla_n + m_n \beta_n + V(\boldsymbol{x}_n)]\Psi. \tag{10.53}$$

α_n represents the velocity operator for the nth particle. Its matrix elements operate only on the spinor index a_n. Similarly for β_n. The α_n have the anticommutation relations

$$[\alpha_n^i, \alpha_n^j]_+ = 2\delta_n^{ij} \qquad \text{(a)} \qquad (10.54)$$

and the commutation relations

$$\left[\alpha_n^i, \alpha_{n'}^j\right]_- = 0 \qquad \text{for } n \neq n'. \qquad \text{(b)} \qquad (10.54)$$

That is to say, the α belonging to one particle commute with those belonging to the others. Similarly

$$[\beta_n, \beta_{n'}]_- = 0 \qquad \text{(c)} \qquad (10.54)$$

and

$$[\beta_n, \alpha_n]_+ = 0. \qquad \text{(d)} \qquad (10.54)$$

Let us now adopt the representation in which all the β_n are diagonal, so that the one-particle wave functions take the form

$$\psi_{\epsilon_n} = \begin{pmatrix} \psi_{1_n} \\ \psi_{2_n} \end{pmatrix},$$

where ψ_{1_n} and ψ_{2_n} are two-component spinors whose indices have been suppressed and where $\epsilon = 1$ refers to the large component and $\epsilon = 2$ refers to the small component. The N-body wave function can then be written as

$$\Psi = \Psi_{\epsilon_1 \ldots \epsilon_N}(\boldsymbol{x}_1 \ldots \boldsymbol{x}_N, t).$$

The Dirac equation will now have 2^N explicit indices representing the large and small components for each particle respectively. The Dirac equation can then be written as

$$\begin{aligned} i\frac{\partial \Psi_{\epsilon_1 \ldots 1_n \ldots \epsilon_N}}{\partial t} &= \sigma_n \cdot \nabla_n \psi_{\epsilon_1 \ldots 2_n \ldots \epsilon_N} \\ &\quad + \left[\sum_{n' \neq n} (\alpha_{n'} \nabla_{n'} + \beta_{n'} m_{n'}) + V + m_n\right] \psi_{\epsilon_1 \ldots 1_n \ldots \epsilon_N}. \end{aligned} \qquad (10.55)$$

$$\begin{aligned} i\frac{\partial \Psi_{\epsilon_1 \ldots 2_n \ldots \epsilon_N}}{\partial t} &= \sigma_n \cdot \nabla_n \psi_{\epsilon_1 \ldots 1_n \ldots \epsilon_N} \\ &\quad + \left[\sum_{n' \neq n} (\alpha_{n'} \nabla_{n'} + \beta_{n'} m_{n'}) + V - m_n\right] \psi_{\epsilon_1 \ldots 2_n \ldots \epsilon_N}. \end{aligned} \qquad (10.56)$$

We now write $\psi \propto \exp[-iEt]$ and assume that all the particles have positive energies. The Pauli wave function will then be approximately $\psi_{1_1 \ldots 1_n \ldots 1_N}$. The total energy will be approximately $E = \sum_n m_n$, in this non-relativistic approximation in which V is assumed to be small compared to $\sum_n m_n$.

For each particle, n, we can solve for $\Psi_{\epsilon_1 \ldots 2_n \ldots \epsilon_N}$ in terms of $\Psi_{\epsilon_1 \ldots 1_n \ldots \epsilon_N}$. Neglecting V for the reasons we have given, we find

$$\Psi_{\epsilon_1 \ldots 2_n \ldots \epsilon_N} \cong -i \frac{\sigma_n \cdot \nabla_n}{2m_n} \Psi_{\epsilon_1 \ldots 1_n \ldots \epsilon_N}. \tag{10.57}$$

If we substitute this into (10.38) it is readily verified that we obtain the Pauli many-body equation. However, what we are especially interested in here is the current of the nth particle which is

$$\begin{aligned} \jmath_n &= \psi^\dagger \alpha_n \psi \\ &= \sum_{\epsilon_1 \ldots \hat{\epsilon}_n \ldots \epsilon_N} \Big[\Psi^\dagger_{\epsilon_1 \ldots 1_n \ldots \epsilon_N} \sigma_n \Psi_{\epsilon_1 \ldots 2_n \ldots \epsilon_N} \\ &\qquad + \Psi^\dagger_{\epsilon_1 \ldots 2_n \ldots \epsilon_N} \sigma_n \Psi_{\epsilon_1 \ldots 1_n \ldots \epsilon_N} \Big]. \end{aligned} \tag{10.58}$$

Let us now use (10.57) in (10.58). We find that (assuming all the particles have the same mass)

$$\begin{aligned} \jmath_n &= \frac{1}{2mi} \sum_\epsilon \Big[\psi^\dagger_{\epsilon_1 \ldots 1_n \ldots \epsilon_N} \sigma_n (\sigma_n \cdot \nabla_n) \psi_{\epsilon_1 \ldots 1_n \ldots \epsilon_N} \\ &\qquad - (\sigma_n \cdot \nabla_n) \psi^\dagger_{\epsilon_1 \ldots 1_n \ldots \epsilon_N} \sigma_n \psi_{\epsilon_1 \ldots 1_n \ldots \epsilon_N} \Big]. \end{aligned} \tag{10.59}$$

This reduces to

$$\begin{aligned} \jmath_n &= \frac{1}{2mi} \sum_\epsilon \Big\{ [\psi^\dagger_{\epsilon_1 \ldots 1_n \ldots \epsilon_N} (\nabla_n \psi_{\epsilon_1 \ldots 1_n \ldots \epsilon_N}) \\ &\qquad - (\nabla_n \psi^\dagger_{\epsilon_1 \ldots 1_n \ldots \epsilon_N}) \psi_{\epsilon_1 \ldots 1_n \ldots \epsilon_N}] \\ &\qquad + [\nabla_n \times \psi^\dagger_{\epsilon_1 \ldots 1_n \ldots \epsilon_N} \sigma_n \psi_{\epsilon_1 \ldots 1_n \ldots \epsilon_N}] \Big\}. \end{aligned} \tag{10.60}$$

As we have suggested before, we can simplify by assuming all the ϵ_n are 1, corresponding to positive energies for each particle. The remaining wave function is then just the Pauli many-body wave function Ψ. We then get

$$\jmath_n = \frac{1}{2mi} [\psi^\dagger (\nabla_n \psi) - (\nabla_n \psi^\dagger) \psi] + \nabla_n \times \frac{(\psi^\dagger \sigma_n \psi)}{2m}. \tag{10.61}$$

From this we conclude that the model (10.61) for the N-body system can be obtained by regarding each particle as having two parts to its velocity

$$\boldsymbol{v}_{n_1} = \frac{(1/2mi)\left[\psi^\dagger(\nabla_n\psi) - (\nabla_n\psi^\dagger)\psi\right]}{|\psi|^2} \tag{10.62}$$

and

$$\boldsymbol{v}_{n_2} = \frac{(1/2m)\nabla_n \times (\psi^\dagger \sigma_n \psi)}{|\psi|^2}. \tag{10.63}$$

Therefore our original picture which adds a circulating current to the velocity of each particle still holds for the many-body system.

We have of course been assuming the interactions are weak enough, so that all the particles will be within the non-relativistic limit. In this approximation the details of the interaction will not significantly change this picture.

10.6 The EPR experiment for two particles of spin one-half

We shall now go on to discuss the EPR experiment for two particles of spin one-half. We shall find that we obtain a consistent interpretation that is in fact very similar to the one that we gave for the case of two particles of spin-one forming a molecule of total spin-zero (chapter 7).

We begin by denoting the coordinates of our two particles by $\boldsymbol{x}_1$ and $\boldsymbol{x}_2$. For the sake of simplicity, we suppose that the two particles are not equivalent. We further assume that each particle is fairly well localised with respective wave packets $f(\boldsymbol{x}_1)$ and $g(\boldsymbol{x}_2)$, the centres of which are separated by a distance a that is much greater than the width of either of the packets. We also suppose that these packets represent s-states about their respective centres.

The wave function for a system with total 'intrinsic' angular momentum zero will be

$$\Psi = f(\boldsymbol{x}_1)g(\boldsymbol{x}_2)\sum_{i,j=-1}^{1}\psi_{ij} \tag{10.64}$$

where

$$\begin{aligned}\psi_{ij} &= 0 && \text{for } i = j \\ &= \frac{1}{\sqrt{2}}\psi_{1,-1} && \text{for } i = 1,\ j = -1 \\ &= -\frac{1}{\sqrt{2}}\psi_{-1,1} && \text{for } i = -1,\ j = 1.\end{aligned} \tag{10.65}$$

ψ_{ij} expresses the fact that the wave function is antisymmetric in the spinor indices of the two particles. For the wave functions that we have assumed, the particles will have no orbital angular momentum. The intrinsic angular momentum of the first particle is

$$\boldsymbol{\Sigma}_1 = \frac{\boldsymbol{r}_1 \times (\nabla_1 \times \boldsymbol{s}_1)}{\rho}, \qquad \text{(a)} \qquad (10.66)$$

and for the second particle it is

$$\boldsymbol{\Sigma}_2 = \frac{\boldsymbol{r}_2 \times (\nabla_2 \times \boldsymbol{s}_2)}{\rho}. \qquad \text{(b)} \qquad (10.66)$$

Clearly the values of $\boldsymbol{\Sigma}_1$ and $\boldsymbol{\Sigma}_2$ are determined by $\boldsymbol{s}_1$ and $\boldsymbol{s}_2$. We readily obtain

$$\boldsymbol{s}_1 = \tfrac{1}{2} \sum_{ijk} \left[\psi_{ij}^{\dagger} \sigma_{ik} \psi_{kj} \right] = 0, \qquad \text{(a)} \qquad (10.67)$$

and likewise

$$\boldsymbol{s}_2 = 0. \qquad \text{(b)} \qquad (10.67)$$

This means that in the initial state, there is no circulating current corresponding to intrinsic moments. We shall discuss the further significance of this in more detail presently. But we may note here that this result is very similar to that obtained for the orbital angular momenta of two atoms of spin-one combined to make a molecule of spin-zero where similarly we found that the particles were not moving.

Let us now suppose that particle number 1 is subjected to a Stern-Gerlach experiment so that its wave function divides into two parts, one moving upwards along the z-axis by a distance d, and the other moving downwards by this distance (we assume also that d is much greater than the width of the wave packets). The wave function then becomes

$$\Psi = \frac{1}{\sqrt{2}} \left[f(z_1 - d) g(z_2) \psi_{1,-1} + f(z_1 + d) g(z_2) \psi_{-1,1} \right]. \qquad (10.68)$$

If particle number 1 enters the packet going upwards, particle number 1 actually is set in circulatory motion in the positive direction around the z-axis, while particle number 2 is likewise set in motion in the opposite direction. And if particle number 1 enters the packet going downwards, the results for both particles will be opposite.

It is clear then that we have explained the correlations in the results of the EPR experiment, not only for the case of the spin measured in the z-direction, but also in an arbitrary direction. This is because the

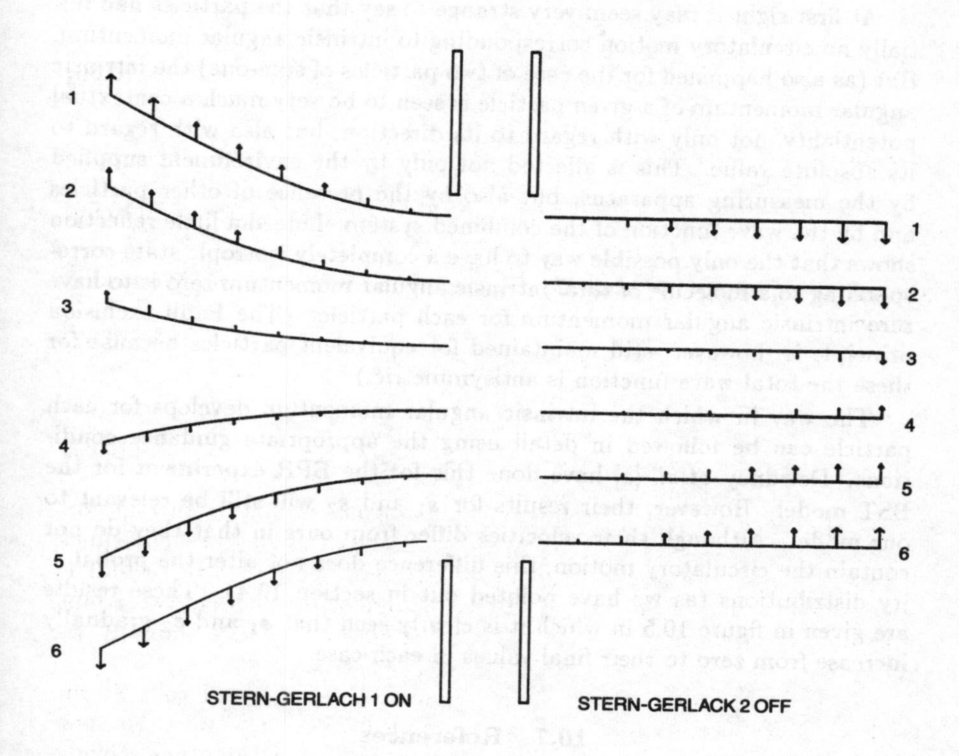

Figure 10.5: Trajectories and correlated spin vector orientations for two particles initially in a singlet state after the impulsive measurement of the z'-component of the spin of particle 1 only

wave function ψ_{ij} is a scalar for rotations, so that the same analysis can be carried out for a Stern-Gerlach experiment performed in an arbitrary direction.

At first sight it may seem very strange to say that the particles had initially no circulatory motion corresponding to intrinsic angular momentum. But (as also happened for the case of two particles of spin-one) the intrinsic angular momentum of a given particle is seen to be very much a contextual potentiality, not only with regard to its direction, but also with regard to its absolute value. This is affected not only by the environment supplied by the measuring apparatus, but also by the presence of other particles and by the wave function of the combined system. Indeed a little reflection shows that the only possible way to have a completely isotropic state corresponding to a molecule of total intrinsic angular momentum zero is to have zero intrinsic angular momentum for each particle. (The Pauli exclusion principle is, however, still maintained for equivalent particles because for these the total wave function is antisymmetric.)

The way in which the intrinsic angular momentum develops for each particle can be followed in detail using the appropriate guidance conditions. Dewdney *et al.* [5] have done this for the EPR experiment for the BST model. However, their results for $\mathbf{s}_1$ and $\mathbf{s}_2$ will still be relevant to our model. Although their velocities differ from ours in that they do not contain the circulatory motion, this difference does not alter the probability distributions (as we have pointed out in section 10.4). These results are given in figure 10.5 in which it is clearly seen that $\mathbf{s}_1$ and $\mathbf{s}_2$ gradually increase from zero to their final values in each case.

10.7 References

1. D. Bohm, R. Schiller and J. Tiomno, *Nuovo Cim. supp.* **1**, 48–66 (1955).

2. D. Bohm and R. Schiller, *Nuovo Cim. supp.* **1**, 67–91 (1955).

3. J. S. Bell, *Speakable and Unspeakable in Quantum Mechanics*, chapter 19, Cambridge University Press, Cambridge, 1987.

4. C. Dewdney, P.R. Holland and A. Kyprianidis, *Phys. Lett.* **119A**, 259–267 (1986).

5. C. Dewdney, P.R. Holland and A. Kyprianidis, *J. Phys.* **A20**, 4717–4732 (1987).

6. L. de Broglie, *Une interprétation causale et non linéaire de la mécanique ondulatoire: la théorie de la double solution*, Gauthiers-Villiars, Paris (1956). English translation, chapter 10, Elsevier, Amsterdam (1960).

7. D. Bohm, *Quantum Theory*, Prentice-Hall, London, 1951. Also available in Dover Publications, New York, 1989.

Chapter 11
The ontological interpretation of boson fields

11.1 Introduction

We are now ready to extend our ontological interpretation to the boson field which in this chapter we consider only from a non-relativistic stand-point. The question of relativistic invariance will be discussed in the next chapter.

We shall begin by giving reasons why it is necessary in doing this to give up the notion of particle as fundamental and to regard our basic 'beables' as the field variables themselves. To understand what this means, we could approximate a continuous field by a very dense distribution of field points, x_n^μ, where n is the nth point. The beables will then be the fields at the points in question (e.g. in the case of a scalar field $\phi(x_n^\mu)$). These are represented by a point in the configuration space $\phi(\ldots x_n^\mu \ldots)$ of all the field beables (which, in the limiting case of continuity, approaches a non-denumerably infinite number of dimensions). This is analogous to a system of infinitely many particles, but the coordinates x_n^μ of the particles now correspond to the values of the field $\phi(x_n^\mu)$ at the specified points at which the field has been assumed to exist. The trajectory of a whole set of particles in their configuration space is then replaced by the trajectory of the whole set of field beables in their configuration space. This means that the concept of particles does not correspond to a basic beable in our interpretation.

How then do we understand the particle-like properties of the bosonic field, e.g. photons or bosons with mass? In this chapter we develop an ontological interpretation which shows systematically how this comes about.

To do this we shall first analyse the field into normal modes and discuss their meaning for excited states of the field, including both Fock states and coherent states. We then discuss the absorption of a quantum of energy by an atom. In this process the field is initially distributed throughout a wide region of space. However, the equations for the field beables are not linear and local wave equations, such as for example, the d'Alembert and

Klein-Gordon equations. Rather they contain an extension of the quantum potential, which we call the super-quantum potential of the total field. This is both nonlinear and nonlocal. Because of this nonlinearity and nonlocality, energy is 'swept in' from the entire wave packet so that it concentrates on the atom as a single quantum of energy $h\nu$. When and whether this will happen depends on the detailed initial conditions of the total set of field beables. But statistically, the probability will be the same as in the usual interpretation. If there are several atoms, these initial conditions will determine which of them will absorb the full quantum of energy. In this way we are able to answer the criticism raised by Renninger [1], who considered a spherical wave and asked how only one atom out of all the atoms on the wave surface comes to be excited. We thus show that while the field beables are distributed throughout space, they generally manifest themselves in a discrete particle-like way (e.g. as if the field were composed of photons or massive bosons).

This explains the wave-particle duality. For it is a wave-like distribution of field beables that determine the discrete particle-like manifestations. This is one of the major differences between our ontological model of particle beables and that of field beables. In the case of the latter there are no particle beables but only dynamic states of the whole set of field beables which suggest certain particle-like features in the overall behaviour.

As an important illustrative example, we consider the photon-like behaviour in a gas arising when a quantum of energy is scattered by the atoms of this gas. Each scattering process gives rise to a subsequent quantum which then goes on to build up a cascade of ionised atoms that constitute what is usually called the 'track of the photon'. In our approach the scattering process is described by the absorption of a full quantum of energy from the overall field in a way that we have already outlined, followed by re-radiation of a quantum of a lower energy. This is then absorbed and re-emitted by a succeeding atom and so on. The typical scattering angle in the Compton effect is of the order of λ/λ_c where λ is the wavelength of the quantum and λ_c is the Compton wavelength of the atom. For high energy quanta, this will be small. Therefore successive collisions will be distributed approximately along a straight line with small random variations. The whole process takes place in a definite time sequence as if a bullet-like object has struck the atoms in a series of collisions leaving ionised particles in its wake. Yet in our interpretation there is no such object.

In this connection we emphasise that photons have never been observed directly but only through the manifestations in particulate matter that are attributed to them. Since our approach explains all these manifestations, we have no need of the further assumption of a bullet-like photon to account for what is actually observed.

In the classical limit, the super-quantum potential becomes negligible in a way similar to that which we have already discussed for the quantum potential in chapter 8. The beables then obey simple linear equations, e.g. the d'Alembert or Klein-Gordon equation. Their Fourier coefficients correspond to simple harmonic oscillators which can, for example, be treated in terms of coherent states that have an essentially classical behaviour of the beables for high quantum numbers.

The above makes it clear that in the classical limit, a bosonic system does not produce significant particle-like manifestations. The nearest we can come to a particle in the classical limit is to make a wave packet in the field beables. In so far as the spread of this packet can be neglected, there is a localised pulse of energy which follows a particle-like path (see, for example, Mandel [2]). But in the classical limit, the restriction to transferring discrete quantities of energy to particulate matter plays no significant role. For the representation of the behaviour of the field as continuous provides a very good approximation to what actually happens.

After thus bringing out the basic concept needed for our ontological interpretation of quantum field theory, we shall extend this interpretation to explain interference phenomena and the Pfleegor-Mandel [3] experiment involving the interference of two separated low-intensity lasers. Finally we treat the EPR experiment along with the Compton effect.

11.2 Reasons for starting with quantum field theories for bosonic fields

In developing an ontological interpretation of bosonic field theories, we have, *a priori*, to consider two possibilities. Firstly we could try to interpret such theories by assuming a set of bosonic particles that would be 'guided' by a quantum field based on a wave function of the whole system of particles. If we could do this, the quantum field theory would not require a treatment basically different from that which we have already given for the many-body system. In the case of the electromagnetic field, for example, such particles would represent photons, while for meson fields of various kinds they would represent the corresponding massive bosons.

But secondly, we could start as we have indeed already suggested in the introduction, by regarding the field quantities themselves as the basic beables. The wave functional of all these beables would then imply new quantum properties for the fields, and the particle-like manifestations would be explained along the lines discussed in the introduction, as dynamic properties of the whole set of field variables.

In this section we shall sketch the principal reasons why the first approach cannot be carried out consistently. We will thus be left with the second approach as the only possibility. And this is, indeed, what will be the content of the remainder of the chapter.

Various authors, de Broglie [4], Cufaro-Petroni *et al.* [5], have proposed, for example, that we begin with the Klein-Gordon scalar equation regarded as the quantum wave equation for a single spin-zero particle. This is

$$\frac{\partial^2 \phi}{\partial t^2} = \nabla^2 \phi + \left(m_0^2 + V\right) \phi. \tag{11.1}$$

(We will again put $\hbar = 1$ throughout this chapter.)

From this it follows that there is a conserved charged current

$$j^\mu = -\frac{1}{2im_0}\left(\phi^* \partial^\mu \phi - \phi \partial^\mu \phi^*\right). \tag{11.2}$$

We could propose the guidance relation

$$\boldsymbol{v} = \frac{\boldsymbol{j}}{j^0}. \tag{11.3}$$

But, as we have already pointed out in chapter 10, section 10.4, in order to be consistent this would require that j^0 be positive definite. However, it is clear that j^0 as given in (11.2) is not, in general, positive definite. Thus if the energy is well defined and positive, j^0 is positive while if it is negative, then j^0 is also negative. It is therefore already clear that this approach cannot be made consistent.

One might perhaps try to get around this difficulty by restricting ϕ to positive energies. But there is no way, in general, to maintain such a restriction, because in principle a potential can always cause transitions from one sign of energy to the other. Therefore we will always be faced with the possibility of a linear combination of positive and negative energies. For example, we can have

$$\phi = \phi_+(x)\mathrm{e}^{-iEt} + \phi_-(x)\mathrm{e}^{+iEt}. \tag{11.4}$$

This gives

$$j^0 = \phi_+^*(x)\phi_+(x) - \phi_-^*(x)\phi_-(x) + \phi_+^*(x)\phi_-(x)\mathrm{e}^{2iEt} + \text{c.c.} \tag{11.5}$$

This can move between positive and negative values. So that even if we began with positive values of j^0 we could not guarantee that j^0 would remain positive indefinitely. Moreover in going from positive to negative values j^0 must pass through zero. This will give rise to infinite velocities which do not make sense physically and which would in addition violate the relativistic requirement that $v \leq c$.

Since we cannot make a consistent particle interpretation of the kind given above for the Klein-Gordon equation, it seems reasonable to conclude that we should adopt the alternative approach of regarding $\phi(x)$ as a field beable. The fact that j^0 can be positive or negative is then understood through the notion that there can be positive and negative charges. This implies the possibility of pair production, so that the number of particles is not conserved.

Evidently in this approach there is no way to regard (11.3) as a guidance condition and the whole attempt to give a particle interpretation to this equation breaks down.

It has been proposed however that there are other conserved quantities which would give positive definite densities, and that these would make it possible to go on with the particle interpretation nevertheless (Wesley [6] and Prosser [7]). What they suggest is that one take the energy density, T^{00}, and flux of energy, T^{0i}, as a starting point. These satisfy the conservation equation

$$\frac{\partial T^{00}}{\partial x^0} + \frac{\partial T^{0i}}{\partial x^i} = 0. \tag{11.6}$$

One would then define the particle velocity as

$$v^i = \frac{T^{0i}}{T^{00}}. \tag{11.7}$$

Thus in the Klein-Gordon equation we have

$$T^{00} = \nabla\phi^* \nabla\phi + \left(m_0^2 + V\right)\phi^*\phi \tag{11.8}$$

and

$$T^{0i} = \frac{1}{2}\left[\frac{\partial\phi^*\partial\phi}{\partial x^i \partial t} - \frac{\partial\phi\partial\phi^*}{\partial x^i \partial t}\right]. \tag{11.9}$$

Since T^{00} is evidently positive definite, it is argued that a particle interpretation of the Klein-Gordon equation can after all be maintained.

A closer inspection shows, however, that such an interpretation would not be consistent with the theory of relativity (although in this chapter we are considering the field theory from a non-relativistic standpoint, it is still necessary that such a theory be applicable relativistically, or else we would contradict a large range of experimental facts).

To show the inconsistency of such a definition of velocity, let us now consider the case of a free particle and a special solution of the wave equation,

$\phi = \exp[i(kx - \omega t)]$. We obtain

$$T^{00} = \left(m_0^2 + k^2\right) \phi^* \phi, \tag{11.10}$$

$$T^{01} = \omega k_1 \phi^* \phi, \tag{11.11}$$

and $T^{02} = T^{03} = 0$.

We can always choose a Lorentz frame in which $k_1 = 0$ and $T^{00} = m_0^2 \phi^* \phi$ while $T^{0i} = 0$. This corresponds, of course, to a particle at rest. If we now transform to a new Lorentz frame so that the particle has velocity $\boldsymbol{v} = (v_1, 0, 0)$, we obtain

$$T'00 = \frac{T^{00}}{1 - v_1^2}. \tag{11.12}$$

If (T^{00}, T^{0i}) is to represent particle density and current, it has to transform as a relativistic four vector, otherwise the integral of the probability will not be the same in every frame and the transformation between frames will not be unitary. This would require that the density in the new frame T'^{00} should be equal to $T^{00}/\sqrt{1 - v^2}$ rather than to $T^{00}/1 - v^2$. We emphasise again that the difficulty arises basically because the charge current has to be a four vector and not a second rank tensor such as $T^{\mu\nu}$.

One can demonstrate even more striking inconsistencies in such an approach by considering the Maxwell field in free space. For this we have the energy density

$$T^{00} = \frac{E^2 + H^2}{8\pi}. \tag{11.13}$$

The energy flux is given by the Poynting vector, so that

$$T^{0i} = \frac{(\boldsymbol{E} \times \boldsymbol{H})^i}{4\pi}. \tag{11.14}$$

Let us consider the special case in which $\boldsymbol{E}$ is parallel to $\boldsymbol{H}$ so that $\boldsymbol{E} \times \boldsymbol{H} = 0$. Assume further $\boldsymbol{E}$ and $\boldsymbol{H}$ are in the direction x^1. This means that, under a Lorentz boost in the direction x^1, we will still have $\boldsymbol{E}'$ parallel to $\boldsymbol{H}'$ and $\boldsymbol{E}' \times \boldsymbol{H}' = 0$. It follows that the velocity is zero in a whole range of Lorentz frames. Clearly it will not make sense to say that a particle is at rest for this whole range of Lorentz frames.

Yet another approach has been proposed for interpreting the Klein-Gordon equation through a particle model (see, for example, Nambu [8]). This is based on the assumption of an invariant parameter τ which represents the order of development of a system. The suggested wave equation is

$$i\frac{\partial \phi}{\partial \tau} = \Box \phi \tag{11.15}$$

where $\Box$ is the d'Alembertian. If we look for solution of the form

$$\phi = \exp[-iK\tau]\psi(x^\mu)$$

we obtain

$$K\phi = \Box\phi. \tag{11.16}$$

Choosing $K = m_0^2$, we see that this reduces to the Klein-Gordon equation. But there is a new feature because we can now obtain solutions with arbitrary values of K and the general solution is

$$\phi = \int f(K)\mathrm{e}^{-iK\tau}\psi(x^\mu)\,\mathrm{d}K. \tag{11.17}$$

This theory obviously has room for a whole range of masses. Indeed if the mass is fixed, no probabilities will depend on τ, because the wave function will be proportional to $\exp[i\alpha m^2\tau]$ where α is a suitable constant. One can obtain a non-stationary state only by a linear combination of terms corresponding to different values of K and therefore of different masses.

In this theory, the probability density is defined as the scalar $\phi^*\phi$ which satisfies the conservation equation

$$\frac{\partial(\phi^*\phi)}{\partial\tau} + \partial_\mu j^\mu = 0, \tag{11.18}$$

where j^μ is given in (11.1). Since $\phi^*\phi$ is positive-definite, we can consistently define the guidance condition

$$v^\mu = \frac{j^\mu}{\phi^*\phi}. \tag{11.19}$$

Writing $\phi = Re^{iS}$ we obtain from (11.15) the generalised Hamilton-Jacobi equation

$$\frac{\partial S}{\partial\tau} + \partial_\mu S\partial^\mu S - \frac{1}{2}\frac{\Box R}{R} = 0, \tag{11.20}$$

where

$$Q = -\frac{1}{2}\frac{\Box R}{R}$$

is the generalised quantum potential.

In order to bring out the meaning of these equations, we first note that $\phi^*\phi$ has to be taken as the probability density in space-time for an event at a point (x^0, x^i). For example, in the usual interpretation it could be regarded as the probability in a measurement of *finding* a particle at the point x^i and at the time x^0, for a certain value of the invariant parameter τ. In an ontological interpretation, it would however be the probability that, at this value of τ, the particle actually is at this point in space-time. The velocity v^μ would then be the rate of change of the coordinates of this space-time point with τ.

This theory evidently goes beyond the simple Klein-Gordon equation because its conceptual basis can only be understood in terms of a particle which may have a range of masses. For if the mass is fixed, the probability is, as we have seen, not dependent on τ, so that there will be no way to discuss the statistics of the development of the system in the parameter τ. For the present we do not know how to determine what should be the actual mass spectrum. To do this would evidently require some extension of the theory involving further new concepts. For example, the particle may have some internal parameters and the quantisation of these might determine a mass spectrum. If this approach were successful, this mass spectrum should be that of the particles that are actually to be found in nature.

It is clear then that the parameter τ introduces a new concept which goes together with the range of masses. The meaning of this concept is that we cease to regard x^0 as determining the time order of development. Rather, x^0 plays a role not very different from that of the space coordinates. Thus at a given value of τ, the wave function will, in general, spread over a range of x^i and x^0. There is no way in general even to be sure x^0 will always increase monotonically with τ, for (as we have seen also in the earlier suggestions regarding the Klein-Gordon equation) v^0 can take on both positive and negative values. Interaction terms (which we have thus far neglected) can cause transitions between positive and negative values of v^0, so that an initial wave function representing a monotonic increase of x^0 with τ can develop parts in which x^0 goes on indefinitely *decreasing* with τ (i.e. tracks parameterised by τ can go backwards in x^0). This is evidently a serious difficulty.

However, further difficulties arise when one considers how one would treat the many-body system. If x_N^μ represents the coordinate of the Nth particle, the wave function will be $\phi(x_1^\mu \ldots x_N^\mu)$. A natural generalisation of the wave equation would be

$$i\frac{\partial\phi}{\partial\tau} = \sum_n \Box_n \phi. \qquad (11.21)$$

Because ϕ spreads out over the coordinates x_n^μ of each particle, it follows that particles considered at a given value of τ will not all have the same value of x^0. Indeed they will in general not even be all located on a simple space-like surface. Rather they will be distributed over a range of x^0 (which is, of course, time as meant in the usual sense). It is not clear how this theory can be made to connect with the usual theory in which all the particles are considered at the same time, or at least on a single space-like surface.

Various authors (see for example Vigier [9]) have suggested that for the two-particle system, this ambiguity can be removed by considering the two particles at the same time in their centre of mass frame. But with many particles it is not clear which frame should be taken as the rest frame. Thus let us divide N particles into two parts with N_1 and N_2 particles respectively. In each part, all the particles could be taken at the same time in their overall centre of mass frame. But since the parts can be chosen arbitrarily, this leaves the whole question undefined. At the very least, this difficulty suggests that such an interpretation cannot be carried out unless the theory is supplemented by further conditions that determine the common time or space-like surface at which the particles are to be taken for a given τ.

There is yet, however, another conceptual difficulty. Because the wave functions ϕ spread out over a range of times for each particle, it follows that the quantum potential for the many-body system $Q = -\frac{1}{2}\sum_n \Box_n R/R$ will directly couple particles at different times. Nonlocality is thus extended from space to time. In one sense this seems reasonable as a way of making the theory relativisitically invariant. Yet the concept of nonlocal action over time introduces new difficulties for understanding causality in its usual sense. And because, as we have seen, parts of the wave function 'travel backward in time' for a long way, this difficulty will not be restricted to short time intervals.

To sum up then we would say that while this idea is an interesting one, we cannot use it confidently on the basis of an ontological interpretation until these difficulties and conceptual ambiguities are cleared up.

11.3 An ontological interpretation of bosonic fields considered from the non-relativistic standpoint

In view of what has been said in the previous section, we feel that the only possibility for an ontological interpretation that is at present viable is to assume that the beables are the fields themselves. In this section we shall illustrate such an interpretation in terms of a real scalar field $\phi(x)$ with zero rest-mass along with its canonical momentum $\pi(x)$. (A more condensed treatment of the electromagnetic field has been given in Bohm [10].) As indicated in the introduction, we shall in this chapter restrict ourselves to a non-relativistic standpoint and in the next chapter we shall discuss how our ideas can be extended to a relativistic context.

We begin with the idea of the configuration space of the total set of field beables. We recall that a good image of the meaning of this set can be obtained by approximating a continuous field by the field variables over a set of discrete but fairly densely distributed points, $\phi(x_n^\mu)$, which will ultimately approach a continuous set. The whole set of beables is then..., $\phi(x_n^\mu)$,... To formulate the theory mathematically we introduce the well-known concept of functional derivative of any function of the whole set of beables $\Psi(\ldots\phi(x_n^\mu)\ldots)$. This is

$$\frac{\delta\Psi}{\delta\phi(x_n^\mu)} = \lim_{\delta\phi\to 0}\left\{\frac{\Psi(\ldots\phi(x_n^\mu)+\delta\phi\ldots)-\Psi(\ldots\phi(x_n^\mu)\ldots)}{\delta\phi}\right\}.$$

If Ψ depends on the derivatives of $\phi(x_n^\mu)$, we may also have quantities like $\delta\Psi/\delta\left(\mathrm{d}\phi/\mathrm{d}x^\mu\right)$. We shall not go into a detailed description here of this general notion of functional derivative (which can be found, for example, in Schiff [11]), but we shall use the ideas from now on.

For a scalar field the classical Lagrangian density is

$$\mathcal{L} = \frac{1}{2}\left[\left(\frac{\partial\phi}{\partial t}\right)^2 - (\nabla\phi)^2\right]. \tag{11.22}$$

In a particle theory the momentum canonically conjugate to x is $\partial\mathcal{L}/\partial\dot{x}$. In a field theory, basically the same concept is used in the sense that the momentum canonically conjugate to $\phi(x^\mu)$ is

$$\pi(x^\mu) = \frac{\delta\mathcal{L}}{\delta\left(\partial\phi/\partial t\right)} = \frac{\partial\phi}{\partial t}. \tag{11.23}$$

It is readily shown from an extension of the Euler-Lagrange equation that this classical field satisfies the wave equation

$$\frac{\partial^2\phi}{\partial t^2} = \nabla^2\phi. \tag{11.24}$$

To quantise a particle theory we introduce the wave function $\psi(\ldots x_n\ldots)$. To quantise the field equation we introduce a corresponding wave functional $\Psi = \Psi(\ldots\phi(x^\mu)\ldots)$. Evidently this wave functional depends on the whole set of values of $\phi(x^\mu)$ at every point in the universe. This then gives $|\Psi|^2$ as the probability of a total given field distribution corresponding to a point in the configuration space of the overall field. In the standard quantum theory the classical number $\pi(x^\mu)$ is replaced by an operator which can be symbolised as

$$\Pi_{op}(x^\mu) = -i\frac{\delta}{\delta\phi(x^\mu)}. \tag{11.25}$$

(This is analogous in the particle theory to $P_{op} = -i(\partial/\partial x^{\mu})$.) $\Pi_{op}(x^{\mu})$ acts on the wave functional $\Psi(\ldots, \phi(x^{\mu}), \ldots)$.

We emphasise again that there are basic similarities between quantum field theory and quantum particle theory (as well as differences, of course). The main similarity is that the field variables at each point play essentially the same role as the particle positions do in the particle theory. This can be made even clearer by considering the Fourier analysis of the field in a cubic box with periodic boundary conditions. The non-denumerable infinity of field variables is replaced by a denumerable infinity of Fourier coefficients ϕ_k with k_x, k_y, k_z restricted to integral multiples of $2\pi/L$ where L is the size of the box. The functional derivative then becomes an ordinary derivative and the similarity to particle quantum mechanics becomes very obvious.

To obtain the ontological interpretation of field theory we do just what was done in the particle theory, that is, we assume the basic set of dynamical variables $\ldots, \phi(x^{\mu}), \ldots$ which we take to represent the beables, have well-defined and continuously changing values. As pointed out in the introduction, the trajectory of the beables corresponds to a single point in the configuration space of the total set of field variables. To determine this trajectory we begin with what we will call the super-Schrödinger equation which is

$$i\frac{\partial \Psi}{\partial t} = H\Psi, \tag{11.26}$$

where H is the Hamiltonian given by

$$H = \frac{1}{2}\int_{\text{All space}} \left[-\frac{\delta^2}{\left(\delta\phi(x,t)\right)^2} + \left(\nabla\phi(x,t)\right)^2\right] \mathrm{d}V. \tag{11.27}$$

We then write $\Psi = R[\ldots\phi(x,t)\ldots]\exp\{iS[\ldots\phi(x,t)\ldots]\}$, and obtain first of all

$$\frac{\partial S}{\partial t} + \frac{1}{2}\int\left[\left(\frac{\delta S}{\delta\phi}\right)^2 + (\nabla\phi)^2\right] \mathrm{d}V + Q = 0, \tag{11.28}$$

where Q is the super-quantum potential which we shall define presently. The above corresponds to the Hamilton-Jacobi equation of a particle system

$$\frac{\partial S}{\partial t} + \frac{1}{2m}\sum\left(\frac{\partial S}{\partial x_i}\right)^2 + V + Q = 0.$$

We also obtain

$$\frac{\partial P}{\partial t} + \int \frac{\delta}{\delta\phi}\left[P\frac{\delta S}{\delta\phi}\right] \mathrm{d}V = 0 \tag{11.29}$$

where $P = |\Psi|^2$ is, as we have already explained, the probability of a given configuration of the total set of field variables. This clearly corresponds in particle theory to the conservation equation

$$\frac{\partial P}{\partial t} + \sum_i \frac{\partial}{\partial \boldsymbol{x}^i}\left(P\frac{\partial S}{\partial \boldsymbol{x}^i}\right) = 0.$$

We can write

$$\Pi(\boldsymbol{x}^\mu) = \frac{\delta S(\ldots\phi(\boldsymbol{x}^\mu)\ldots)}{\delta\phi(\boldsymbol{x}^\mu)}. \tag{11.30}$$

This is an extension of the guidance condition for the particle case, which is

$$\boldsymbol{p} = \nabla S. \tag{11.31}$$

The super-quantum potential will be

$$Q = -\frac{1}{2}\int \left\{ \frac{\left[\delta^2/(\delta\phi(\boldsymbol{x}^\mu))^2 R(\ldots\phi(\boldsymbol{x}^\mu)\ldots)\right]}{R(\ldots\phi(\boldsymbol{x}^\mu)\ldots)} \right\} \mathrm{d}V \tag{11.32}$$

and this is comparable in the many-particle system to

$$Q_{MP} = -\frac{1}{2m}\sum_i \frac{\nabla_i^2 R(\ldots\boldsymbol{x}_i,\ldots,t)}{R(\ldots\boldsymbol{x}_i,\ldots,t)}. \tag{11.33}$$

From the extended Hamilton-Jacobi equation (11.28) we obtain the 'field velocity'

$$\frac{\partial\phi}{\partial t} = \frac{\delta S}{\delta\phi}. \tag{11.34}$$

Then equation (11.29) clearly represents the conservation of probability. The remaining equation of motion is obtained from (11.28) as

$$\frac{\partial\pi}{\partial t} = -\frac{\delta H}{\delta\phi} = \nabla^2\phi - \frac{\delta Q}{\delta\phi}. \tag{11.35}$$

Putting (11.23) into the above we obtain

$$\frac{\partial^2\phi}{\partial t^2} = \nabla^2\phi - \frac{\delta Q}{\delta\phi}. \tag{11.36}$$

In line with the particle case given in chapter 7, the classical limit is approached whenever Q becomes negligible. In this case, equation (11.36) reduces to the ordinary wave equation. However, in the quantum domain there is an extra term $-\delta Q/\delta\phi$. In general, this is a non-linear and non-local function of ϕ. As we shall see, it profoundly alters the meaning of the equation and leads to radically new implications. We emphasise again, however, that in the classical limit none of these will be significant.

11.4 Analysis into normal modes and the ground state of the field

In order to discuss the ground state and excited states of the field, it will be convenient to go to a representation in terms of normal modes. To simplify the problem we may, as we have already suggested, consider a cubic box of side L and assume periodic boundary conditions. We take L to be finite, but much larger than any dimension of physical interest, i.e., in principle, as large as we please. The normal modes will then be trigonometric waves and an arbitrary field can be expressed as

$$\phi(\boldsymbol{x},t) = \sqrt{\frac{2}{V}}\sum_{\boldsymbol{k}}{}' \left[a_{\boldsymbol{k}}(t)\cos(\boldsymbol{k}\cdot\boldsymbol{x}) + b_{\boldsymbol{k}}(t)\sin(\boldsymbol{k}\cdot\boldsymbol{x})\right], \tag{11.37}$$

where

$$\boldsymbol{k} = k_x, k_y, k_z \quad \text{with} \quad k_x = \frac{2\pi n_x}{L}; \quad k_y = \frac{2\pi n_y}{L}; \quad k_z = \frac{2\pi n_z}{L}$$

and where n_x, n_y, n_z are integers and V is the volume of the box. Because $-\boldsymbol{k}$ and $\boldsymbol{k}$ cover the same functions we should sum only over half the possible values of $\boldsymbol{k}$, e.g. n_x and n_y cover all the integers positive, negative and zero, while n_z is restricted to values greater than or equal to zero. Whenever the sum is taken with the above restrictions we shall write it as above $\sum_{\boldsymbol{k}}{}'$.

We also introduce $\pi(\boldsymbol{x},t)$, the momentum canonically conjugate to $\phi(\boldsymbol{x},t)$, which satisfies the Poisson bracket relations

$$\{\phi(\boldsymbol{x},t), \pi(\boldsymbol{x}',t)\} = \delta(\boldsymbol{x}-\boldsymbol{x}'). \tag{11.38}$$

We Fourier analyse $\pi(\boldsymbol{x},t)$ writing

$$\pi(\boldsymbol{x},t) = \sqrt{\frac{2}{V}}\sum_{\boldsymbol{k}}{}' \left[\alpha_{\boldsymbol{k}}(t)\cos(\boldsymbol{k}\cdot\boldsymbol{x}) + \beta_{\boldsymbol{k}}(t)\sin(\boldsymbol{k}\cdot\boldsymbol{x})\right]. \tag{11.39}$$

With the aid of (11.38) we then obtain the Poisson bracket relations

$$\{a_{\boldsymbol{k}}, \alpha_{\boldsymbol{k}'}\} = \{b_{\boldsymbol{k}}, \beta_{\boldsymbol{k}'}\} = \delta_{\boldsymbol{k}'\boldsymbol{k}},$$

with all other Poisson brackets equal to zero. This means that a_k, α_k and b_k, β_k form a set of canonical variables.

From (11.27) we obtain the Hamiltonian

$$H = \frac{1}{2}\int \left[\pi^2(\boldsymbol{x}) + (\nabla\phi(\boldsymbol{x}))^2\right] \mathrm{d}V, \tag{11.40}$$

which in terms of Fourier analysis becomes

$$H = \frac{1}{2}\sum_{\boldsymbol{k}}{}' \left[(\alpha_{\boldsymbol{k}})^2 + k^2(a_{\boldsymbol{k}})^2 + (\beta_{\boldsymbol{k}})^2 + k^2(b_{\boldsymbol{k}})^2\right]. \tag{11.41}$$

It is convenient to go to the complex representation of the normal modes. Writing

$$q_{\boldsymbol{k}} = \frac{1}{\sqrt{2}}(a_{\boldsymbol{k}} - ib_{\boldsymbol{k}}), \qquad \pi_{\boldsymbol{k}} = \frac{1}{\sqrt{2}}(\alpha_{\boldsymbol{k}} - i\beta_{\boldsymbol{k}}) \tag{11.42}$$

we obtain

$$\phi(\boldsymbol{x}, t) = \frac{1}{\sqrt{V}}\sum_{\boldsymbol{k}} q_{\boldsymbol{k}}(t)\exp(i\boldsymbol{k}\cdot\boldsymbol{x}) \tag{11.43}$$

and

$$\pi(\boldsymbol{x}, t) = \frac{1}{\sqrt{V}}\sum_{\boldsymbol{k}} \pi_{\boldsymbol{k}}(t)\exp(i\boldsymbol{k}\cdot\boldsymbol{x}). \tag{11.44}$$

From the above we see that the reality of ϕ implies that $q_{\boldsymbol{k}} = q^*_{-\boldsymbol{k}}$ and that $\pi_{\boldsymbol{k}} = \pi^*_{-\boldsymbol{k}}$.

We then obtain the Poisson bracket relations

$$\{q_{\boldsymbol{k}}, \pi^*_{\boldsymbol{k}'}\} = \{q^*_{\boldsymbol{k}}, \pi_{\boldsymbol{k}'}\} = \delta_{\boldsymbol{k}\boldsymbol{k}'}$$

with all other Poisson brackets being zero. And the Hamiltonian becomes

$$H = \sum_{\boldsymbol{k}}{}' \left[\pi^*_{\boldsymbol{k}}\pi_{\boldsymbol{k}} + k^2 q^*_{\boldsymbol{k}} q_{\boldsymbol{k}}\right]. \tag{11.45}$$

We can easily verify that the classical equations of motion are

$$\dot{q}_k = \frac{\partial H}{\partial \pi^*_{\boldsymbol{k}}} = \pi_{\boldsymbol{k}}, \qquad \dot{\pi} = -\frac{\partial H}{\partial q^*_{\boldsymbol{k}}} = -k^2 q_{\boldsymbol{k}},$$

so that

$$\ddot{q}_{\boldsymbol{k}} + k^2 q_{\boldsymbol{k}} = 0.$$

We now go to the quantum theory. We represent the wave functional in terms of the $q_{\boldsymbol{k}}$ and $q^*_{\boldsymbol{k}} = q_{-\boldsymbol{k}}$. We are restricting ourselves to half the values of k as before. We then write

$$\Psi = \Psi(\ldots q_{\boldsymbol{k}} \ldots; q^*_{\boldsymbol{k}} \ldots; t).$$

Because the system reduces to a set of independent harmonic oscillators, the wave functional can be expressed as a linear combination of the product functions,

$$\Psi_0 = \psi\left(q_{\boldsymbol{k}_1}, q^*_{\boldsymbol{k}_1}\right) \psi\left(q_{\boldsymbol{k}_2}, q^*_{\boldsymbol{k}_2}\right) \ldots \psi\left(q_{\boldsymbol{k}_i}, q^*_{\boldsymbol{k}_i}\right) \ldots \tag{11.46}$$

Each $\psi\left(q_{\boldsymbol{k}}, q^*_{\boldsymbol{k}}\right)$ satisfies an independent Schrödinger equation representing a harmonic oscillator,

$$i\frac{\partial \psi}{\partial t}\left(q_{\boldsymbol{k}}, q^*_{\boldsymbol{k}}\right) = \left[\frac{-\partial^2}{\partial q_{\boldsymbol{k}} \partial q^*_{\boldsymbol{k}}} + k^2 q^*_{\boldsymbol{k}} q_{\boldsymbol{k}}\right] \psi\left(q_{\boldsymbol{k}}, q^*_{\boldsymbol{k}}\right). \tag{11.47}$$

Writing $\psi\left(q_{\boldsymbol{k}}, q^*_{\boldsymbol{k}}\right) = R_{\boldsymbol{k}} \exp(iS_{\boldsymbol{k}})$ we obtain the equivalent Hamilton-Jacobi equation

$$\frac{\partial S_{\boldsymbol{k}}}{\partial t} + \left(\frac{\partial S_{\boldsymbol{k}}}{\partial q_{\boldsymbol{k}}}\right)\left(\frac{\partial S_{\boldsymbol{k}}}{\partial q^*_{\boldsymbol{k}}}\right) + k^2 q^*_{\boldsymbol{k}} q_{\boldsymbol{k}} - \frac{i}{R_{\boldsymbol{k}}} \frac{\partial^2 R_{\boldsymbol{k}}}{\partial q_{\boldsymbol{k}} \partial q^*_{\boldsymbol{k}}} = 0. \tag{11.48}$$

The above result is similar to that obtained in the causal interpretation of a particle undergoing harmonic oscillation. The ground state of equation (11.47) then represents the zero point energy. Its wave function is

$$\psi_0\left(q_{\boldsymbol{k}} q^*_{\boldsymbol{k}}\right) = \exp\left[-k q^*_{\boldsymbol{k}} q_{\boldsymbol{k}}\right]. \tag{11.49}$$

The ground state of the whole set of oscillators is then

$$\Psi_0 = \exp\left[-\sum_{\boldsymbol{k}}{}' k q^*_{\boldsymbol{k}} q_{\boldsymbol{k}}\right]. \tag{11.50}$$

The quantum potential for this state is

$$Q_0 = \sum_{\boldsymbol{k}}{}' \left(k - k^2 q^*_{\boldsymbol{k}} q_{\boldsymbol{k}}\right). \tag{11.51}$$

Writing

$$\frac{\partial S}{\partial t} = -E,$$

we get

$$E = \sum_{\boldsymbol{k}}{}' \left[\left(\frac{\partial S}{\partial q_{\boldsymbol{k}}} \right) \left(\frac{\partial S}{\partial q^*_{\boldsymbol{k}}} \right) + k \right]. \tag{11.52}$$

Since the wave function is real, $\partial S/\partial q_{\boldsymbol{k}} = \partial S/\partial q^*_{\boldsymbol{k}} = 0$ and we obtain

$$E = \sum_{k}{}' k. \tag{11.53}$$

This is, evidently, the usual expression for the zero point energy (note that each $\boldsymbol{k}$ corresponds to two real oscillators a_k and b_k so that the usual factor $\frac{1}{2}$ is not present).

We shall now discuss the meaning of the wave function for the ground state, equation (11.50). Firstly, let us note that although the quantum potential will, in general, be nonlocal, in the special case of the ground state it reduces to a *local* function. In fact, it cancels the term $\sum_{\boldsymbol{k}}{}' k^2 q^*_{\boldsymbol{k}} q_{\boldsymbol{k}}$, which, in the space representation, corresponds to $\nabla\phi^* \nabla\phi$. This is evidently a local function (as we shall see later it is only in the excited states that nonlocality appears, although in the classical limit it disappears again).

Let us now consider the significance of the probability distribution. (Wheeler [12] has considered this problem for the electromagnetic field and has obtained results similar to those that will be given here. His analysis is, however, developed in less detail and, of course, does not consider the role of the quantum potential.) We begin by expressing the q_k in terms of $\phi(\boldsymbol{x})$, i.e.

$$q_k = \frac{1}{\sqrt{V}} \int \exp[-i\boldsymbol{k} \cdot \boldsymbol{x}]\phi(\boldsymbol{x})\, \mathrm{d}V. \tag{11.54}$$

(We have suppressed the coordinate t which will not be relevant in this context.) This yields

$$\Psi_0 = \exp\left[-\iint \phi(\boldsymbol{x})\phi(\boldsymbol{x}')f(\boldsymbol{x}' - \boldsymbol{x})\, \mathrm{d}V\, \mathrm{d}V' \right], \tag{11.55}$$

where

$$f(\boldsymbol{x}' - \boldsymbol{x}) = \frac{1}{V} \sum_{k}{}' k \exp\left[i\boldsymbol{k} \cdot (\boldsymbol{x}' - \boldsymbol{x})\right].$$

To turn this sum into an integral, we introduce the density of $\boldsymbol{k}$-vector which is $V/(2\pi)^3$ and we obtain

$$f(\boldsymbol{x}' - \boldsymbol{x}) = \frac{1}{(2\pi)^3} \int k \exp\left[i\boldsymbol{k} \cdot (\boldsymbol{x}' - \boldsymbol{x})\right] \mathrm{d}\boldsymbol{k}. \tag{11.56}$$

This is not a properly defined function but we can evaluate it by taking a limit in which there is a cut-off in $\boldsymbol{k}$-space. We can represent such a cut-off by multiplying the integrand by a factor $e^{-\lambda k}$ and allowing λ to become as small as we please. These integrations are then straightforward and the result is

$$f(r) = \frac{1}{\pi^2} \frac{r^2 - 3\lambda^2}{(\lambda^2 + r^2)^3} \tag{11.57}$$

where $r = |\boldsymbol{x}' - \boldsymbol{x}|$.

The probability function is then

$$P = |\Psi_0|^2 = \left| \exp\left[-\iint \frac{\phi(\boldsymbol{x})\phi(\boldsymbol{x}')}{\pi^2} \frac{r^2 - 3\lambda^2}{(\lambda^2 + r^2)^3} \mathrm{d}V \, \mathrm{d}V' \right] \right|^2 . \tag{11.58}$$

This probability is largest when $\phi(\boldsymbol{x})$ and $\phi(\boldsymbol{x}')$ have different signs. Therefore it is quite clear that for a given value of $\phi(\boldsymbol{x})$ it is highly improbable that a neighbouring field $\phi(\boldsymbol{x}')$ will have the same sign within a distance of the order of λ. (Recall that λ is as small as we please.) This point can be brought out more clearly by going to spherical polar coordinates where $r = |\boldsymbol{x}' - \boldsymbol{x}|$. The contribution to the probability coming from a field $\phi(\boldsymbol{x})$ which is on a shell of radius r will be

$$\left| \exp\left[-\iint \frac{\phi(\boldsymbol{x})\phi(\boldsymbol{x} + \boldsymbol{r})}{\pi^2} \frac{r^2 - 3\lambda^2}{(r^2 + \lambda^2)^2} r^2 \, \mathrm{d}r \, \mathrm{d}\Omega \right] \right|^2 .$$

Clearly this probability can become very small for small λ and small r, when $\phi(\boldsymbol{x})$ and $\phi(\boldsymbol{x} + \boldsymbol{r})$ have the same sign. It is therefore most probable that the field $\phi(\boldsymbol{x})$ is highly discontinuous, the sharpness of the discontinuity being dependent on the cut-off radius λ. Such a discontinuity implies that the most probable spatial form of the field will be chaotic, i.e. in the sense of modern chaos theory [13]. However, this chaotic variation will be limited, because the variation of the q_k will be the order of $1/\sqrt{k}$ as can be seen from equation (11.50).

Because $\partial S/\partial q_k = \pi_k = \dot{q}_k = 0$, it follows that the field is static. This result is surprising as one generally thinks that the zero point fluctuations of the vacuum correspond to some kind of chaotic *dynamical* behaviour. We can, indeed, obtain such a dynamical behaviour by introducing the further assumption that the field is taking part in a random fluctuation in which the mean value of $\dot{\pi}_k$ is zero. This would be along the lines suggested in our stochastic particle interpretation given in chapter 9. However, for the purposes of the present treatment, we shall confine ourselves to the initial deterministic model, which we emphasise is completely self-consistent.

The ground state energy which we ordinarily ascribe to the dynamic behaviour will now be attributed to the super-quantum potential. This is analogous to the particle case in a stationary state, with $p = 0$ and $E = V + Q$, or alternatively $Q = E - V$. In the classical case, we have $T = E - V$. So Q is in some sense playing a role analogous to T. For example, in an excited state, the energy of the quantum potential enters into the quantum transitions in the same way that the kinetic energy would in the classical limit. And if, as we have suggested above, the energy of the quantum potential is accounted for by further random motions, then the analogy to our ordinary way of thinking will be very close.

Whether we take the static or dynamic model, the ground state is not covariant because it defines a favoured frame in which either $\dot{q}_k = 0$ or its mean value is zero. Will this be consistent with relativity? We will discuss this question in the next chapter.

11.5 The excited state of the field

There is a considerable range of possible forms for the excited state of the field. One of these is the well-known set of Fock states in which the number of quanta and the energy are well defined. Another is the set of coherent states [14], which are especially suited to the discussion of the time dependence of the field and its approach to the classical limit. We will discuss these in detail in the next section. For the case where the mean quantum number is low, both kinds of states lead to the same results for most purposes. However, as we shall see, for states with high quantum numbers there may be significant differences between the two types of state.

Let us begin with a treatment of the Fock state. A typical wave function for the excitation of a normal mode q_k with a definite $\boldsymbol{k}$-vector is

$$\Psi = q_{\boldsymbol{k}} \exp\left[-\sum k q^*_{\boldsymbol{k}} q_{\boldsymbol{k}}\right] = q_{\boldsymbol{k}} \Psi_0. \tag{11.59}$$

The probability function is

$$P_{\boldsymbol{k}} = q^*_{\boldsymbol{k}} q_{\boldsymbol{k}} |\Psi_0|^2. \tag{11.60}$$

What this means is that the actual field in the excited state will in general not look very different from that of the ground state. As a function of $\boldsymbol{x}$ we will have

$$P_{\boldsymbol{k}} = \iint \phi(\boldsymbol{x})\phi(\boldsymbol{x}') \exp\left[i\boldsymbol{k} \cdot (\boldsymbol{x}' - \boldsymbol{x})\right] \mathrm{d}V\, \mathrm{d}V' \, |\Psi_0|^2. \tag{11.61}$$

The factor $|\Psi_0|^2$ will still represent a generally chaotic variation. However, the integral in front implies a tendency for waves with vector $\boldsymbol{k}$ to be emphasised relative to the original chaotic distribution. From the actual value of the field one could therefore not say for certain whether the state is excited or not because the same field configuration could exist in either state. Indeed, the degree of excitation is determined only with the aid of the super-quantum potential which involves the state of the whole as discussed in chapter 4 in the treatment of the many-body system.

A wave functional having only a single normal mode excited is an extreme abstraction. A more realistic representation of the state would involve something like a wave packet. To represent such a packet we need a range of wave vectors, f_k. In this case the wave packet will take the form

$$\Psi = \sum_{\boldsymbol{k}}{}' f_{\boldsymbol{k}} q_{\boldsymbol{k}} \Psi_0. \tag{11.62}$$

In the above equation, the sum is over all $\boldsymbol{k}$ and we do not make the restriction that $f_{-\boldsymbol{k}} = f^*_{\boldsymbol{k}}$ because the wave functional is, in general, complex even though $f(\boldsymbol{x})$ is real. Indeed considering that $q_{-\boldsymbol{k}} = q^*_{\boldsymbol{k}}$, we can also write

$$\Psi = \sum_{\boldsymbol{k}}{}' \left[f_{\boldsymbol{k}} q_{\boldsymbol{k}} + f_{-\boldsymbol{k}} q^*_{\boldsymbol{k}} \right] \Psi_0 \tag{11.63}$$

where the $\sum_{\boldsymbol{k}}{}'$ indicates summation over a suitable half of the total set of values of $\boldsymbol{k}$ as discussed before. For the applications made in this chapter we may assume that $f_0 = 0$, i.e. the space average of the field is zero.

To obtain the time dependence of Ψ, we note that the state $q_{\boldsymbol{k}} \Psi_0$ and $q^*_{\boldsymbol{k}} \Psi_0$ both have the same energy k, so that they oscillate as $\exp(-ikt)$. In this case the wave function becomes

$$\Psi(t) = \sum_{\boldsymbol{k}}{}' \left[f_{\boldsymbol{k}} q_{\boldsymbol{k}} \mathrm{e}^{-ikt} + f_{-\boldsymbol{k}} q^*_{\boldsymbol{k}} \mathrm{e}^{-ikt} \right] \Psi_0.$$

It is clear from the above that the $q_{\boldsymbol{k}}$ terms correspond to running waves in the direction $+z$, while the $q^*_{\boldsymbol{k}}$ terms correspond to running waves in the $-z$-direction.

To simplify the discussion let us form a wave packet running only in the $+z$-direction. This will be sufficient to illustrate the general meaning of the super-quantum potential for these wave packets. We obtain

$$\Psi(t) = \int F(\boldsymbol{x}, t) \phi(\boldsymbol{x}) \, \mathrm{d}V \, \Psi_0 \tag{11.64}$$

where

$$F(\boldsymbol{x}, t) = \frac{1}{\sqrt{V}} \sum_{\boldsymbol{k}}{}' f_{\boldsymbol{k}} \exp\left[-i\boldsymbol{k} \cdot \boldsymbol{x} - ikt \right]. \tag{11.65}$$

To interpret this, it is convenient to write, in accordance with equation (11.37),

$$\phi(\boldsymbol{x},t)=\sqrt{\frac{2}{V}}\sum_{\boldsymbol{k}}{}' \left[a_{\boldsymbol{k}}(t)\cos(\boldsymbol{k}\cdot\boldsymbol{x})+b_{\boldsymbol{k}}(t)\sin(\boldsymbol{k}\cdot\boldsymbol{x})\right],$$

and define

$$\phi_c=\sqrt{\frac{2}{V}}\sum{}' a_{\boldsymbol{k}}(t)\cos(\boldsymbol{k}\cdot\boldsymbol{x})$$

and

$$\phi_s=\sqrt{\frac{2}{V}}\sum{}' b_{\boldsymbol{k}}(t)\sin(\boldsymbol{k}\cdot\boldsymbol{x})$$

with $\phi=\phi_c+\phi_s$. We note that ϕ_c and ϕ_s are orthogonal. Similarly, we split $F(\boldsymbol{x})$ into two parts F_c and F_s with

$$F_c=\frac{1}{\sqrt{V}}\sum{}' f_{\boldsymbol{k}}\exp[-ikt]\cos(\boldsymbol{k}\cdot\boldsymbol{x})$$

and

$$F_s=-\frac{i}{\sqrt{V}}\sum{}' f_{\boldsymbol{k}}\exp[-ikt]\sin(\boldsymbol{k}\cdot\boldsymbol{x}).$$

We further note that F_s is orthogonal to ϕ_c and F_c to ϕ_s. Using these orthogonality relations we then obtain

$$\Psi(t)=\int\left[F_c(\boldsymbol{x},t)\phi_c(\boldsymbol{x})+F_s(\boldsymbol{x},t)\phi_s(\boldsymbol{x})\right]\,\mathrm{d}V\,\Psi_0. \tag{11.66}$$

This wave function will be large in absolute magnitude for fields in which $\phi_c=\alpha_cF_c$ and $\phi_s=\alpha_sF_s$, where α_c and α_s are real and positive proportionality factors. This follows because the total integrand is then a non-negative function of both integrands. Fields that are not proportional in this way will tend to produce a smaller integral because of cancellation, and this is especially true for the fields that are orthogonal to those described above.

Introducing $\Gamma(\boldsymbol{x})$ to represent the part of the field that is orthogonal to $F_c(\boldsymbol{x},t)$ and $F_s(\boldsymbol{x},t)$ we can write

$$\phi(\boldsymbol{x},t)=\alpha_cF_c(\boldsymbol{x},t)+\alpha_sF_s(\boldsymbol{x},t)+\Gamma(\boldsymbol{x},t). \tag{11.67}$$

The most probable state will then be one in which α_c and α_s are as large as possible (but let us recall that within the Gaussian function, Ψ_0, will tend to limit the variation of the overall field).

It is therefore most likely that the field $\phi(\boldsymbol{x},t)$ will have the form of a wave packet corresponding to $\alpha_c F_c(\boldsymbol{x},t)+\alpha_s F_s(\boldsymbol{x},t)$. The functions $\Gamma(\boldsymbol{x},t)$ which are orthogonal to $F_c(\boldsymbol{x},t)$ and $F_s(\boldsymbol{x},t)$ will then not affect the factor in front of Ψ_0. This means that their variation will be the same as it is in the ground state. Therefore the generally chaotic variation of the field as a whole will be modified by a statistical tendency to vary around an average which has the form of a wave packet as described above.

We shall now show that inside this wave packet the super-quantum potential introduces nonlocal connections between fields at different points separated by a finite distance (unlike what happens in the ground state). To obtain the quantum potential, we must first write down the absolute value of the wave function in the $q_{\boldsymbol{k}}$ representation. Recalling that we are choosing $f_{-\boldsymbol{k}}=0$ in equation (11.63), we write

$$\Psi=\sum_{\boldsymbol{k}}{}' f_{\boldsymbol{k}} q_{\boldsymbol{k}} \exp[-ikt]\Psi_0.$$

Then writing

$$g=\sum_{\boldsymbol{k}}{}' f_{\boldsymbol{k}} q_{\boldsymbol{k}} \exp[-ikt]$$

we have

$$R=\sqrt{\Psi^*\Psi}=\sqrt{gg^*}\Psi_0. \tag{11.68}$$

The quantum potential is

$$Q=-\sum_{\boldsymbol{k}}{}' \frac{\partial^2 R}{\partial q_{\boldsymbol{k}}^* \partial q_{\boldsymbol{k}}} \Big/ R.$$

Let us now evaluate the change of quantum potential, ΔQ, from the ground state. This is

$$\Delta Q=-\frac{1}{4}\sum_{\boldsymbol{k}}{}' \frac{f_{\boldsymbol{k}} f_{\boldsymbol{k}}^*}{g^* g}+\frac{1}{2}\sum_{\boldsymbol{k}}{}' \frac{k f_{\boldsymbol{k}} q_{\boldsymbol{k}} \exp[-ikt]}{g}+\text{c.c.} \tag{11.69}$$

In a wave packet which has only a small range of wave vectors, the factor $\boldsymbol{k}$ on the right-hand side reduces to a fixed number $\boldsymbol{k}_0$, while the remaining factors reduce to unity. This term then gives the extra energy of the whole field above the ground state (recalling that only one quantum has been excited). More generally, this term varies with the time in a rather complex way, but it will be sufficient for our purposes to consider only those wave packets in which the spread of $\boldsymbol{k}$ makes a negligible contribution to this term.

What will be of interest to us is the remaining term. When the $q_{\boldsymbol{k}}$ are expressed in terms of $\phi(\boldsymbol{x})$ through equation (11.54), the quantum potential then reduces to

$$\Delta Q = -\frac{1}{4}\sum_{\boldsymbol{k}}{}' \frac{f_{\boldsymbol{k}} f_{\boldsymbol{k}}^*}{\sqrt{\int F(\boldsymbol{x},t)\phi(\boldsymbol{x})\,\mathrm{d}V}\sqrt{\int F^*(\boldsymbol{x}',t)\phi(\boldsymbol{x}')\,\mathrm{d}V'}}. \tag{11.70}$$

Clearly this term implies a nonlocal interaction between $\phi(\boldsymbol{x})$ at one point and the $\phi(\boldsymbol{x}')$ at other points for which the integrand is appreciable. Writing $Q = \Delta Q + Q_0$, where Q_0 is the quantum potential of the ground state given in equation (11.55) we can write the field equation (11.36) as

$$\frac{\partial^2\phi}{\partial t^2} = \nabla^2\phi - \frac{\delta\Delta Q}{\delta\phi} - \frac{\delta Q_0}{\delta\phi}. \tag{11.71}$$

Using equation (11.55) and expressing Q in terms of $\phi(\boldsymbol{x})$ by Fourier analysis we obtain

$$\begin{aligned}\frac{\partial^2\phi}{\partial t^2} &= \frac{1}{8}\left(\sum_{\boldsymbol{k}}{}' f_{\boldsymbol{k}} f_{\boldsymbol{k}}^*\right)\\ &\quad\times \frac{f(\boldsymbol{x},t)}{\left(\int F(\boldsymbol{x},t)\phi(\boldsymbol{x})\,\mathrm{d}V\right)^{3/2}\left(\int F^*(\boldsymbol{x}',t)\phi(\boldsymbol{x}')\,\mathrm{d}V'\right)^{1/2}} + \text{c.c.}\end{aligned} \tag{11.72}$$

Recall that in the ground state, the field is static because the effect of the quantum potential cancels out $\nabla^2\phi$ in the field equation. But now in the excited state there is a further term which causes the wave packet to move. As happens with the quantum potential itself, the field equation is nonlocal as well as nonlinear. This nonlocality represents instantaneous connection of the field at different points in space. However, it is significant only over the extent of the wave packet. In the usual interpretation the spread of the wave packet corresponds to a region within which, according to the uncertainty principle, nothing at all can be said as to what is happening. Therefore the causal interpretation attributes nonlocality only to those situations in which, in the usual interpretation, no well-defined properties at all can be attributed.

It is important to note that a wave functional of the form $\Psi = q_k \exp[-ikt]\Psi_0$ does not correspond to the usual picture of an oscillation. This can be seen from equation (11.72) because the term $\nabla^2\phi$ is absent. This result follows essentially because, as explained in a previous chapter, stationary wave functions generally correspond to a static situation in contradiction to our intuitive expectations of a dynamic state of movement.

11.6 Coherent states and the classical limit

In the previous section we discussed how Fock states are to be understood in terms of our ontological interpretation and we found that although our results are consistent, they do not correspond to our ordinary physical intuition as to what to expect under these conditions. In order to obtain wave functions giving results that are closer to such intuitive notions as well as to give a treatment that applies as we approach the classical limit, we may use coherent oscillator wave functions (generally called coherent states in the usual interpretation [14]). To illustrate what is meant by a coherent oscillator wave function, let us consider a particle with coordinate $\boldsymbol{x}$ undergoing harmonic oscillation. The Schrödinger equation is

$$i\frac{\partial\psi}{\partial t} = -\frac{1}{2m}\frac{\partial^2\psi}{\partial \boldsymbol{x}^2} + \frac{k}{2}\boldsymbol{x}^2\psi.$$

Introducing $\xi = \sqrt{m\omega}\, x$, where $\omega = \sqrt{k/m}$, we obtain

$$i\frac{\partial\psi}{\partial t} = -\frac{\omega}{2}\left[-\frac{\partial^2}{\partial\xi^2} + \xi^2\right]\psi.$$

The ground state wave function is

$$\psi_0 \propto \exp\left[-\frac{\xi^2}{2}\right].$$

One can show by direct differentiation that a solution for Schrödinger's equation is obtained by writing

$$\psi \propto \exp\left[-\tfrac{1}{2}(\xi - \gamma e^{-i\omega t})^2\right], \tag{11.73}$$

where γ is a complex number which we can write as $\gamma = |\gamma| e^{-i\theta}$. We can rewrite the above as

$$\begin{aligned}\psi \;\propto\; & \exp\left[-\tfrac{1}{2}\left(\xi - |\gamma|\cos(\omega t + \theta)\right)^2\right] \exp\left[i|\gamma|\sin(\omega t + \theta)\xi\right] \\ & \times \exp\left[-i|\gamma|^2 \sin(\omega t + \theta)\cos(\omega t + \theta) + \tfrac{1}{2}|\gamma|^2 \sin^2(\omega t + \theta)\right].\end{aligned} \tag{11.74}$$

The latter exponential is a factor that can be absorbed into the normalising coefficient. This solution, therefore, represents a Gaussian wave packet whose centre oscillates harmonically with amplitude $|\gamma|$ and initial phase θ. The mean momentum (in these units) at any time, t, is $|\gamma|\sin(\omega t+\theta)$.

Evidently the packet oscillates as a unit. Wave functions of this kind which are, of course, not stationary are what we have called coherent wave functions. They represent the nearest approximation to the classical behaviour of an oscillator.

If $a^\dagger$ represents the creation operator for a quantum (so that $a^\dagger a = N$ is the number operator) then one can show [14] that

$$\psi = \exp\left[-\tfrac{1}{2}|\gamma|^2)\right]\sum_{n=0}^{\infty}\frac{a^{\dagger n}\gamma^n\psi_0}{n!}. \tag{11.75}$$

The probability for the nth state of excitation is

$$P_n = \frac{\exp\left[-|\gamma|^2\right]|\gamma|^{2n}}{n!}.$$

This is a Poisson distribution and the most probable value of n is $n = |\gamma|^2$. The root-mean-deviation is $\delta_n \cong \sqrt{n}$ or $\delta_n/n \cong 1/\sqrt{n}$. For large values of $|\gamma|$, although the number of quanta is indefinite, the fractional deviation of this number from the mean approaches zero. This behaviour further shows the classical significance of coherent wave functions of high $|\gamma|$, because their energy becomes nearly determinate in the sense that the fractional deviation is small, while the absolute magnitude of the indeterminacy in energy becomes large enough so that one quantum more or less makes no significant difference. Therefore, in interactions with other systems, the oscillator with a coherent state of high mean energy behaves nearly like a classically well-defined object.

To further show up the classical significance of such wave functions, we evaluate the quantum potential and compare it with the classical potential. The quantum potential (in these units) will be

$$Q = -\tfrac{1}{2}\left[\xi - |\gamma|\cos(\omega t+\theta)\right]^2 + \tfrac{1}{2}, \tag{11.76}$$

but because of the Gaussian probability function, the range of fluctuation of particle position and therefore of the quantum potential (which is essentially proportional to the argument of the Gaussian) will be of the order of 1 (in these units). Evidently, as the absolute value of $|\gamma|$ becomes large, the total energy (which is kinetic + potential + quantum potential) will be correspondingly large and therefore much greater than the quantum potential. Since the quantum potential can be neglected, the dynamics will then be essentially that of classical mechanics.

Of course, the coherent state represents classical motion under conditions in which, in the usual interpretation, we would say that x and p are defined as accurately as possible within the limits of the uncertainty principle. However, more general wave functions are clearly possible which approach the classical limit in the sense that the quantum potential can be neglected but which correspond to much greater uncertainty in x and p.

It is evident that we can discuss coherent wave functions for the radiation oscillators in a straightforward way. For low degrees of excitation the coherent wave function of an oscillator given by equation (11.75) can be approximated as

$$\Psi \propto (1 + |\gamma| a^{\dagger})\Psi_0. \tag{11.77}$$

This represents a time-dependent linear combination of the ground state and the first excited state. For most purposes, such as calculating probability of transitions, the results obtained from such a state are essentially the same as those obtained from the corresponding Fock state which is not time-dependent and in which the number of quanta is well defined. The only difference is that the mean field will oscillate harmonically in the coherent state while it does not do so in the Fock state.

The most interesting case to consider is that of a wave packet in which the mean value of ϕ actually moves with the form of this packet even as we approach the limit of high quantum numbers. To obtain this we go back to equation (11.73), writing for the wave function of an oscillator

$$\psi \propto \exp\left[-\frac{\xi^2}{2} + \xi\gamma e^{-i\omega t}\right].$$

It will be convenient here to return to expressing the field ϕ in terms of a_k and b_k, the real and imaginary parts of q_k. This field is

$$\phi(\boldsymbol{x}, t) = \sqrt{\frac{2}{V}} \sum_k{}' \left[a_k(t)\cos(\boldsymbol{k}\cdot\boldsymbol{x}) + b_k(t)\sin(\boldsymbol{k}\cdot\boldsymbol{x})\right].$$

We want a packet in this field whose average behaviour is given by

$$a_k = |r_k| \cos[\omega_k t + \delta_k]$$

and

$$b_k = |s_k| \cos[\omega_k t + \epsilon_k].$$

The appropriate overall wave packet will have to be made up of a product of states

$$\Psi \propto \exp\left[-\sum_k k\frac{(a_k^2 + b_k^2)}{2}\right] \exp\left[r_k \exp\left(i(\delta_k + \omega_k t)\right) a_k\right]$$
$$\times \exp\left[s_k \exp\left(i(\epsilon_k + \omega_k t)\right) b_k\right]. \tag{11.78}$$

For large r_k and s_k this will represent a very nearly well-defined state of movement as a wave packet (corresponding, for example, to the oscillations of an electromagnetic field in a resonant cavity). In this state the average field will have a packet form and there will be random Gaussian deviations from this packet which become negligible when the mean amplitude is great enough. This means that, as pointed out in the introduction, the particle-like properties fade away and we are left with the field beables as an essentially complete description of the system. (Essentially because the super-quantum potential produces negligible effects.)

11.7 The concept of a photon

Thus far in discussing wave packets we have not encountered any of the discrete quantised properties of the electromagnetic field. We shall now show how these come in by discussing the process of absorption of energy from the field by an atom which goes from the ground state to an ionised state. If the field is in an excited state with a single quantum of energy then it will interact with the atom. During this interaction, a very complex wave functional of the field together with the atom will develop. The resulting super-quantum potential can change dramatically to bring in bifurcation points. When the process is complete the whole system of field plus particle will then either remain in the initial state or will enter into a new state in which a quantum of energy has gone into the atom, while the field is left in the ground state. As we have already suggested in the introduction, the super-quantum potential, being nonlocal, sweeps up energy in this way from the whole wave packet and brings it to the atom. The discrete quantised manifestations of the field are thus explained in terms of a theory which nevertheless allows the fields to vary more or less continuously over a wide range of values.

The treatment is very similar to that of the Auger-like effect which we discussed in a previous chapter. We begin with the initial wave functional of the field, which for mathematical simplicity we treat as a scalar field. This is given by equation (11.62), which represents a packet of Fock states with one quantum of excitation,

$$\Psi_0^f = \sum_{\boldsymbol{k}} f_{\boldsymbol{k}} q_{\boldsymbol{k}} \exp\left[-\sum_{\boldsymbol{k}'}{}' k' q_{\boldsymbol{k}'}^* q_{\boldsymbol{k}'}\right].$$

The initial wave function of the electron in an atom is $\psi_0(\xi - \xi_0)$ where ξ_0 represents the coordinates of the centre of the atom. The combined initial wave functional is

$$\Psi_0 = \sum_{\boldsymbol{k}} f_{\boldsymbol{k}} q_{\boldsymbol{k}} \psi_0(\xi - \xi_0) \exp\left[-\sum_{\boldsymbol{k}'}{}' k' q^*_{\boldsymbol{k}'} q_{\boldsymbol{k}'}\right].$$

We assume an interaction Hamiltonian

$$H_I = \lambda \int \phi(x)\delta(x-\xi)\mathrm{d}^3 x = \lambda\phi(\xi) = \lambda \sum_{\boldsymbol{k}} q_{\boldsymbol{k}} \exp[i\boldsymbol{k}\cdot\xi]. \qquad (11.79)$$

What this means is that the particle interacts locally, i.e. only at the central point of the atom.

As happened in the Auger-like effect, the interaction Hamiltonian will introduce further terms into the overall wave function. The coefficient $q_{\boldsymbol{k}}$ is the sum of a creation operator and an annihilation operator. In this case, only the annihilation operator will be significant and it will produce the ground state of the field. The quantum of energy will then go into the particle at ξ. We assume that this energy is considerably more than enough to ionise the atom. The final wave function of the electron can be written as $\psi_f = \sum_n C_n \psi_n(\xi, t)$ where the $\psi_n(\xi, t)$ represent packets going in various directions. Perturbation theory then enables us to compute C_n as function of the time. The final wave function for the whole system is

$$\Psi = \Psi_0^f \psi_0(\xi - \xi_0) + \sum_n \int_0^t \alpha_n(t', t)\psi_n(\xi, t - t')\,\mathrm{d}t' \Psi_0^f \qquad (11.80)$$

where $\alpha_n(t', t)$ can be calculated using ordinary time-dependent perturbation theory along the lines explained in chapter 6.

The coefficient $\alpha_n(t', t)$ is proportional to V_{n0}, which is the matrix element between the initial state ψ_0 and one of the final states ψ_n of the electron. In more detail this is

$$V_{n0} = \lambda \int \psi_0^*(\xi') \sum_{\boldsymbol{k}} f_{\boldsymbol{k}} \exp\left[i\boldsymbol{k}\cdot(\xi' + \xi_0)\right] \psi_n(\xi')\,\mathrm{d}V', \qquad (11.81)$$

with $\xi' = \xi - \xi_0$. (The term $\sum_{\boldsymbol{k}} f_{\boldsymbol{k}} \exp\left[i\boldsymbol{k}\cdot(\xi' + \xi_0)\right]$ has arisen by taking the matrix element of the field between its initial excited state and final ground state.)

What equation (11.80) means is that during a small time interval dt, a contribution to the wave function will be produced which will be described by the integrand. This contribution is multiplied by the ground state of the field.

The packet functions $\psi_n(\xi, t - t')$ correspond to particles that move in a direction represented by n and are set in motion at $t = t'$. These packets will all separate so that they cease to overlap both with each other and with the initial wave function y_0.

Returning to the ontological interpretation as applied to the electron, we can say as shown in chapter 6 that the corresponding particle will eventually either enter one of the outgoing packets or it will remain in the initial state still bound to the atom. After it enters one of these channels, it will remain there indefinitely. Moreover, because the packets do not overlap with each other or with the initial wave functional, the quantum potential in each channel will be the same as if the other channels did not exist. In other words, as explained in chapter 6, all the effects of the collapse of a wave function will be present without any actual collapse having occurred. Furthermore, as also explained in chapter 6, when the electron goes on to interact with the thermal environment, the probability that these packets could ever come to interfere again will be negligible.

The effective wave function of the combined system then reduces to

$$\Psi_{\text{eff}} = \psi_{n_f}(\xi, t - t')\Psi_0^f \tag{11.82}$$

where n_f signifies the packet actually occupied by the particle. This represents a situation in which the particle now carries away a discrete quantum of energy, while the field is left in its ground state.

As we have indeed already indicated, during the period of transition the quantum potential will be very complex and will contain many bifurcation points, implying a highly unstable movement of the whole system. According to the initial conditions of the particle and of the total field, the whole system will in some cases enter channels corresponding to a complete transition in which a whole quantum of energy has been absorbed by the particle, while in other cases no absorption will have taken place. These initial conditions are very complex because they include not only the particle position, but also the initial configuration of the entire set of field beables which are, as we recall, basically chaotic in their distribution. A very slight change in the initial field configuration could, in many cases, change the result from one outcome to something very different, for example from absorption of a quantum to non-absorption. However, when the probability of transition is worked out using the assumptions that the initial conditions have a distribution of field and particle variables corresponding to $P = |\psi|^2$, then the usual probability of transition will be obtained.

This probability is proportional to $|V_{n0}|^2$ where V_{n0} is the matrix element given in equation (11.81). It will be useful to transform this equation by first introducing the function

$$K(\xi' + \xi_0) = \sum_{\boldsymbol{k}} f_{\boldsymbol{k}} \exp\left[i\boldsymbol{k} \cdot (\xi' + \xi_0)\right].$$

This is essentially the spatial form of the mean wave packet as determined by the Fourier coefficients $f_{\boldsymbol{k}}$, which appeared in the original wave functional of the field (11.62). It is clear that the wave functional of the field determines a function $K(\xi)$ which is the nearest that one can come to what the classical field would be in the corresponding classical situation. We may then expand $\exp[i\boldsymbol{k} \cdot \xi'] = 1 + i\boldsymbol{k} \cdot \xi'$ because ψ_0 is generally appreciable only in regions considerably smaller than the typical wavelength in the packet. Then noting that $\psi_0(\xi')$ and $\psi_n(\xi', 0)$ are orthogonal where $\psi_n(\xi', 0)$ is the value of ψ_n at $t - t' = 0$, we obtain

$$\begin{aligned} V_{n0} &= \lambda \int \psi_0^*(\xi') \sum_{\boldsymbol{k}} i\boldsymbol{k} \cdot \xi' f_{\boldsymbol{k}} \exp[i\boldsymbol{k} \cdot \xi_0] \psi_n(\xi', 0)\, \mathrm{d}V' \\ &= \lambda \int \psi_0^*(\xi') \xi' \psi_n(\xi', 0)\, \mathrm{d}V' \cdot \nabla K(\xi_0). \end{aligned} \tag{11.83}$$

This means that the matrix element depends on the mean field K through its gradient. The fact that it depends on the gradient has come about because we are working with a scalar theory. With a vector theory it would not depend on the gradient in this way. However, this difference is not of fundamental significance because it means only that the probability of transition depends on the gradient of the mean field function (so that, for example, the maxima will occur at points of steepest gradient rather than at points of greatest intensity of the wave).

When the ground state is spherically symmetric, the probability of transition, P, will also be spherically symmetric so that

$$P \propto |\nabla K|^2. \tag{11.84}$$

We thus complete the demonstration that, although the field is essentially continuous, the possibility of transfer of energies is discrete and quantised according to the usual rules. Evidently, similar results will be obtained for emission, excitation, etc.

By means of a calculation similar to that leading to equation (11.72) it can be shown that the energy distributed throughout the wave packet is swept up through the nonlocal and nonlinear effects of the quantum potential. The fact that the amount of energy transferred is 'quantised' can be understood by noting that in the long run the effective wave function has to take the form of an excited state in which the energy differs by a definite and discrete amount from the ground state.

Thus far we have considered an excited field incident on a single atom. What would happen if there were several such atoms in different positions that could be reached by the wave packet? (This is essentially the situation considered by Renninger [1], but simplified by taking a plane wave as the limit of a spherical wave of infinite radius.) Is it possible then that we would still end up, as experimental fact requires, with a single quantum of energy transferred to only one of these atoms while the others remain unexcited?

To treat this problem we would need to introduce not only the wave functional of the field, but also the wave functions $\phi_A(x_1), \phi_B(x_2), \phi_C(x_3)$ of the various particles. If we did this however, the perturbation theory could still be applied as in the one-particle case. The state in which a quantum had been absorbed would contain a linear combination of free states of the various particles. But this would correspond to having several possible channels of this kind. Our treatment would still go through and the particle would end up in one of these channels. The channel in which the quantum is absorbed is determined by the precise initial conditions of the particles and of the set of field beables as a whole (recalling that the whole process will, in general, be unstable and chaotic). However, the probability of each outcome will still be the same as in the usual interpretation.

Finally, in section 11.2, we showed that it is not possible to make an ontological interpretation of the field in which there would be a well-defined particle trajectory. How then is it possible to understand the observed fact that a γ-ray, for example, passing through a gas, not only leaves a trail of ionised particles looking like a track, but also shows an observable development in time which suggests a bullet-like object, i.e. a photon, striking one atom after the other as it moves along? To understand this we must first recall that, as has already been pointed out in the introduction, this bullet-like object is never itself observed. All that can ever be observed are the manifestations of the field in particulate matter (e.g. by ionising gas atoms or producing the photo-electric effect, etc.). So it cannot be concluded that experiments have proved the existence of 'particles' of light.

The way is therefore open for a different explanation such as that which we are proposing here. This is that basically what is happening is that the electromagnetic field is undergoing inelastic scattering involving definite quanta of field energy in the way we have described earlier. We need not go into detail here, but we shall merely recall that this scattering is brought about, roughly speaking, by the absorption of a quantum of energy which is swept up into the atom and by the re-emission of a quantum of a slightly lower energy moving in a slightly different direction. If the initial energy of the quantum is large compared with that absorbed by the atom, then a series of such inelastic scatterings will determine a distribution of ions resembling a jaggered track.

In order to bring out the time development of this track one could, for example, have an initial wave packet that is narrow compared with the distance between the atoms. It would then follow that one scattering process after another would take place, as if there had been a bullet-like particle responsible for these events. However, even if the wave packet were broad, a similar result would follow. For, as shown in chapter 5, a transition process can take place in a time much less than that required for the wave packet to pass a given point. When such a process takes place, e.g. the ionisation of an atom, the subsequent behaviour follows at a relatively well-defined time in a way that is independent of the width of the wave packet.

We conclude that even though our ontological interpretation does not imply any boson particles such as photons, it can still explain all the experimental behaviour that has been attributed to such 'particles' in the usual interpretation.

11.8 Interference experiments

11.8.1 The treatment of interference

We are now ready to discuss how interference is to be treated in the ontological interpretation of the quantum theory. We shall do this by assuming that quanta of the field are detected by atoms that absorb discrete units of energy in the way described in the previous two sections. We shall see that in an interference experiment with two slits, for example, the probability of absorption will be proportional to what would be thought of classically as the field intensity at the point where the atom is located. Of course, this probability will vary in just such a way as to explain the observed interference patterns.

We shall then go on to discuss the Pfleegor-Mandel experiment [3] in which two independent laser sources are excited in such a way that there is only one quantum of energy in the whole system. (A more detailed discussion of this topic can be found in Paul [15].) If the two lasers can interfere in a certain region of space, this gives rise to a paradox when we try to think of where the photon that is absorbed 'comes from'. In our treatment this situation presents no serious problem. The energy is distributed throughout the space between the sources, while in the process of absorption, it is swept into the single atom by the action of the nonlocal, nonlinear quantum potential.

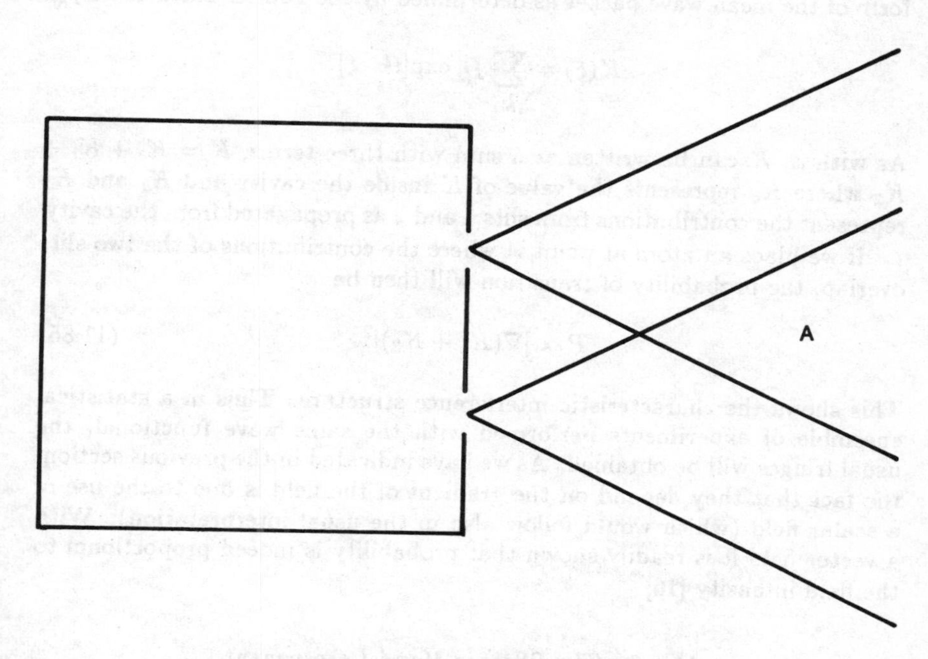

Figure 11.1: Two-slit interference

11.8.2 The two-slit interference pattern

Let us now return to a consideration of the interference in the two-slit experiment. To simplify the discussion we shall assume a resonant cavity with one of its normal modes excited to its first quantum state. Let us suppose that two small holes are made in the cavity close to each other, as shown in figure 11.1. We assume that this will make a negligible change to the normal modes inside the cavity. In the classical description, waves will radiate out from the holes 1 and 2, and will interfere in the region A. We will treat this by representing the field inside the cavity by ϕ_c and the field outside by ϕ_0. It will be convenient to treat ϕ_0 as the sum ϕ_1 and ϕ_2 of the contributions from slits 1 and 2 respectively. The effective normal mode is then $\phi = \phi_c + \phi_1 + \phi_2$. Of course, this is an approximation that will be valid if the holes are small enough so that the overall wave function will not change appreciably during the course of an experiment.

The probability of transition is not determined by ϕ alone but depends on K in the way given by (11.84). As we have said earlier, K is the spatial form of the mean wave packet as determined by the Fourier coefficients $f_{\boldsymbol{k}}$,

$$K(\xi) = \sum_{\boldsymbol{k}} f_{\boldsymbol{k}} \exp[i\boldsymbol{k} \cdot \xi].$$

As with ϕ, K can be written as a sum with three terms, $K = K_c + K_1 + K_2$ where K_c represents the value of K inside the cavity and K_1 and K_2 represent the contributions from slits 1 and 2 as propagated from the cavity.

If we place an atom at point A where the contributions of the two slits overlap, the probability of transition will then be

$$P \propto |\nabla(K_1 + K_2)|^2. \tag{11.85}$$

This shows the characteristic interference structure. Thus in a statistical ensemble of experiments performed with the same wave functional, the usual fringes will be obtained. As we have indicated in the previous section, the fact that they depend on the gradient of the field is due to the use of a scalar field (which would follow also in the usual interpretation). With a vector field it is readily shown that probability is indeed proportional to the field intensity [16].

11.8.3 The Pfleegor-Mandel experiment

Let us now go on to the Pfleegor-Mandel experiment [3]. We shall treat this in terms of two resonant cavities C_1 and C_2, each with a small hole in it. The beams coming out of the holes will cross at A as shown in figure 11.2. Each cavity will have normal modes denoted respectively by the functions K_1 and K_2. As in the interference experiment, we divide each of these into a part that is inside and a part that is outside the cavity, $K_1 = K_{I_1} + K_{O_1}$ and $K_2 = K_{I_2} + K_{O_2}$. Suppose that one of these lasers is excited to the first excited Fock state while the other is left in the ground state. (It is easy to show that a similar result for this experiment would be obtained for coherent states as well.) According to equation (11.59) the overall wave functional is then

$$\Psi_1 = \sum_{\boldsymbol{k}} f_{1\boldsymbol{k}} q_{\boldsymbol{k}} \Psi_0$$

where Ψ_0 is the vacuum state and where $f_{1\boldsymbol{k}}$ is the Fourier coefficient of K_1. But if the second laser is excited, then the overall wave function is

$$\Psi_2 = \sum_{\boldsymbol{k}} f_{2\boldsymbol{k}} q_{\boldsymbol{k}} \Psi_0.$$

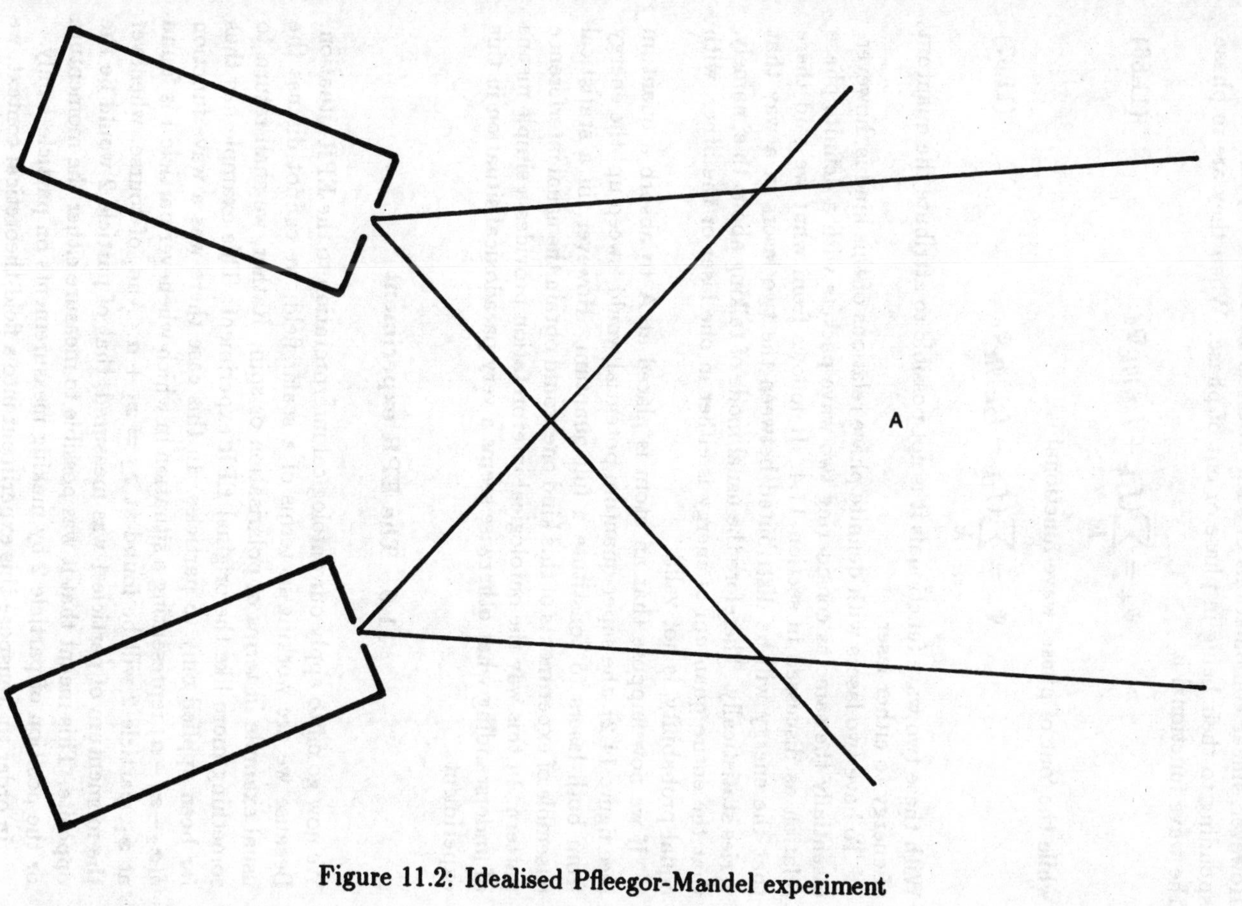

Figure 11.2: Idealised Pfleegor-Mandel experiment

However, linear combinations of such wave functionals are possible corresponding to their being in phase or out of phase. When they are in phase the wave functional is

$$\Psi_+ = \sum_{\boldsymbol{k}} (f_{1\boldsymbol{k}} + f_{2\boldsymbol{k}}) q_{\boldsymbol{k}} \Psi_0 \tag{11.86}$$

while the 'out of phase' wave functional is

$$\Psi_- = \sum_{\boldsymbol{k}} (f_{1\boldsymbol{k}} - f_{2\boldsymbol{k}}) q_{\boldsymbol{k}} \Psi_0. \tag{11.87}$$

With these two wave functionals it is not possible to attribute the quantum of energy to either laser.

To have two lasers with definite phase relations of this kind is, however, essentially the same as considering two wave packets with a definite phase relation as discussed in section 11.4. It follows from what we said there that the energy will be distributed between the two lasers in a way that varies statistically. Therefore the usual mode of talking about this, namely, that the entire quantum of energy is either in one laser or the other, with equal probability, is not valid.

If we now suppose that an atom is placed at **A** to absorb a quantum (see figure 11.2), the super-quantum potential would 'sweep up' the energy from both lasers to constitute a full quantum. However, in a statistical ensemble of experiments of this kind one would obtain the usual interference pattern. In this way the ontological interpretation provides a simple means of understanding what otherwise seems a very paradoxical situation in this experiment.

11.9 The EPR experiment

We now go on to apply our ontological interpretation to the EPR situation. Because we are working in terms of a scalar field, we cannot discuss the usual example in terms of polarisation or spin. Rather, we shall return to something more like the original EPR experiment. This example has thus far been applied only to particles. In this case there was a wave function $\delta(\boldsymbol{x}_2 - \boldsymbol{x}_1 - \boldsymbol{a})$ representing a situation in which whenever particle 1 is found at $\boldsymbol{x}_1$, particle 2 will be found at $\boldsymbol{x}_2 = \boldsymbol{x}_1 + \boldsymbol{a}$. And, of course, whenever the momentum of particle 1 was measured, that of particle 2 would be the opposite. This meant that it was possible to measure either the momentum or the position of particle 2 by making measurements on particle 1 only.

In order to transpose this experiment into a field-theoretical context, we first replace $\delta(\boldsymbol{x}_2 - \boldsymbol{x}_1 - \boldsymbol{a})$ by a function $S(\boldsymbol{x}_2 - \boldsymbol{x}_1 - \boldsymbol{a})$ which is sharply peaked but regular, and can be Fourier analysed as

$$S = \sum_{\boldsymbol{k}} g_{\boldsymbol{k}} \exp[i\boldsymbol{k} \cdot (\boldsymbol{x}_2 - \boldsymbol{x}_1 - \boldsymbol{a})].$$

If the width of the function S is considerably less than $\boldsymbol{a}$, then clearly the conditions of the experiment are not significantly altered but we can avoid the mathematical difficulties that arise from the δ-function.

The experimental situation that we envisage is to have two normal modes of the field, each excited with opposite momentum, but in a correlated way and with the distance between the respective regions of excitation of these normal modes approximately equal to $\boldsymbol{a}$. The detection of the position of these normal modes will be again by means of an atom that absorbs one of the quanta near the position of the atom ξ_0 (with the initial wave function of the electron that absorbs the quanta being $\psi_0(\xi - \xi_0)$). What we have to do is to show what happens to the other normal mode when this process of position detection takes place on the first normal mode.

The initial wave functional of the combined system of field and particle will then be

$$\psi = \sum_{\boldsymbol{k}} g_{\boldsymbol{k}} \exp[-i\boldsymbol{k} \cdot \boldsymbol{a})] q_{\boldsymbol{k}} q_{-\boldsymbol{k}} \Psi_0 \psi_0(\xi - \xi_0) \qquad (11.88)$$

or

$$\psi = \iint S(\boldsymbol{x}_2 - \boldsymbol{x}_1 - \boldsymbol{a}) \phi(\boldsymbol{x}_2) \phi(\boldsymbol{x}_1) \Psi_0 \psi_0(\xi - \xi_0)\, \mathrm{d}V_1\, \mathrm{d}V_2. \qquad (11.89)$$

Let us now consider detection of the position of the normal modes in more detail. We can see from equation (11.89) that there will be a statistical tendency that whenever the field ϕ is excited at $\boldsymbol{x}_1$, there is a corresponding excitation at $\boldsymbol{x}_2 = \boldsymbol{x}_1 + \boldsymbol{a}$ (and this is, of course, just the experimental situation we wish to consider). To treat the process of detection of position, we introduce the Hamiltonian for interaction between the field and the atom which we represent as

$$H_I = \lambda \sum_{\boldsymbol{k}} q_{\boldsymbol{k}} \exp[i\boldsymbol{k} \cdot \xi].$$

This Hamiltonian will lead to the absorption of a quantum of energy associated with, say, $q_{\boldsymbol{k}}$. Because the two quanta of energy are oppositely correlated in momenta, this will leave the total field with only a single quantum associated with $q_{-\boldsymbol{k}} = q^*_{\boldsymbol{k}}$, and it will introduce a factor into the wave functional $\exp(-i\boldsymbol{k} \cdot \xi)$.

However, as shown by equation (11.80), this term will also bring about a transition of the particle wave function from $\psi_0(\xi - \xi_0)$ to a final wave function which includes terms such as $\psi_n(\xi, t)$. Each ψ_n determines a separate channel, and eventually the electron enters one of them, ψ_{n_f}, after which the other channels may be dropped from consideration. The final wave function of field plus particle will then be

$$\Psi_f = \sum_{\boldsymbol{k}} g_{\boldsymbol{k}} \exp[-i\boldsymbol{k} \cdot (\xi + \boldsymbol{a})] q_{-\boldsymbol{k}} \Psi_0 \psi_{n_f}(\xi, t). \tag{11.90}$$

(Recall $q_{-\boldsymbol{k}} = q^*_{\boldsymbol{k}}$ when ϕ is real.) We are left with

$$\Psi_f = \int S(\boldsymbol{x} - \xi - \boldsymbol{a}) \phi(\boldsymbol{x}) \Psi_0 \psi_{n_f}(\xi, t)\, \mathrm{d}V. \tag{11.91}$$

This result is the same as the one we would obtain using the usual quantum theory, which effectively contains the assumption of the collapse of the wave function to the final state. However, in the ontological interpretation there has been no collapse, but an equivalent result has been brought about by the action of the super-quantum potential. The field now fluctuates in a region near $\boldsymbol{x} = \xi + \boldsymbol{a}$. The absorption of the quantum of energy by the atom at ξ has therefore given a more definite shape and form to the remaining field at $\boldsymbol{x} = \xi + \boldsymbol{a}$. This is, of course, a nonlinear and nonlocal effect of the super-quantum potential.

To complete the discussion of the EPR situation, we now need an experiment in which the momentum is determined rather than the position. One way of measuring momentum of a quantum is through the Compton effect. For example, if the frequency of the quantum in one of the normal modes is ν, we can measure this frequency by scattering the quantum from a very weakly bound electron. The quantum will be absorbed and a new quantum will be emitted with frequency ν' and with an angle, θ, relative to the original wave vector. The electron absorbs the energy difference $\nu - \nu'$ and scatters at an angle ϕ with momentum $\boldsymbol{p}$.

Using conservation of energy and momentum, we obtain by means of a well-known calculation

$$\nu = \frac{p\,(1 - p^2/4m^2)}{2\cos\phi - p/m}. \tag{11.92}$$

By measuring $\boldsymbol{p}$ and ϕ for the recoil electron, we can therefore find the value of ν. And according to the reasoning behind the EPR effect we expect the second normal mode to have the same frequency but the opposite momentum. To make clear how this comes about, let us represent the initial state of the electron by $\psi_0(\xi - \xi_0)$. This will be very poorly localised so that the range of momenta will be small. The limiting accuracy of the measurement will be determined by this range of momenta. To treat this problem in detail would involve time-dependent perturbation theory for the scattering process which includes the absorption of the original quantum and emission of the new quantum in another direction. However, for our purposes here we do not have to go into all the details. It is sufficient to take into account the fact that the initial wave function of the electron, $\psi_0(\xi - \xi_0)$, goes into a linear combination of packet functions, $\psi_n(\xi, t)$, representing the recoil electron. Each of these packets will constitute a channel and the electron will enter one of them so that thereafter the other channels can be neglected. We know from a more detailed mathematical treatment that these packets are very nearly plane waves with a range of wave numbers k corresponding roughly to the range in $\psi_0(\xi - \xi_0)$.

Let us denote the creation operator initially associated with the first normal mode, by $q_{\boldsymbol{k}}$, while the corresponding scattered quantum will be associated with $q_{\boldsymbol{k}'}$ (which has a different direction from $q_{\boldsymbol{k}}$ as well as a different energy). If the recoil electron determines that the quantum originally in the first normal mode had frequency ν (and therefore momentum $k = \nu$), it follows that the final wave function will be effectively reduced to

$$\psi_f \propto g_{\boldsymbol{k}'} \exp[-i\boldsymbol{k}' \cdot \boldsymbol{a}] q_{\boldsymbol{k}'} q_{-\boldsymbol{\nu}} \Psi_0 \psi_n(\xi t). \tag{11.93}$$

This result shows that after the Compton effect has taken place, not only does the scattered quantum go into a definite state associated with $q_{\boldsymbol{k}'}$, but also that the quantum in the second normal mode goes into a definite state associated with $q_{-\boldsymbol{\nu}}$. In other words the measurement of the properties of the recoil electron determines that the second normal mode transforms into a wave-like form correlated to the results of this measurement even though, according to the usual interpretation, there is no connection between the first normal mode and the second. But, of course, in our interpretation this is brought about by the super-quantum potential, and thus the explanation is basically similar to that given for the EPR situation with particles.

11.10 Appendix: the destruction of interference by radiation

In chapter 8, section 3, we explained how radiation falling on a large object, such as a planet in space, can eliminate the possibility of interference between contributions of different parts of its wave front. In section 4 of chapter 8 we actually proved that this would happen when a distribution of particles falls on such an object, but we deferred the proof that this would happen for radiation until now.

The situation we are discussing is shown in figure 8.2 (chapter 8) in which radiation is assumed to be falling on the planet. We consider two possible positions of the planet, A and B, and the shadows S_{A_1} and S_{B_1} which would be cast from the respective positions. In section 4 of chapter 8 we saw that particles would cast similar shadows and we proved that if there is a large enough number of such particles, the overlap between the total wave function for the planet at A and for the planet at B will become negligible so that interference in a region O (see figure 8.1) will be destroyed.

A very similar treatment can be given for radiation. Let us suppose there is a beam of radiation incident in the z-direction constituted of quanta having some range of k-vectors. A Fock wave function for a single quantum of definite wave vector $\boldsymbol{k}_s$ can be written as

$$\psi_i = q_{\boldsymbol{k}_s}\Psi_0 \tag{11.94}$$

where Ψ_0 is the ground state wave function for all the quanta as given in equation (11.49)

$$\Psi_0 = \prod_k \exp[-kq_k^* q_k]. \tag{11.95}$$

We are assuming that the planet completely absorbs the quantum if it falls on it, otherwise we suppose that nothing happens. We now write

$$q_{\boldsymbol{k}_s} = \int \mathrm{d}\boldsymbol{x} \exp[i\boldsymbol{k}_s \cdot \boldsymbol{x}]\phi(\boldsymbol{x})$$

where $\phi(\boldsymbol{x})$ is the field at $\boldsymbol{x}$. The contribution to the net wave function when the planet is at A can be written as

$$\Psi_1 = \int \mathrm{d}\boldsymbol{x} \exp[i\boldsymbol{k}_s \cdot \boldsymbol{x}]\, \{\phi(\boldsymbol{x})S_{\mathrm{A}}(\boldsymbol{y}-\boldsymbol{x}) + 1 - S_{\mathrm{A}}(\boldsymbol{y}-\boldsymbol{x})\}\, \Psi_0 \Phi_{\mathrm{A}}(\boldsymbol{y})$$

where S_{A} is zero in the shadow of the planet and is unity otherwise, while $\Phi_{\mathrm{A}}(\boldsymbol{y})$ is the planetary wave function for the region A. The above expresses the fact that the excited state of the field wave function will only occur outside the shadow of the planet while in the shadow we have only the ground state.

The corresponding contribution when the planet is at B is

$$\Psi_2 = \int \mathrm{d}\boldsymbol{x} \exp[i\boldsymbol{k}_s \cdot \boldsymbol{x}]\, \{\phi(\boldsymbol{x})S_{\mathrm{B}}(\boldsymbol{y}-\boldsymbol{x}) + 1 - S_{\mathrm{B}}(\boldsymbol{y}-\boldsymbol{x})\}\, \Psi_0 \Phi_{\mathrm{B}}(\boldsymbol{y}).$$

We must now estimate the overlap between Ψ_1 and Ψ_2. When we treated the case of particle bombardment in chapter 8, section 4, this was simple as there was no overlap of the trajectories corresponding to the two cylindrical shadows S_{A_1} and S_{B_1}. With fields the situation is a bit more complicated because what we have to consider is not an overlap of positions, but rather of field variables between the excited state wave function and the ground state wave function. Because the Gaussian ground state has a finite spread of field variables, there is some overlap between this wave function and the first excited state wave function corresponding to the single quantum. Nevertheless this is evidently less than the overlap of the first excited state with itself. So there will be some decrease of overlap and will give rise to a fraction F of overlap which will be less that given by F in chapter 8, section 4 for the case of particles.

So we may write

$$F' = \epsilon F = \frac{\epsilon a^3}{\alpha L^3}$$

where ϵ may be estimated to be of the order of one-half. From here on the proof can be carried through as in chapter 8, section 4. Therefore when we consider N quanta, we will find that the overlap is decreased by a factor $\exp[-NF] = \exp[-N\epsilon a^3/\alpha L^3]$.

We conclude that when there are sufficient quanta falling on the planet, the interference between planetary wave functions coming from different parts of the wave front can be neglected.

11.11 References

1. M. Renninger, *Zeit. Phys.* **158**, 417–421 (1960).
2. L. Mandel, *Phys. Rev.* **144**, 1071–1077 (1966).
3. R. L. Pfleegor and L. Mandel, *Phys. Rev.* **159**, 1084 (1967) and *J. Opt. Soc. Am.* **58**, 946 (1968).
4. L. de Broglie, *Une interprétation causale et non linéaire de la mécanique ondulatoire: la théorie de la double solution*, Gauthiers-Villiars, Paris (1956). English translation: Elsevier, Amsterdam (1960).
5. N. Cufaro-Petroni, C. Dewdney, P. Holland, T. Kiprianidis and J-P. Vigier, *Phys. Rev.* **32D**, 1375–1383 (1985).
6. J. P. Wesley, *Found. Phys.* **14**, 155 (1984).
7. R. D. Prosser, *Int. J. Theor. Phys.* **15**, 169 (1976).
8. Y. Nambu, *Prog. Theor. Phys.* **5**, 82–94 (1950).

9. J-P. Vigier, 'Nonlocal Quantum Potential Interpretation of Relativistic Action at a Distance in Many-body Problems', in *Open Questions in Quantum Physics*, ed. G. Tarozzi and A. van der Merwe, Reidel, Dordrecht, 1985.

10. D. Bohm, *Phys. Rev.* **85**, 180–193 (1952).

11. L. I. Schiff, *Quantum Mechanics*, chapter 8, sec. 45, McGraw-Hill, New York, 1955.

12. J. A. Wheeler, in *Battelle Rencontres*, ed. C. M. DeWitt and J. A. Wheeler, Benjamin, New York, 1968, 255.

13. Hao Ban-Lin, *Chaos*, World Publications, Singapore, 1984.

14. J. Glauber, *Phys. Rev.* **130**, 2529, and **131**, 2766 (1963).

15. H. Paul, *Rev. Mod. Phys.* **58**, 209–231 (1986).

16. P. N. Kaloyerou, 'Investigations of the Quantum Potential in the Relativistic Domain', Ph.D. Thesis, London, 1985.

Chapter 12
On the relativistic invariance of our ontological interpretation

12.1 Introduction

In this chapter we shall examine the question of how far Lorentz invariance of our ontological interpretation can be maintained.

We shall see that it is indeed possible to provide a Lorentz invariant interpretation of the one-body Dirac equation. For the many-body system we find that it is still possible to obtain a Lorentz invariant description of the manifest world of ordinary large scale experience which we introduced in chapter 7. In addition we show that all statistical predictions of the quantum theory are Lorentz invariant in our interpretation. This means that our approach is consistent with Lorentz invariance in all experiments that are thus far possible.

When this question is pursued further however, it is found that we cannot maintain a Lorentz invariant interpretation of the quantum nonlocal connection of distant systems. This is, of course, not surprising. Indeed we show that there has to be a unique frame in which these nonlocal connections are instantaneous. A similar result is also shown to hold for field theories. These likewise give Lorentz invariant results in the manifest world of ordinary experience and for the statistical predictions of the quantum theory. But where individual quantum processes are concerned, our ontological interpretation requires a unique frame of the kind we have described both for field theories and particle theories.

We discuss the meaning of this preferred frame and show that the idea is not only perfectly consistent, but also fits in with an important tradition regarding the way in which new levels of reality (e.g. atoms) are introduced in physics to explain older levels (e.g. continuous matter) on a qualitatively new basis.

12.2 The one-particle Dirac equation

We shall begin with the one-particle Dirac equation and show that it is possible with this equation to obtain an ontological interpretation that is invariant in content to a Lorentz transformation.

We have already given the essence of this interpretation in chapter 10. To recapitulate briefly, it follows from the Dirac equation that there is a conserved current density whose time component is positive definite. This is

$$\begin{aligned} \boldsymbol{j} &= \psi^\dagger \boldsymbol{\alpha} \psi \\ j^0 &= \psi^\dagger \psi \end{aligned} \tag{12.1}$$

with

$$\frac{\partial j^0}{\partial t} + \nabla \cdot \boldsymbol{j} = 0. \tag{12.2}$$

We define the velocity as

$$\boldsymbol{v} = \frac{\psi^\dagger \boldsymbol{\alpha} \psi}{|\psi|^2}. \tag{12.3}$$

We start with the identity

$$(\psi^\dagger \beta \gamma_\mu \psi)(\psi^\dagger \beta \gamma^\mu \psi) = (\psi^\dagger \beta \psi)^2 + (\psi^\dagger (\alpha \beta \gamma^5) \psi)^2 (\psi^\dagger \psi)^2 - (\psi^\dagger \alpha \psi)^2. \tag{12.4}$$

This reduces to

$$(\psi^\dagger \psi)^2 - \sum_{\boldsymbol{k}} (\psi^\dagger \alpha_{\boldsymbol{k}} \psi) = (\psi^\dagger \beta \psi)^2 + (\psi^\dagger 2 \beta \gamma^5 \psi)^2. \tag{12.5}$$

The left-hand side of this equation is just $\rho^2 - \sum j_i^2$ and it is evident that the right-hand side is positive definite so that $|v| < 1$.

If we denote the density $\rho = j^0$, we obtain for the conservation equation

$$\frac{\partial \rho}{\partial t} + \nabla \cdot (\rho \boldsymbol{v}) = 0. \tag{12.6}$$

Therefore the velocity (12.3) may be taken as determining the guidance relation for a particle. If we assume a probability density, $P = \rho$, for any time, then equation (12.6) guarantees that it will hold for all time.

As we explained in chapter 10, it is not necessary to have a simple relationship for the acceleration analogous to the one in the non-relativistic theory which expresses the latter as the sum of two terms, one corresponding to the classical potential and the other to the quantum potential. Even without such a relationship the path of the particle is completely determined by equation (12.3) and the whole theory will eventually go through by an extension of the approach given in chapter 10.

It is evident that this theory is Lorentz invariant in content because the Dirac equation is covariant, and because from this it follows that the relationship (12.6) is also covariant. It is important, however, to ask what would happen if we assumed, as was done in chapter 9, that the probability density P was not initially equal to ρ. Could covariance still be maintained under these conditions?

To discuss this question, we first note that in chapter 9 we have given two ways in which the probability distribution might approach ρ. Firstly, we might have a complex system consisting of many particles in interaction, e.g. electrons in a metal. We showed that in such a case, the probability P would eventually approach ρ. To be sure, the situation that brings about this approach is not covariant. For example, the electrons in the metal will determine a process that is uniquely related to the centre of mass frame of the metal. Yet once the distribution $P = \rho$ has been brought about, the result will be covariant and will remain so. Thus electrons boiling out of the metal and accelerated to high velocities will still have a covariant distribution with $P = \rho$ because this distribution was established before they boiled out.

The second way of bringing about the distribution $P = \rho$ is to assume an underlying stochastic process. No one has yet found a way to give a coherent description of a covariant stochastic process, although attempts have been made (Lehr and Park [1], Fenech and Vigier [2]). The principal difficulty is that the probability of a given velocity would have to be uniform on a time-like hyperbola. This would mean not only that almost all the particles will be going very close to the speed of light, but also that there is no way to normalise the probability distribution (essentially because the Lorentz group is not compact).

There is no need, however, to have such a Lorentz invariant stochastic process. Indeed, as long as we have a stochastic process that leads to $P = \rho$ in any particular frame, this will be enough. For it follows that from then on we will have a theory that will have a Lorentz invariant content. Thus, for example, we could introduce equations for osmotic and diffusion velocity along the lines of equations (9.38) and (9.39) to give

$$\frac{\partial \rho}{\partial t} + \nabla \cdot \left[P \left(\boldsymbol{v} + D \frac{\nabla \rho}{\rho} \right) \right] - D \nabla^2 P = 0 \tag{12.7}$$

where $\boldsymbol{v} = \psi^\dagger \boldsymbol{\alpha} \psi / |\psi|^2$ and the osmotic velocity is evidently $D(\nabla \rho / \rho)$. This equation would hold only in a certain frame and would evidently imply a non-covariant process of approach to equilibrium. Nevertheless along the lines that we have described, it would lead to an equilibrium distribution that is Lorentz invariant. This would be enough to bring about a covariant ontological interpretation of the one-particle Dirac equation. (The osmotic velocity $\boldsymbol{u}_0 = D(\nabla\rho/\rho)$ may be infinite where ρ approaches zero and this would imply speeds greater than that of light. We shall however discuss this point in section 12.6 where we show that it leads to no inconsistencies.)

12.3 The ontological interpretation of the Dirac equation for a many-body system

In chapter 10 we have given the main features of the extension of the ontological interpretation for Dirac particles to a many-body system. We begin by summing up what was done there. Firstly we define the wave function for the many-body system to be $\Psi = \Psi_{i_1 \ldots i_n \ldots}(\boldsymbol{x}_1 \ldots \boldsymbol{x}_n \ldots, t)$. Note that we are, at least for the moment, taking all the particles at the same time t. We introduce the operators, α_n, which operate only on the four indices, i_n, belonging to the nth particle (similarly for β_n).The Dirac equation for the N-body system is then

$$i\frac{\partial\Psi}{\partial t} = -i\left\{\sum_n \alpha_n\left[\nabla_n - i\frac{e}{c}\boldsymbol{A}_n(\boldsymbol{x}_n, t)\right] + \sum_n (\beta_n m + [\Phi_n(\boldsymbol{x}_n, t)])\right\}\Psi \tag{12.8}$$

where Φ and $\boldsymbol{A}$ represent the electromagnetic potentials. From this it is readily shown that

$$\frac{\partial\Psi^\dagger\Psi}{\partial t} + \sum \nabla_n \Psi^\dagger \alpha_n \Psi = 0 \tag{12.9}$$

where in the above we sum over all the indices of all the particles.

Defining the density $\rho(\boldsymbol{x}_1 \ldots \boldsymbol{x}_n \ldots) = |\Psi|^2$ in the configuration space of all the particles and the guidance relation for the velocity of the nth particle, $\boldsymbol{v}_n$,

$$\boldsymbol{v}_n = \Psi^\dagger \alpha_n \Psi/\rho, \tag{12.10}$$

we obtain the conservation equation in configuration space

$$\frac{\partial\rho}{\partial t} + \sum \nabla_n \cdot \rho\boldsymbol{v}_n = 0. \tag{12.11}$$

If $P = \rho$ initially, it follows that this will hold for all time. Along the lines that we have given in previous cases, we can see that the above provides a consistent ontological interpretation of the many-body Dirac equation.

This consistency is evident in the particular frame that we have chosen for the representation of the beables of the system, which are $\boldsymbol{x}_n$ and $\boldsymbol{v}_n$.

The question then arises, however, as to whether this interpretation has a covariant content, i.e. one that has the same meaning in every Lorentz frame. Thus, for example, the many-body Dirac equation (12.8) considers all the particles at the same time and it is not immediately evident that one could make a Lorentz transformation in which all the particles would be considered at the same value of the transformed time $t' = (t - vx)/\sqrt{1 - v^2}$. Nor is it evident that the $\boldsymbol{v}_n$ defined in equation (12.10) would be obtainable from some four vector v_n^μ. And finally one can ask what is to be done with the negative energy states.

Let us first consider the meaning of the negative energy states. This is usually treated in terms of fermionic creation and annihilation operators satisfying anticommutating relations, which, of course, describe particles that satisfy the exclusion principle. The most obvious anticommutating relations for the operators $\Psi^\dagger(x)$ and $\Psi(x)$ would be

$$[\Psi^\dagger(x), \Psi(x')]_+ = \delta(x - x'). \tag{12.12}$$

If we Fourier analyse Ψ as

$$\Psi = \sum \left[a_{\boldsymbol{k}_+} e^{i\boldsymbol{k}\cdot\boldsymbol{x}} + a_{\boldsymbol{k}_-} e^{-i\boldsymbol{k}\cdot\boldsymbol{x}}\right], \tag{12.13}$$

this leads to the anticommutation relations

$$\begin{aligned}
\left[a^\dagger_{\boldsymbol{k}'_+}, a_{\boldsymbol{k}_+}\right]_+ &= \delta^{(3)}(\boldsymbol{k}_+ - \boldsymbol{k}'_+) \\
\left[a^\dagger_{\boldsymbol{k}'_-}, a_{\boldsymbol{k}_-}\right]_+ &= \delta^{(3)}(\boldsymbol{k}_- - \boldsymbol{k}'_-) \\
\left[a^\dagger_{\boldsymbol{k}_+}, a_{\boldsymbol{k}_-}\right]_+ &= 0
\end{aligned} \tag{12.14}$$

while all the $a_{\boldsymbol{k}}$ and the $a^\dagger_{\boldsymbol{k}}$ anticommute among themselves.

The state of no particles would be one containing neither particles with positive nor negative energy. Therefore it would be possible in general for interactions to produce particles with negative as well as positive energies. To avoid this, standard treatments in field theory effectively interchange annihilation and creation operators. That is to say they write

$$a_{\boldsymbol{k}_-} = b^\dagger_{\boldsymbol{k}} \qquad \text{and} \qquad a^\dagger_{\boldsymbol{k}_-} = b_{\boldsymbol{k}}$$

where $b_{\boldsymbol{k}}$ and $b^\dagger_{\boldsymbol{k}}$ are the creation and annihilation operators for the corresponding antiparticles with positive energy, so that we can write

$$\Psi = \sum a_{\boldsymbol{k}} e^{i\boldsymbol{k}\cdot\boldsymbol{x}} + b^\dagger_{\boldsymbol{k}} e^{-i\boldsymbol{k}\cdot\boldsymbol{x}}.$$

In this way we avoid having particles with negative energy.

Although the approach described above is most convenient for field theory, it must be emphasised that it is equivalent in all its results to the procedure adopted by Dirac, who originally worked with a purely particle point of view. What he proposed was that in the vacuum, all the negative energy states were filled. Because the particles satisfy the exclusion principle, the transition of particles to the negative energy states could therefore never occur. However, when a particle went to a positive energy state, it would leave a hole that acted like an antiparticle. So in effect a positive and negative pair would be created in such a process (more or less as happens in a semi-conductor where likewise a transition from one band to another creates a pair of positive and negative carriers).

In order that the exclusion principle be satisfied, it is necessary that the wave function for all the particles be antisymmetric. Once this condition is satisfied, then Dirac's approach and the usual approach in terms of fermionic operators are essentially equivalent. (Indeed the two differ only by a unitary transformation in which the $a_{\boldsymbol{k}_-}$ are replaced by $b^{\dagger}_{\boldsymbol{k}}$ and the $a^{\dagger}_{\boldsymbol{k}_-}$ by $b_{\boldsymbol{k}}$.)

It is important to start with the Dirac approach if we wish to obtain an ontological interpretation of the Dirac equation, because fermionic fields cannot be given an ontological interpretation in terms of beables in the same way that it is possible with boson fields. For the fermionic field operators have only two possible states and cannot be put in correspondence with field beables that would change continuously. (On the other hand bosonic operators with their infinity of states can be represented in terms of continuous fields.) All that we have to do then is to assume a position $\boldsymbol{x}_n$ for each fermionic particle and to make the total wave function $\Psi_{i_1 \ldots i_n \ldots}(\boldsymbol{x}_1 \ldots \boldsymbol{x}_n \ldots, t)$ antisymmetric. Then, as we have already pointed out, provided that we assume the vacuum corresponds to having all the negative energy states 'filled', we will be able to guarantee that there are no transitions to negative energy states. The interpretation we have given here will thus go through in a straightforward way.

12.4 Lorentz invariance of the many-body Dirac equation

Let us now go on to consider the question of whether the many-body Dirac equation is Lorentz invariant in spite of the fact that it uses a common time for all the particles. To show that it is Lorentz invariant we shall begin by making use of the fundamental relationship that we have indicated in section 12.3 between the field theoretic formulation for fermions and the particle formulation for the case that the total number of particles is well defined. We make use of the fact that the field formulation has been shown in standard treatments to be invariant not only to Lorentz transformations but also to any change of the space-like surface on which the initial values of the wave function are given. (For details see Schweber [3].) From the equivalence of the field and particle formulations, it will be seen to follow that the Dirac many-body equation is also Lorentz invariant.

Let us begin by assuming that the fermionic field contains N particles. The basic orthonormal set can be formed from linear combinations of wave functions

$$\psi^{\dagger}_{i_1}(x_1)\psi^{\dagger}_{i_2}(x_2)\ldots\psi^{\dagger}_{i_N}(x_N)\Psi_0, \tag{12.15}$$

where $\psi^{\dagger}_{i_1}(x_1)$ is the operator for the creation of a particle at the point x_i, and Ψ_0 is the wave function representing the situation with no particle at any point. The general wave function can be written as a liner combination of this orthonormal set

$$\Psi = \int\!\!\int \sum \phi_{i_1 \ldots i_N}(x_1 \ldots x_N)\psi^{\dagger}_{i_1}(x_1)\psi^{\dagger}_{i_2}(x_2)\ldots\psi^{\dagger}_{i_N}(x_N)\Psi_0\,\mathrm{d}x_1\ldots\mathrm{d}x_N. \tag{12.16}$$

Since the operators $\psi^{\dagger}(x_j)$ anticommute, it follows that the ϕ are antisymmetric functions of the x_j and the indices i_j. Indeed, as is well known, the ϕ are just the ordinary antisymmetric wave functions for the N-particle system.

In terms of the field approach we can write the wave equation as

$$i\frac{\partial\Psi}{\partial t} = H\Psi, \tag{12.17}$$

where the Hamiltonian is

$$H = \int \frac{\mathrm{d}^3x}{2i}\left[\psi^{\dagger}\boldsymbol{\alpha}\cdot(\boldsymbol{\nabla} - ie\boldsymbol{A}(x))\psi - \psi\boldsymbol{\alpha}\cdot(\boldsymbol{\nabla} + ie\boldsymbol{A}(x))\psi^{\dagger}\right]. \tag{12.18}$$

This Hamiltonian operates on the creation operators in equation (12.16). However, by integration by parts we can bring about an equivalent operation on the ϕ. It then follows that

$$\begin{aligned} i\frac{\partial}{\partial t}\phi_{j_1\ldots j_N}(x_1\ldots x_N) &= \sum_{j_1\ldots j_N}\sum_{n}\left[\tfrac{1}{i}(\alpha_n\cdot\nabla_n - e\boldsymbol{A}(x_n))\right. \\ &\quad \left. + m\beta_n\right]\phi_{j_1\ldots j_N}(x_1\ldots x_N). \end{aligned} \tag{12.19}$$

Since the operator formalism has been demonstrated to be covariant, it follows that the above form of the Dirac equation can be obtained in any frame. As is well known, the transformation between frames in field theory is unitary. From this it follows that at least as far as probabilities are concerned, the many-body Dirac equation is Lorentz invariant both in form and in content.

12.5 The multiple time formalism

Although we have adequately proved the invariance of the Dirac many-body equation, it will be instructive to consider an alternative approach through the many-time formalism [4] because the latter can provide further insight into meaning of this invariance.

Let us begin by considering the Dirac equation for N particles, each of mass m, which is

$$i\frac{\partial\psi}{\partial t} = \sum H_n\psi, \tag{12.20}$$

where

$$H_n = \boldsymbol{\alpha}\cdot\left[\frac{\boldsymbol{\nabla}_n}{i} - e\boldsymbol{A}(\boldsymbol{x}_n, t)\right] + m\beta_n. \tag{12.21}$$

When the equation is in this form we can only calculate probabilities for obtaining certain results (e.g. the values of the $\boldsymbol{x}_n$) when measurements are made at the same time t. But suppose that we wish, for example, to measure the positions of the particles $\boldsymbol{x}_n$ at different times τ_n. The multiple time formalism provides a way of calculating the probability of such results.

To do this we begin by postulating the set of N wave equations

$$i\frac{\partial\chi}{\partial\tau_n} = H_n\chi. \tag{12.22}$$

These equations will all be integrable provided the Hamiltonians H_N and $H_{N'}$ all commute. But this will follow provided that the various points $(\boldsymbol{x}_1\tau_1, \boldsymbol{x}_2\tau_2$, etc.) are on a space-like surface, so that they are outside each other's light cones. In this case one can start with $\chi = \chi(\boldsymbol{x}_1, \ldots, \boldsymbol{x}_N, t)$ with $t = \tau_1 = \tau_2 = \cdots = \tau_N$, and integrate the equations (12.22) to obtain

$$\chi = \chi(\boldsymbol{x}_1, \tau_1; \boldsymbol{x}_2, \tau_2; \ldots \boldsymbol{x}_N, \tau_N).$$

Bloch has proposed an interpretation for the meaning of χ (see Wentzel [4]). This is that $|\chi|^2$ gives the net probability density for obtaining all these results together. It must be remembered however that Bloch is referring to the usual interpretation and not to our interpretation.

To see what Bloch's interpretation means, suppose that we consider an ordered series of times $\tau_1 < \tau_2 < \cdots < \tau_N$. If, for example, we measure

$\boldsymbol{x}_1$ at τ_1 then according to the theory given in chapter 2, which deals with the usual interpretation, the wave function after the measurement of the position of the first particle will 'collapse' to

$$\phi(\xi - \xi_{1r})\delta(\boldsymbol{x}_1 - \boldsymbol{x}_{1r})\chi(\boldsymbol{x}_1, \tau_1; \boldsymbol{x}_2, \tau_2; \ldots \boldsymbol{x}_N, \tau_N),$$

where $\phi(\xi - \xi_{1r})$ represents the wave function of the measuring apparatus whose wave packet centres at ξ_{1r} which corresponds to obtaining the value $\boldsymbol{x}_1 = \boldsymbol{x}_{1r}$ (where $\boldsymbol{x}_{1r}$ is the rth possible result of the measurement). The important thing to notice here is that after this measurement, there will be no interference with parts of the wave function initially corresponding to values of $\boldsymbol{x}_{1r}$ different from the result actually obtained, $\boldsymbol{x}_{1r}$. A similar description will hold for subsequent measurement of $\boldsymbol{x}_2, \boldsymbol{x}_3$, etc. and therefore, as Bloch proposed, $|\chi|^2$ will give the probability that if $\boldsymbol{x}_1 = \boldsymbol{x}_{1r}$ in the result of the first measurement, then $\boldsymbol{x}_2 = \boldsymbol{x}_{2s}$ will be the result of the second measurement, and so on.

The consistency of this formalism is further demonstrated by showing that propagation through a time Δt is the same as that obtained by propagating each of the individual equations (12.22) through $\Delta\tau_n = \Delta t$. Thus we obtain for the total change of χ

$$i\Delta\chi = \left(\sum H_n\right)\chi\Delta\tau$$

and this is just what we would obtain from the single time formalism.

Let us now consider Lorentz invariance of this formalism. Each of the equations (12.22) is evidently Lorentz invariant. If we simply Lorentz transform each of the equations from a wave function χ in which all of the τ_n are equated to a common time, t, we must then, in addition, consider each of the particles at a common time t' such that

$$\tau_n = \frac{t' + v\boldsymbol{x}'_n}{\sqrt{1 - v^2}}$$

as shown in figure 12.1. It is evident from this that by suitable choices of τ_n we can always have a wave function considered at a common transformed time t'. The Lorentz invariance of the formalism is thus evident.

12.6 On the question of Lorentz invariance of the ontological interpretation of the Dirac equation

Let us now raise the question as to what extent the ontological interpretation of the many-body Dirac equation is invariant in content to a Lorentz

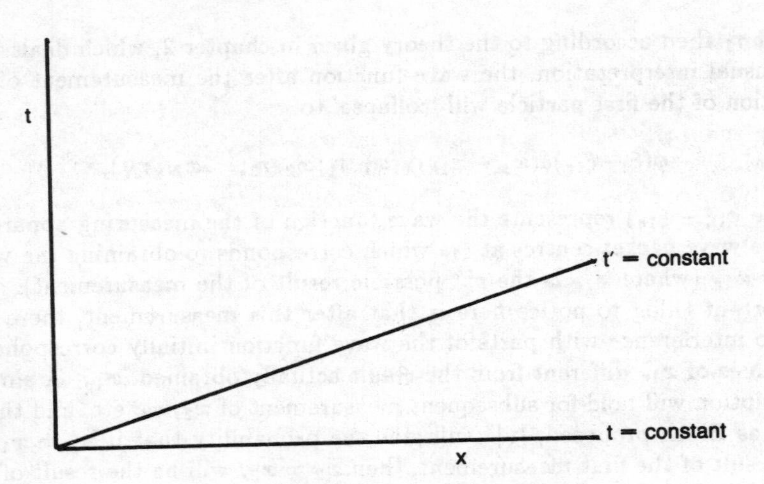

Figure 12.1: Space-time frames

transformation. To simplify the discussion, let us consider a two-body system. In so far as the wave function of the combined system can be written as a product $\psi = \phi_a(\boldsymbol{x}_1, t_1)\phi_b(\boldsymbol{x}_2, t_2)$ it is clear that the velocities (12.10) of the two particles will take the form

$$\begin{aligned} \boldsymbol{v}_1 &= \phi_a^\dagger \boldsymbol{\alpha} \phi_a / \phi_a^\dagger \phi_a, \qquad \text{and} \\ \boldsymbol{v}_2 &= \phi_b^\dagger \boldsymbol{\alpha} \phi_b / \phi_b^\dagger \phi_b. \end{aligned} \tag{12.23}$$

The density in configuration space will be $\rho_a(\boldsymbol{x}_1)\rho_b(\boldsymbol{x}_2)$, where

$$\rho_a(\boldsymbol{x}_1) = \phi_a^\dagger(\boldsymbol{x}_1)\phi_a(\boldsymbol{x}_1)$$

and

$$\rho_b(\boldsymbol{x}_2) = \phi_b^\dagger(\boldsymbol{x}_2)\phi_b(\boldsymbol{x}_2).$$

The particles thus behave independently, and, from our previous discussion of the one-particle case, it follows that this system is Lorentz invariant.

If there is a linear combination of wave functions,

$$\psi = A\phi_a(\boldsymbol{x}_1, t)\phi_b(\boldsymbol{x}_2, t) + B\phi_c(\boldsymbol{x}_1, t)\phi_d(\boldsymbol{x}_2, t), \tag{12.24}$$

then, in general, the two systems will cease to be independent. However, for the special case in which $\phi_c(\boldsymbol{x}_1, t)\phi_d(\boldsymbol{x}_2, t)$ does not overlap $\phi_a(\boldsymbol{x}_1, t)\phi_b(\boldsymbol{x}_2, t)$, the interference terms will vanish and the net probability density in the configuration space of the two particles, $|\psi|^2$, will divide into two distinct channels of the general kind discussed in chapters 5 and 6. In this case the particles will be in one of the channels, while the potentially active information in the other channel will have no actual effect. It is clear then that once again the two particles will be independent.

More generally, however, there will be a nonlocal connection between the two particles, as can be seen, for example, from the expression for the probability density in configuration space

$$\begin{aligned} |\Psi|^2 &= |A|^2|\phi_a(\boldsymbol{x}_1,t)|^2|\phi_b(\boldsymbol{x}_2,t)|^2 + |B|^2|\phi_c(\boldsymbol{x}_1,t)|^2|\phi_d(\boldsymbol{x}_2,t)|^2 \\ &\quad + A^*B\phi_a^\dagger(\boldsymbol{x}_1,t)\phi_b^\dagger(\boldsymbol{x}_2,t)\phi_c(\boldsymbol{x}_1,t)\phi_d(\boldsymbol{x}_2,t) + \text{c.c.}, \end{aligned} \tag{12.25}$$

and for the current density of the first particle in configuration space

$$\begin{aligned} \boldsymbol{j}_1 &= |A|^2\phi^\dagger{}_{,a}(\boldsymbol{x}_1,t)\boldsymbol{\alpha}_1\phi_a(\boldsymbol{x}_1,t)|\phi_b(\boldsymbol{x}_2,t)|^2 \\ &\quad + |B|^2\phi_c^\dagger(\boldsymbol{x}_1,t)\boldsymbol{\alpha}_2\phi_c(\boldsymbol{x}_1,t)|\phi_d(\boldsymbol{x}_2,t)|^2 \\ &\quad + A^*B\phi_a^\dagger(\boldsymbol{x}_1,t)\boldsymbol{\alpha}_1\phi_c(\boldsymbol{x}_1,t)\phi_b^\dagger(\boldsymbol{x}_2,t)\boldsymbol{\alpha}_2\phi_d(\boldsymbol{x}_2,t) + \text{c.c.}, \end{aligned} \tag{12.26}$$

with a similar expression for the second particle.

It is clear from the above that the velocities $\boldsymbol{v}_1 = \boldsymbol{j}_1/\rho$ and $\boldsymbol{v}_2 = \boldsymbol{j}_2/\rho$ are no longer independent. Rather, the velocity of the first particle, for example, will depend on the location of the second particle, which may in general be far away. And this, of course, implies the nonlocal connection that we have mentioned above. When we go to the non-relativistic limit, this will lead, in the way explained in chapter 10, to the Pauli equation, and eventually in the limit in which the effects of 'spin' can be neglected, we obtain the Schrödinger equation. In the latter case, a nonlocal connection in the velocities of the particles implies a nonlocal quantum potential that accounts for the mutually related accelerations that are entailed by the guidance conditions.

In the limit in which such nonlocal connections can be neglected, it follows from what we have said earlier in this section that the theory will be Lorentz invariant. In this way we see that what we have in chapter 8 called the manifest world, i.e. the world of ordinary large scale classically describable experience, will be invariant. It is in this world that quantum properties will have to manifest themselves, whether through measurements or through equivalent processes that occur naturally (e.g. as an ion going through a cloud of gas in a naturally occurring electric field will produce large scale cascades of charged particles in essentially the same way as happens in ion chambers constructed by human beings).

In order to agree with the experimental facts available, it is necessary to show that all statistical results of measurement-manifestation processes are

Lorentz invariant. In doing this, it will be convenient first of all to point out how Lorentz invariance arises in the usual interpretation. What happens there is that the wave function undergoes a unitary transformation, U, when we go from one Lorentz frame to another (where U depends on the Hamiltonian). Therefore operators such as $\boldsymbol{x}'_n = U\boldsymbol{x}_n U^{-1}$ do not in general commute with $\boldsymbol{x}_n$ (even though the $\boldsymbol{x}_n$ all commute with each other, as do the $\boldsymbol{x}'_n$). In spite of this however, because the wave equations are the same in every frame, measurement theory will go through in a covariant way.

It has been shown, however, that the ontological interpretation gives the same statistical results for all measurement-manifestation processes as does the usual interpretation. This means that our interpretation will have a covariant content for all statistical consequences of the measurement-manifestation process. But the only quantum experiments that can be performed thus far are of this statistical nature. So, in this domain, our ontological interpretation is also Lorentz invariant in content. (In chapter 14 we discuss the possibility of going beyond the present quantum mechanical statistical results.)

It is evident, however, that from the above discussion, it does not follow that the ontological interpretation for an individual system in terms of a particle with a well-defined trajectory is a Lorentz invariant concept. Indeed it would be extremely surprising to obtain a Lorentz invariant theory of particles that were connected nonlocally. To illustrate the typical difficulties involved in this notion, consider two separated particles, A and B, both at rest in the laboratory frame as shown in figure 12.2.

If there is a nonlocal connection of the kind implied by our guidance conditions, then it follows that, for example, point a and point b instantaneously affect each other. But if the theory is covariant, there should be similar instantaneous connections in every Lorentz frame. Therefore a and b' as well as b' and a' will also be in such an immediate connection. It would then be possible for A acting at a to affect its own past a' by affecting b'. This would evidently imply a paradoxical situation (e.g. as in the case of a person who killed his own father before he was conceived and in doing so annihilated himself so that he was incapable of carrying out the assumed action).

Of course, until we are able to carry out experiments on individual processes that are more accurate than the limits set by the uncertainty principle, it will not be possible to use quantum nonlocality for the purpose of sending signals. For as shown in chapter 7, in statistical measurements the EPR correlations between distant particles do not make it possible for signals to be sent from one particle to another. More generally any attempt to send a signal by influencing one of a pair of particles under EPR correlation will encounter the difficulties arising from the irreducibly

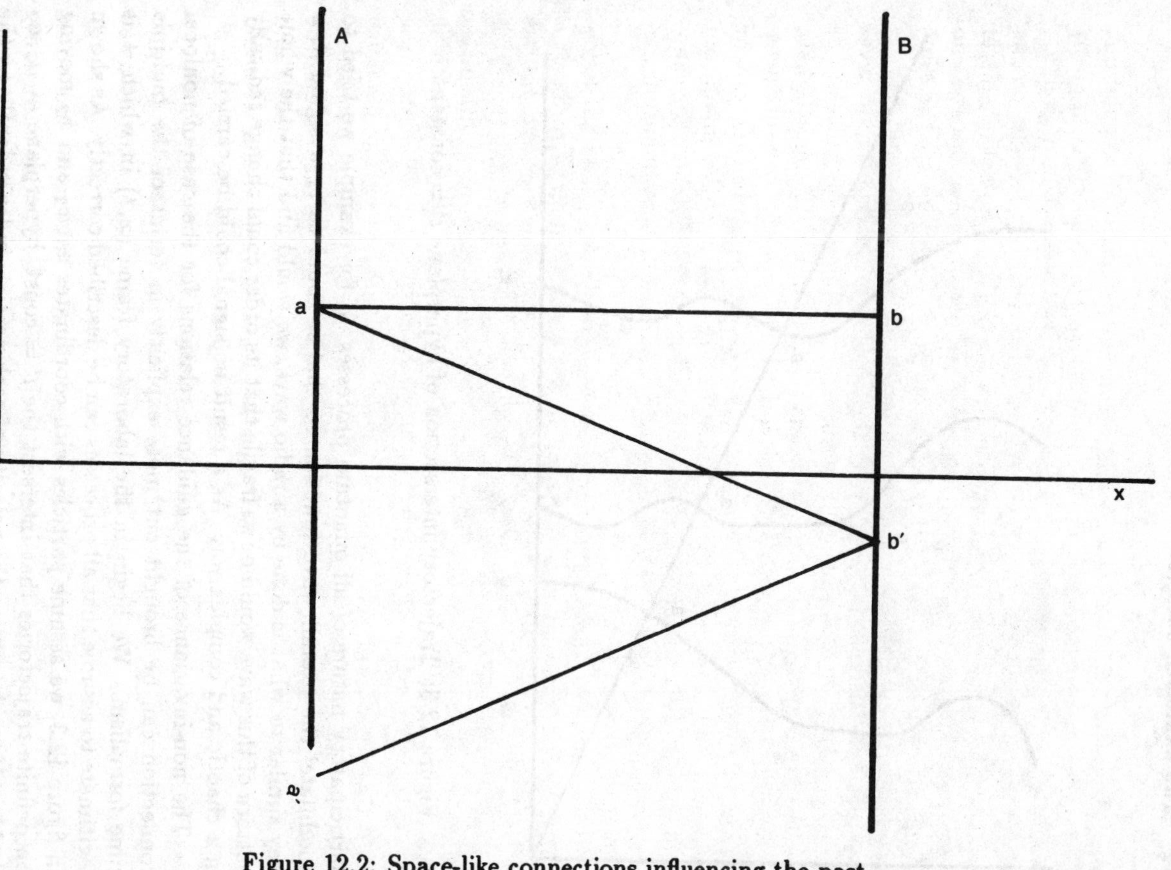

Figure 12.2: Space-like connections influencing the past

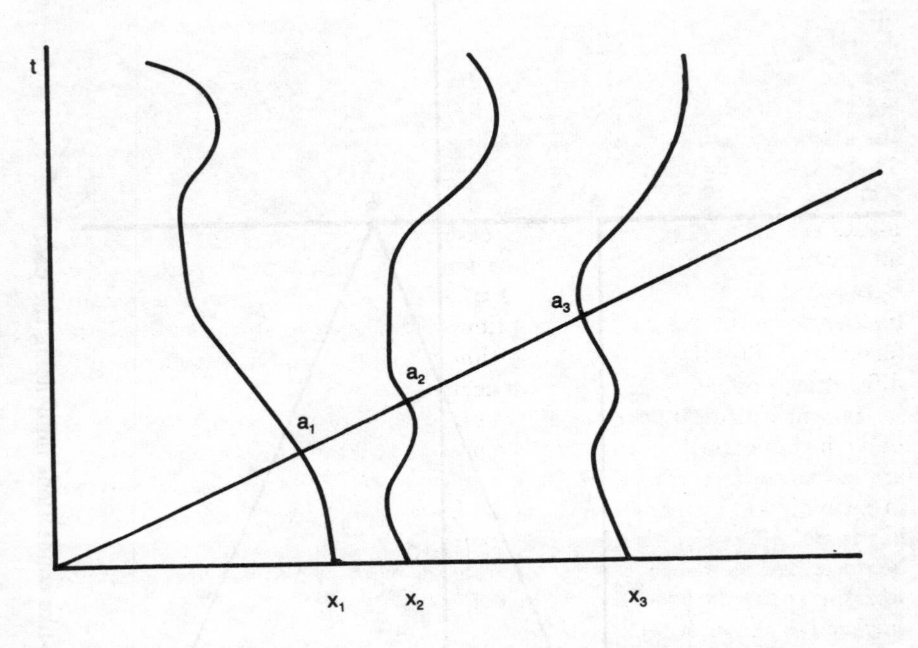

Figure 12.3: Trajectory intersection of hyperplane $t' =$ constant

participatory nature of all quantum processes. If for example we tried to 'modulate' the overall wave function so that it could carry a signal in a way similar to what is done by a radio wave, we would find that the whole pattern of this wave would be so fragile that its order could change radically in a chaotic and complex way. As a result no signal could be carried.

The non-invariance of the guidance relations for the case of nonlocal connection can be brought out more explicitly in terms of the multiple time formalism. We begin in the laboratory frame $(\boldsymbol{x}, t)$ in which it is legitimate to assume that all processes can be described correctly. As shown in figure 12.3, we assume particles with coordinates $\boldsymbol{x}_1, \boldsymbol{x}_2$ and $\boldsymbol{x}_3$ moving on definite trajectories that intersect the $t' =$ const. hyperplane at $\boldsymbol{a}_1, \boldsymbol{a}_2$ and $\boldsymbol{a}_3$. If we determine the velocity of each particle independently by the multiple time equations (12.22), the results would indeed be relativistically invariant.

But as we have seen in the previous section, this will give us the statistical behaviour of a system for which $\boldsymbol{x}_1$ has been measured at τ_1, $\boldsymbol{x}_2$ at a later time τ_2, and so on. Although our results were obtained from measurement theory in the usual interpretation, they will still follow in our interpretation. That is to say, after the measurement of $\boldsymbol{x}_1$ is over and has yielded a certain result $\boldsymbol{x}_{1r}$, the subsequent behaviour of $\boldsymbol{x}_2$, $\boldsymbol{x}_3$ etc. will no longer be affected by the parts of the wave function corresponding to $\boldsymbol{x}_{1s}$ not equal to $\boldsymbol{x}_{1r}$ (that is for positions which are not realised in the actual individual process of interaction with the measuring apparatus). Therefore the wave function will not be the right one for calculating the guidance relations that actually operate on the particles when no measurements are made. For in this latter case the wave packets corresponding to all possible values of $\boldsymbol{x}_{1r}$ can come to interfere with the passage of time. Because of nonlocal connections of $\boldsymbol{x}_1$ with $\boldsymbol{x}_2$, $\boldsymbol{x}_3$, etc. this will introduce further terms in the guidance relations not contained in the multiple time formalism. Therefore the multiple time formalism cannot get us out of the difficulties implied by nonlocal connection.

Indeed it is clear that the entire calculation of the particle velocity will be ambiguous until we specify the frame in which the nonlocal connections are instantaneous. The concept of a particle guided in a nonlocal way will, in general, not be Lorentz invariant. As we have already pointed out, one has therefore to assume some definite frame in which the connections are to be described as instantaneous, while in other frames they are described as working either backwards or forwards in time, although still on a space-like surface (as shown in figure 12.2).

If it turns out that the laboratory frame is not the one in which the connections are instantaneous, then it might seem at first sight as if the present could affect the past. But because the effect is only on a space-like surface, it follows that there will necessarily be a frame in which the nonlocal connections act only between points that are at the same time in this frame. Briefly, what this means is that there is always a unique frame in which the nonlocal connections operate instantaneously. In this frame there is no intrinsic logical difficulty about having nonlocal connections. The behaviour of these connections in other frames will then always be obtained by transforming these results from the special frame in which the connections are instantaneous.

From the above it follows that even though the underlying theory requires a preferred frame, all the statistical results of the quantum theory will still be covariant (as will, of course, the manifest world of large scale experience). The basic idea behind this is that every measurement process is determined by an operator. The covariance of the wave equation implies that to any operator O, there is a corresponding operator, O', which is the Lorentz transformation of O. In general, O and O' do not commute, but this need not concern us since we are interested here only in measurement processes that are occurring in a particular Lorentz frame. The theory of measurement processes that we have given, along with the demonstration that the probability distribution $P = \rho$ is covariant, then ensures that the statistical results of these measurement processes will have a Lorentz invariant content.

Various authors have however tried to retain covariance of the underlying theory by assuming nonlocal connections that are Lorentz invariant. For example, Droz-Vincent [5], and Vigier [6] suggested that for a two-particle system, one would assume that the nonlocal connections were instantaneous in their centre of mass frame. Clearly this is a covariant notion. But the typical difficulty arises when we consider three or more particles. As we pointed out in chapter 11, section 11.2, where we discussed particle alternatives to the bosonic field theories, the centre of mass frame of a pair of particles in a many-body system is ambiguous, because each particle may be considered in connection with any one of the remaining particles in defining such a frame. Even if we consider all the particles of the given system regarded as classical, the concept is ambiguous because the latter may always interact with yet another system. Thus we do not feel that any overall theory of this kind can consistently define a covariant relationship of particles that are nonlocally connected.

To sum up then, we have found that it is necessary to assume a basically non-Lorentz invariant theory for the individual particles. Nevertheless this theory becomes Lorentz invariant where nonlocal connections can be neglected. From this we conclude that the manifest level of ordinary large scale experience will be covariant in its behaviour. We have also shown that the statistical laws of the quantum theory are covariant. We emphasise again that there is Lorentz invariance in all of the domains of particle theory that have thus far been investigated experimentally, but these do not necessarily invalidate our assumptions concerning the underlying level in which the order of succession is unique.

12.7 On the Lorentz invariance of the ontological interpretation of boson field theories

The question of Lorentz invariance in the ontological interpretation of boson field theories is not basically different from that in fermion many-particle theories. To bring out this essential similarity, we note that in the fermionic theory, the beables are particle positions governed by guidance conditions determined by the wave function (see equation 12.3), while in boson theories, the beables are the fields whose movement is governed by guidance conditions determined from the wave functional (see equation 11.3). The similarity of the two theories can be made even more evident by using the model discussed in chapter 11, section 11.1, in which field variables are considered only at a set of discrete, but rather densely distributed points in space-time. The field beables then correspond fairly closely to the particle beables.

It is clear that as long as we stick to a certain Lorentz frame, this interpretation is coherent. But if we try to go from one Lorentz frame to another, we then encounter difficulties similar to those met in fermionic particle theories. For as we have already pointed out in chapter 11, there will in general be nonlocal connections amongst the field beables. And as we have seen in this chapter, this means that a consistent description of these connections can be obtained only if we assume that there is a unique frame in which they are instantaneous. Also, if fermionic and bosonic systems are to interact, it is clearly necessary that this frame be the same for both of them.

As was done with fermions, we can show that in the large scale limit of ordinary experience, bosonic fields can approach a simple classical behaviour in which nonlocal connections become negligible. We have indeed already shown in chapter 11 that harmonic oscillators excited to high quantum numbers can have wave functions (e.g. those corresponding to coherent states) in which the fractional fluctuation in the energy of excitation becomes very small, so that this energy can be regarded as very nearly well defined. Nevertheless, at the same time the total spread in the number of quanta becomes very large. And as Glauber [7] has shown, for such a wave function one can define an approximate meaning to the phase of an oscillator. Therefore the oscillator can be regarded as moving very nearly in a classical orbit. If we then put together all the equivalent oscillators corresponding to the Fourier coefficients of the field, we obtain in the large scale limit of ordinary experience something that approximates to the behaviour of a classical field. Because interactions that exchange a few quanta with the environment have no significant effect on the field as a whole, it follows that the latter can be regarded as independently existent. What this means is that one can ignore the basic quantum mechanical participation which is involved in a measurement, or more generally, in any manifestation in the large scale world of ordinary experience, whether this takes place naturally or is induced by human beings. (Chew and Stapp [8] have in fact given a more detailed treatment of this question for the case of massless fields, but they start from the point of view of the S-matrix rather than the usual quantum mechanics.)

As happens with fermions, this approximately independent part of the field in the manifest world will also have a negligible contribution from nonlocal quantum connections (essentially because the super-quantum potential is negligible for such high quantum states). However, it must be recalled that this will generally hold only for massless bosons. However, with massive bosons the energies required for such a classical situation are so high that these do not play a significant role in the manifest world of ordinary experience.

We can then go on to consider processes in which the quantum laws become important (e.g. scattering, ionisation, etc.) but as we have shown throughout this book, the statistical distribution of manifestations of these processes in the ordinary large scale world of experience will be the same in our interpretation as in the usual interpretations. Since the latter is Lorentz invariant, it follows that our approach will also lead to Lorentz invariant results for all statistical measurement-manifestation processes.

As with fermions, it is only in the individual processes in which the quantum laws are important that the theory ceases to be covariant and requires the assumption of a unique Lorentz frame. (It is interesting that Chew and Stapp [8] also have such a frame which is implied by their concept of a unique universal time.)

12.8 On the meaning of non-Lorentz invariance of processes involving individual beables

As we have seen in the previous section, our ontological interpretation leads to Lorentz invariant results in the manifest world of ordinary experience and also in the statistical predictions of quantum theory. These two fields cover all the experimental knowledge that we have so far. It is only in connection with individual processes involving the beables that Lorentz invariance may break down. But as long as the quantum theory is valid, there is no way to demonstrate this non-Lorentz invariance experimentally.

Nevertheless, it is understandable that there may be strong objections to giving up Lorentz invariance even at the level of individual processes involving the beables. For relativity theory (both special and general) has had a very wide range of successful applications. Moreover, the beauty and symmetry of its mathematical form have not only been attractive to physicists, but have led them to have great confidence in the ultimate truth of the theory.

We have to be careful, however, not to assume that a theory has an absolute and unlimited validity just because it has agreed with a very wide range of experiments and because its form is aesthetically pleasing. From these reasons it does not follow that the theory of relativity is an absolute truth. In fact there is evidently no way to prove that any theory whatsoever is an absolute truth no matter how well verified it may be. Of course, as Popper [9] has pointed out, a good scientific theory should, at least in principle, be refutable and the theory of relativity clearly satisfies this requirement. But if we tacitly proceed in such a way as to take this theory as unchallengeable, we are, in effect, denying the applicability of Popper's requirement. However, such challenges may take many forms not only those involving experimental tests, but also those involving theoretical criticisms of the consistency of the theory in question with itself and with other generally accepted theories. We could then say that our ontological interpretation helps bring out a fundamental inconsistency between relativity and quantum theory, centred on the question of nonlocality. (Recall here that other interpretations of the quantum theory also imply nonlocality, though not in such a clear way, as shown in chapter 7 and as we shall show in further detail in chapter 13.)

What this suggests is that either we accept nonlocality, in which case relativity is not fully adequate in the quantum domain, or we reject nonlocality, in which case quantum theory is not fully adequate in the relativistic domain. Of course, it is also possible that both theories may break down somewhere in the domain in which they meet. We are therefore led to consider the possibility that either relativity or the quantum theory, or both, may prove to be false when extended beyond certain domains. In chapter 14 we shall discuss the possibility that it is the quantum theory that breaks down (though we shall not enquire specifically into whether the new theory would be compatible with relativity). In the present chapter we are investigating the possibility that it is special relativity that breaks down, while quantum mechanics remains valid.

The most essential feature of special relativity is Lorentz invariance. We already know that Lorentz invariance must be limited at large distances where the effects of curvature are appreciable. We can also give an argument showing that Lorentz invariance should be limited at short distances (probably at the order of the Planck length). To do this let us consider a vector, v^μ, of ordinary length and regard it as made up of many smaller vectors, δv_i^μ, which are on a straight line such that

$$v^\mu = \sum_i \delta v_i^\mu .$$

Suppose that we make a Lorentz transformation. Strictly speaking this cannot be defined without giving the metric tensor $g_{\mu\nu}$. But at very short distances this tensor has large quantum fluctuations and does not reduce to its simple diagonal form. The Lorentz transformation, $\epsilon^{\mu\alpha}$, will then give us

$$v'^\mu = \sum_i \epsilon^{\mu\alpha} g_{\alpha\beta}(x^\lambda) \delta v_i^\beta .$$

It is clear that what was a simple line vector is transformed into a rather foggy distribution of constituent elements which are not well defined. So the whole idea of a simple Lorentz transformation of one line into another breaks down. The idea of Lorentz invariance only becomes relevant over intermediate distances for which the fluctuations of $g_{\mu\nu}$ will average out to something relatively small, implying that $g_{\mu\nu}$ can be taken as effectively constant.

All of this suggests that the Lorentz group may not be as fundamental as we have generally assumed. If it is not valid in the domain of very large distances and it is not valid in the domain of very small distances, it may also cease to be valid in yet other domains. The analysis given in this chapter suggests that even at intermediate distances it may cease to be valid for *individual processes* involving the beables wherever quantum nonlocality is important.

We are thus led to assume something similar to a unique universal frame, or more accurately, to a universal order of succession. In certain ways such a theory would be reminiscent of the Lorentz-type ether theory within which there were large scale objects with structures undergoing processes that would change with velocity in such a way as to bring about Lorentz invariance in terms of frames defined through these structures.

An explanation of this kind of theory is given in Bohm [10]. Here one appeals to the notion that matter is constituted of atoms consisting of negatively charged electrons and positively charged nuclei. The forces between atoms, responsible for binding them into molecules, and ultimately into macroscopic solid objects, are assumed to originate in the attractive forces between electrons and the nuclei, and the repulsive forces between electron and electrons as well as between nuclei and nuclei. Consider, for example, a crystal lattice. The places where the electrical forces come to a balance would then determine the distance, d, between the neighbouring atoms. Therefore, in the last analysis the size of such a crystal containing a specified number of atomic steps in any given direction is determined in this way.

Lorentz assumed that the electrical forces were states of stress and strain in the ether. From Maxwell's equations, assumed to hold in the reference frame in which the ether was at rest, it was possible to calculate the electromagnetic field surrounding the charge particle. For a particle at rest in the ether, it followed that this field was derivable from a potential, ϕ, which was a spherically symmetric function of the distance r from the charge, i.e. $\phi = q/r$, where q is the charge of the particle. When a similar calculation was performed for a charge moving through the ether, it was found that the force field was no longer spherically symmetric. Rather, its symmetry became that of an ellipsoid of revolution, having unchanged diameters in the directions perpendicular to the velocity, but shortened in the direction of motion by the factor $\sqrt{1-(v^2/c^2)}$. This shortening is evidently an effect of the movement of the electron through the ether.

Because the electrical potential due to all the atoms in the crystal is just the sum of the potentials due to each particle out of which it is constituted, it follows that the whole pattern of equipotentials is contracted in the direction of motion and left unaltered in the perpendicular directions, in just the same way as happens with a single electron. Now the equilibrium positions of the atoms are at points of minimum potential. It follows then that when the pattern of equipotentials is contracted in the direction of motion, there will be a corresponding contraction of the whole crystal in the same direction. A measuring rod made of such a crystal will therefore be shortened by the factor $\sqrt{1-(v^2/c^2)}$. But if the rod is perpendicular to the direction of motion its length will not be altered.

In the reference given above, it is also shown that because the field intensity is increased by the contraction, it follows that the electromagnetic inertia of a charged particle is also increased. The net result is that any physical process involving the crystal (e.g. vibrations) is slowed down by the factor $\sqrt{1-(v^2/c^2)}$. Therefore if oscillatory processes are used to measure time, we obtain the well-known formula

$$T=\frac{T_0}{\sqrt{1-(v^2/c^2)}},$$

where T_0 is the period for a crystal at rest in the ether and T is the period for a crystal in motion through the ether. Finally it can be shown that two such processes equivalent to clocks will get out of phase when they are separated.

When all three of these effects are put together, we obtain the Lorentz transformation between length and time as measured by a moving system and those measured by a system at rest in the ether. It follows from this sort of argument that Lorentz invariance is perfectly compatible with the assumption of an absolute frame such as that of the ether. Therefore there is no way from the experimental facts alone to prove there is no absolute frame determining a universal order of succession.

However, for various reasons which were mainly either aesthetic or philosophical, such a notion of a universal time order was rejected early in the twentieth century. For example, the ether theories available at the time assumed very complex and implausible mechanisms to explain the behaviour of fields and physicists found these assumptions unattractive. In addition, the ether theory in the form presented by Lorentz was such that it was impossible to detect its presence by any experiments that could be contemplated at that time. Therefore at the very least such theories were felt not to be very relevant, while by many they were regarded as violating a philosophical principle that was widely accepted in the positivistic atmosphere of the time, i.e. that all concepts contained in any theory ought to be observable.

Nevertheless physics developed in such a way that some features of the ether theory were retained. Thus Einstein [11] in his general theory attributed properties to empty space in the form of the gravitational tensor, $g_{\mu\nu}$. Einstein therefore later came to say that his field theories were theories of a relativistic ether that did not involve detailed descriptions of any mechanisms. It was essential for Einstein that such theories should be local and thus he was able to retain Lorentz invariance, at least in regions that were not too large (strictly speaking only in infinitesimal domains described by tangent spaces). But now we have additional reasons to limit Lorentz invariance implying that it does not hold even in the small. And nonlocality further suggests that even in intermediate domains, Lorentz invariance no longer holds for individual processes involving the beables, but is valid only in the manifest domain and in statistical applications of the quantum theory.

The notion of having a domain in which all phenomena become essentially independent of the preferred frame is one that is actually quite familiar in other contexts. For example, consider a cubic crystal which evidently determines a preferred frame, yet large scale properties such as sound waves of long wavelength are isotropic and show essentially no trace of the frame. Similarly we may get Lorentz invariance for long waves even when there is a structure determining a preferred frame.

We have used the ether theory to illustrate how it is possible to retain an absolute time order and yet retain Lorentz invariance. But the same result can be obtained in a much more general way without detailed and implausible assumptions about a material space-filling substratum of the kind to which Einstein objected. For example, one could suppose that in addition to the known types of field there was a new kind of field which would determine a space-like surface along which nonlocal effects would be propagated instantaneously. At present we can say very little about this field, but one could surmise that this space-like surface would be close to a hyperplane of constant time as determined in a certain Lorentz frame. A good candidate for such a frame could be obtained by considering at each point in space-time, the line connecting it to the presumed origin of the universe. This would determine a unique time order for the neighbourhood of that point around which one would expect isotropic properties in space. We may plausibly conjecture that this frame would be the one in which the 3K background radiation in space has an isotropic distribution.

This unique frame would not only make possible a coherent account of nonlocal connections, but could also be significant in other ways. For example in this frame there will evidently be no limit to the possible speed of the particles. Thus, in the stochastic form of the ontological interpretation of the Dirac equation it is possible, as we have seen in connection with equation (12.7), to have random movements that are faster than light. In the unique frame that we have introduced, there evidently is no reason why such movements should not take place. (Recall however that a non-Lorentz invariant process of this kind can give a Lorentz invariant equilibrium distribution $\boldsymbol{j} = \psi^\dagger \boldsymbol{\alpha} \psi, j^0 = \psi^\dagger \psi$.)

This brings us to the second objection, i.e. that the absolute frame is unobservable. Certainly if we restrict ourselves to the assumptions that have been made thus far, this would be a valid criticism. However, in terms of our ontological approach it is possible to alter these assumptions in such a way that the absolute frame would be observable, while negligible changes would be produced in the domain of experiments that have been possible thus far.

A great many such assumptions are possible, but we shall limit ourselves here to a simple illustration. Our ontological approach allows us, for example, to consider the possibility that the current quantum mechanics is an approximation to a deeper theory (as we shall explain in more detail in chapter 14). Let us assume then that the long range connections of distant systems are not truly nonlocal, as is implied by the quantum theory, but that they are actually carried in the preferred frame at a speed that is finite, but very much greater than that of light. For measurements made at levels of accuracy thus far available, the results will be very close to those predicted by the present quantum theory. But if we can make measurements in periods shorter than those required for the transmission of quantum connections between particles, the correlations predicted by quantum theory will vanish. In effect we would thus be explaining quantum nonlocality as an outcome of a deeper kind of non-Lorentz invariant locality. This however is relevant only in a domain beyond that in which current quantum theory is adequate.

If we consider such modifications of the theory, it then becomes possible to contemplate an experimental test that would reveal the preferred frame. In essence this test would consist of doing an EPR experiment in which the relative time of detection of the two particles was extremely accurately determined. In principle the Bell inequality would then no longer be violated for there would be no time for the disturbance of one particle to propagate to the other before the measurement was made on it. And from this it follows that the latter particle would no longer go into a corresponding state of close correlation with the first one. Of course this will require a measurement of extremely high accuracy because the speed of transmission of the quantum connection is assumed to be very much greater than that of light. It is clear that the accuracy required is far beyond that of the Aspect experiment [12].

In certain ways this test is reminiscent of the Michelson-Morley experiment, but it differs in the crucial respect that what is at issue here is not the measurement of the speed of light, but rather a measurement of the immensely greater speed of propagation of our assumed quantum connection of distant particles. Nevertheless as in the Michelson-Morley experiment, we would have to take into account the speed of the earth relative to the preferred frame. Thus one would have to make measurements in different directions and at different times. If one detected a change between these measurements, one would be able to demonstrate the existence of a preferred frame thus making the latter in principle observable.

However, if such changes were found, this would indicate a failure of both quantum mechanics and relativity, which would be much more significant than the mere observation of the speed of the preferred frame. The meaning of such a result would be that we can eliminate quantum nonlocality provided that we assume a deeper level of reality in which the basic laws are neither those of quantum theory nor of relativity (which latter come out as suitable limiting cases and approximations).

Of course, we cannot at present say any more about what these experiments are that could falsify quantum mechanics and relativity. To say more we would have to develop our ideas in greater detail. We may give here the analogy of the atomic theory which was originally proposed in ancient Greece by Democritus about 2500 years ago. At that time there was very little to guide assumptions about the atoms (e.g. it was proposed that they were held together by 'hooks' and that there were atoms of all sorts, such as atoms of intelligence, etc.). In any case it turns out that the experiments available at the time were much too crude to have given any significant information about atoms. It took 2000 years for ideas and experiments to develop far enough to permit some progress in this regard. Nevertheless the idea of atoms was relevant during all this time. What they indicated was that continuity at the large scale could be regarded as an appearance underlying which was the assumed discrete atomic reality. In any domain in which the notion of continuity was adequate atoms could not be observed. However, it was not assumed that atoms were intrinsically unobservable. Indeed the very notion of the atomic theory helped people to be alert to the fact that there might be observable consequences on the large scale which could be explained through these atoms and some of which would not be compatible with the assumption of continuity. Such consequences would provide observable evidence of the atomic constitution of matter.

Similarly with regard to the special frame implied by 'nonlocal' quantum connections and the deeper fields that would propagate these connections locally, we do not propose that these are intrinsically unobservable. We merely say that in the statistical and manifest domain in which the current quantum theory and relativity are valid, these new properties cannot be observed. Just as the observation of atoms became possible where continuity broke down, so the observation of the new properties would become possible where quantum theory and relativity break down. But of course it may take some time before we are able to pursue this matter experimentally (and hopefully fewer than 2000 years!).

12.9 References

1. W. J. Lehr and J.L. Park, *J. Math. Phys.* **18**, 1235–1240 (1977).
2. C. Fenech and J-P. Vigier, *C. R. Acad. Sc. Paris* **293**, 249–252 (1981).
3. S. S. Schweber, *An Introduction to Relativistic Quantum Field Theory*, Harper and Row, New York, 1961.
4. G. Wentzel, *Quantum Theory of Fields*, Interscience, New York, 1949, 138–152.
5. P. Droz-Vincent, *Phys. Rev.* **19D**, 702–706 (1979).
6. J-P. Vigier, 'Nonlocal Quantum Potential Interpretations of Relativistic Action at a Distance in Many-body Problems', in *Open Questions in Quantum Physics*, ed. G. Tarozzi and A. van der Merwe, Reidel, Dordrecht, 1985.
7. R. J. Glauber, *Phys. Rev.* **130**, 2529, and **131**, 2766 (1963).
8. G. Chew and H. P. Stapp, *Found. Phys.* **18**, 809–831 (1988).
9. K. R. Popper, *The Logic of Scientific Discovery*, Hutchinson, London, 1968.
10. D.Bohm, *The Special Theory of Relativity*, Addison-Wesley, New York, 1989.
11. A. Einstein, in *Scientific Papers Presented to M. Born on his retirement*, Oliver and Boyd, London, 1953.
12. A. Aspect, J. Dalibard and G. Roger, *Phys. Rev. Lett.* **44**, 1804–1807 (1982).

Chapter 13
On the many-worlds interpretation

13.1 Introduction

Over a period of time there have developed a number of other ontological interpretations of the quantum theory. In the next chapter we shall discuss some of these, but here we shall give a more detailed account of one of these which generally goes by the name of the many-worlds interpretation. We are giving this interpretation special attention because it has come to be fairly widely accepted, especially by those who work in general relativity and cosmology and who therefore feel a need to regard the universe as existing in itself whether observed or not.

We must emphasise from the very outset, however, that there is no single generally accepted version of the many-worlds interpretation. Indeed this interpretation has been constantly developing as a result of attempts to clarify its meaning and to remove what are perceived as its inadequacies. In this process a great many ideas have been explored which do not fully agree with each other, and it seems that, as yet, there is no final version of this interpretation that is accepted by all of its protagonists. (For a comprehensive account of this development see Ben-Dov [1]. Further detailed discussions will be found in Ballentine [2], Bell [3], Cooper and van Vechten [4], d'Espagnat [5], Primas [6], Stapp [7], Zeh [8].) The key difficulty stems from the fact that, as we shall see, the two main protagonists of this interpretation, Everett [9] and DeWitt [10], differ significantly in their basic formulations, and from this difference further divisions arise from those who have adopted various approaches to deal with the difficulties that these two did not resolve.

13.2 Everett's approach to the many-worlds interpretation

We shall begin with a discussion of Everett's approach to the many-worlds interpretation [9], which emphasises the inclusion of observers as purely physical systems within the theory. It seems that in doing this Everett intends to deny dualistic notions in which mind, for example, might be regarded as a kind of spiritual reality beyond that of the material universe and separate from it. (This denial would also include Wigner's notion of a collapse of the wave function as produced by the interaction between matter and such a mind [11].)

Everett effectively assumes that the universe as a whole including all observers exists objectively and is described completely by a vector in Hilbert space. Within this universe he wants to attribute subjective states to the observers which are in perfect correspondence with some aspects of the material universe and therefore of Hilbert space. (This sort of idea has been widely called psycho-physical parallelism.) He feels that "it will suffice for his purposes that observers possess memories, i.e. parts of a relatively permanent nature whose states are in correspondence with the past experiences of the observer" [12]. What this means is that, as in a computer whose memories are contained in the state of a disc, some aspect of the physical state of the observer, presumably within his brain, serves as the basis of his memories. Deductions can then in principle be made about the subjective experience of the observer by examining the contents of the memory.

Everett proposes that we can extend this model of an observer by considering him to behave like an automatic machine possessing sensory apparatus coupled to registering devices capable of recording past sensory data as well as machine configurations. He further supposes that the present actions of the machine are determined, not only by its present sensory data, but by the content of its memory as well. Such a machine will be capable of performing a sequence of observations (measurements), and furthermore it will be able to determine its future behaviour on the basis of past results. Everett feels that for such a machine we are justified in using phrases such as "the machine has perceived A" or "the machine is aware of A" if the occurrence of A is recorded in its memory.

We can further illustrate Everett's ideas in terms of the example of the measurement of the spin of a particle. In treating this question, Everett makes use of the von Neumann theory of measurement which is, of course, also what we have used in our own approach. If the initial wave function of the spin is

$$\psi = \alpha\psi_+(s) + \beta\psi_-(s) \tag{13.1}$$

and the initial wave function of the 'pointer' of the apparatus is a wave packet $\phi_0(y)$, then after the measurement interaction is over, the combined system is left with the wave function

$$\psi_f = \alpha\phi_+(y)\psi_+(s) + \beta\phi_-(y)\psi_-(s), \tag{13.2}$$

where $\phi_+(y)$ and $\phi_-(y)$ are wave packets corresponding respectively to positive and negative results of the spin measurement. The essential further step of Everett is to introduce wave functions $\chi_+(M)$ and $\chi_-(M)$, where χ_+ and χ_- represent states of the brain (or computer) that correspond respectively to memories of positive and negative results of this measurement. The overall wave function is then

$$\lambda = \alpha\chi_+(M)\phi_+(y)\psi_+(s) + \beta\chi_-(M)\phi_-(y)\psi_-(s). \tag{13.3}$$

In equation (13.2) we had a 'quantum state' in which the spin and the observing apparatus were entangled in the sense suggested by Schrödinger. Similarly equation (13.3) describes a state in which the memory within the observer has further been entangled in a similar way. Therefore in a fundamental sense the observer does not have a definite memory of what is the outcome of the measurement (unless either α or β is zero).

In order to give the memory a definite state, we would have to have another observation in which the content of this memory was observed. But this would introduce yet another memory of the second observer and the same difficulty would be repeated at one stage removed. It is clear therefore that if we add no further concepts to the quantum theory (e.g. such as well-defined particle positions as we have done and if we do not allow collapse) then there is no way in what Everett has said thus far, to describe the fact that a definite result is actually obtained in each case.

In order to proceed further, some additional assumptions are evidently needed. Everett does not state these assumptions clearly and indeed puts them in such a tacit way that one can readily fail to notice them. However, his essential assumption is implicit in the statement given on page 68 of reference [12]. We shall give here a paraphrase of this statement which we feel brings out more directly what we think that Everett wants to say.

In order to avoid the need for bringing about a definite result for the measurement by means of a collapse of the wave function in equation (13.3), Everett assumes that each part of this wave function corresponds to a definite state of awareness of the content of the observer's memory. He says that in one sense, there is only one total awareness belonging to such an observer. Nevertheless he says that (for the case of spin measurement) this single total awareness simultaneously can be regarded as two-fold, and that each of the two partial awarenesses is not aware of the other (nor of course of the whole).

It is evident that in a series of measurements, the number of partial awarenesses must multiply indefinitely. There are correspondingly many possible branches consisting of such sequences of partial awarenesses. He assumes that *one* of these branches represents the experience of a particular person. The other branches do not represent merely ***possible*** experiences that this person might have had (as they would, for example, in a corresponding treatment in terms of our own interpretation). Rather they represent distinctly different sequences of experiments which are actually recorded in other awarenesses similar to, but different from, this particular one. We emphasise again that all these different sequences of awareness are not aware of each other, nor of the whole.

Everett assumes that the whole universe including all observers exists objectively and is described, as we have already pointed out, completely and perfectly by a vector in Hilbert space. One may explain what this means loosely by saying that in some sense the universe is regarded as a multidimensional reality. But what Everett is doing, is to make a theory that relates the universe to various points of view that are contained within it. What we experience of the universe is a single subsequence of these points of view. Or, as Squires [13] has put it, what we are dealing with is not a many-universe theory, but a theory of one universe with many viewpoints. Each point of view establishes a relationship between a state of awareness and the state of some other part of the universe containing the observing instrument and the observed object.

The existence of this sort of relationship has been described by Everett as a *relative state* of the system corresponding to a particular state of the observer (or more accurately, to the state of his awareness). Everett and those who follow him seem to regard this notion of relative state as a fundamentally new concept comparable perhaps to Einstein's regarding the meaning of a position coordinate as being relative to some material system which serves as a frame of reference. Thus in Everett's language we can say that any part of the total vector in Hilbert space has a perceptible meaning only in relationship to 'frames' constituted of the memory of the observer. As there are many permissible coordinate frames in relativity so there are many observer 'frames' in quantum theory.

As an example of Everett's interpretation, let us consider again equation (13.3). In this case the wave function of a partial awareness corresponding to a state of memory of spin $+\frac{1}{2}$ will be multiplied by wave functions describing the state of the apparatus and of the spin itself also corresponding to spin $+\frac{1}{2}$. From there on this memory will develop further awarenesses which fit in with this state. It will however not be aware of the development of other components of the Hilbert space vector corresponding to the observing instrument and observed object. So on one branch the awareness is as if the spin had 'collapsed' to $-\frac{1}{2}$.

Everett gives some arguments aiming to show the consistency of this approach. For example, if a second observer interacts with this system after the first observation, the overall wave function will become

$$\alpha\chi_+(M_2)\chi_+(M_1)\phi_+(\boldsymbol{y}_1)\phi_+(\boldsymbol{y}_2)\psi_+(s) + \beta\chi_-(M_2)\chi_-(M_1)\phi_-(\boldsymbol{y}_1)\phi_-(\boldsymbol{y}_2)\psi_-(s)$$

where $\boldsymbol{y}_1$ and $\boldsymbol{y}_2$ represent the coordinates of the first and second apparatuses respectively, while M_1 and M_2 represent the memories of the corresponding observers. From Everett's assumption, it follows that the memories of the two observers, the states of their apparatuses, and the value of the spin are completely correlated. Therefore if, for example, the first observer remembers spin $+\frac{1}{2}$, this means not only that the spin will act as if it were $+\frac{1}{2}$ in all his subsequent experiences, but that the second observer will also remember the spin was $+\frac{1}{2}$ and experience it as such from then on. In this way Everett explains how different observers can come to agree about the implications of their observations. Everett gives further arguments aiming to show in more detail how all the observed results of the quantum theory would follow from analyses of this nature.

It is implicit in Everett's approach (and as we shall see, this is made explicit in DeWitt's approach [10]) that the main purpose of this interpretation is to introduce the minimum number of concepts into the theory (for example, Everett would regard our notion of particle variables as further assumptions that are not needed).

It seems to be especially important for Everett to put all the concepts in his theory into correspondence with the basic mathematical structure of the quantum theory, i.e. in terms of Hilbert space. Thus he not only assumes that the physical universe can be described completely in terms of Hilbert space, but he seems to imply that the same is true for mind, which he regards as being in essence just awareness and memory as he has defined them. It must be emphasised however that this in itself is a highly speculative assumption with very little evidence behind it. Everyone agrees that we know very little about mind or conscious awareness, but many would claim that it must include much more than just memory. Even if we accept the as yet unproven assumption that memory can be explained in the way that Everett does, it does not follow that this could be done for the whole of mind which includes attention, sensitivity to incoherence, all sorts of subtle feeling and thoughts and creative imagination as well as much more. We will discuss this in more detail in chapter 15, where we, too, consider the possibility of including the observer in our overall view of the universe. However, we feel that the entire framework of the current quantum theory is too limited to do this properly and we propose extensions of the theory that might make a more coherent account of the observer possible.

Perhaps we could put Everett's hypothesis about the mind into a more favourable light by regarding it as just a convenient simplifying assumption that makes it possible for him to get on with the interpretation of the quantum theory. But even if we do this, we find that there is a further speculative hypothesis of a more physical nature. This is that it is possible to put some quality, such as awareness, into a precise and accurate correspondence with some *part* or *component* of the total vector in Hilbert space. Thus for example, returning to equation (13.3), it would be consistent to say that the memory is well defined when either $\alpha = 0$ or $\beta = 0$. For in these cases, the memory would be determined by an eigenvector of what may be called a memory operator. Otherwise the memory has to be ambiguous, just as other properties, such as momentum and position, have to be ambiguous when the Hilbert space vector representing the state of the system is not an eigenvector of the corresponding operators. But Everett is further assuming that there is a property called awareness (which he regards as essentially physical) that can be put into accurate correspondence with some part of the total Hilbert space vector, even when the latter is not an eigenvector of the memory. If a similar statement were made about an ordinary physical quantity, this would imply a violation of the uncertainty principle. Or to put it differently, Everett is assuming that the mind can be aware of its own state of memory without measuring the latter, therefore without disturbing the latter, and hence without introducing any ambiguity. It is clear then that Everett has assumed a kind of property going far beyond what is allowed in the ordinary quantum mechanics. In effect he has introduced a basically speculative theory of mind, along with an equally speculative theory as to how this can be related to a component of a total vector in Hilbert space.

Everett's approach may well be interesting to pursue further, but it should not be presented as having a smaller number of hypotheses than other interpretations. Even if we do not object to the assumption of a large number of awarenesses undetectable to us or to each other, we must still notice that we are introducing large and unverifiable assumptions about mind and attributing new properties to Hilbert space which go beyond those which follow from ordinary quantum mechanics.

If we accept Everett's approach in the above spirit, we can then go on to ask whether it is consistent in full detail and whether it deals adequately with all the problems raised in such an approach. We shall in the subsequent sections outline briefly what are the main questions involved and what are the difficulties that arise in doing this.

13.3 Comparison of Everett's approach to that of DeWitt

In this section we shall develop DeWitt's approach to the many-worlds interpretation and compare it with that of Everett. In doing this we shall see that there are some very important differences between these two approaches. However, the fact that these differences exist appears not to be widely known or recognised, and indeed many authors including DeWitt himself essentially identify the two. The fact that both are commonly given the same name (though Everett did not use the term 'many-worlds') helps further to draw attention away from the importance of these differences.

DeWitt's motivation for his approach is clearly stated in his paper [10]. He felt dissatisfied with all the 'metaphysical' assumptions contained in previously available interpretations. He then proposed that we "forget all these" and begin afresh [14]. In particular his intentions were stated as the following:

1. To take the mathematical formalism of quantum mechanics as it stands without adding anything to it.
2. To deny the existence of a separate classical reality.
3. To assert that the state vector never collapses.

He then asserts that it is possible to achieve these intentions and that this was proved by Everett [9] with the encouragement of Wheeler [15] and that this has subsequently been elaborated by Graham [16]. As with Everett, DeWitt starts with the assumption that the mathematical concept of Hilbert space provides the whole of the basic conceptual structure underlying quantum theory. He then says that only one additional postulate is needed to give meaning to the mathematics, namely, the postulate of complexity: "The world must be sufficiently complicated that it can be decomposed into systems and apparatuses" [14]. From this it follows that "the mathematical formalism of the quantum theory is capable of yielding its own interpretation" (and all this without external metaphysics or mathematics).

DeWitt evidently agrees with the implication of Everett's approach in assuming that the universe is faithfully represented by a vector in Hilbert space. But of course, as he pointed out, this vector must be immensely more complex than the sort that appears in equation (13.3). DeWitt then asserts that the universe itself is splitting into a stupendous number of branches, all resulting from the measurement-like interactions between its myriads of components. Such interactions take place, not only in measurement, but also in natural processes of all kinds which likewise bring about a continual splitting of the universe.

Here we encounter a serious difficulty. For DeWitt clearly attributes his theory to Everett in a number of places. Yet as we have seen there is no mention of splitting universes in Everett's work. Rather there is one universe with many awarenesses that are not aware of each other. To be sure Everett includes memory which is the content of these awarenesses as part of the total universe. Nevertheless he specifically requires that there be relatively permanent memories without which he could not formulate his notion of awareness. Of course, memory implies a very complex structure, but not every complex structure is a memory. So at the very least, DeWitt should have said that Everett requires a kind of complexity that would provide the basis of memory, e.g. a brain or at least a super computer (as indeed we have seen from Everett's comparison of awareness to the function of such a machine). But even this would not be enough; Everett also says that no very significant change takes place when an observer becomes aware of a certain result. It is not the universe that splits, but it is just awareness as a whole that divides into many parts that are not aware of each other. We repeat again what Squires [13] has said, i.e. it is not a theory of many universes, but a theory of many viewpoints of one universe. Everett's aim is thus not mainly to explain the universe, but rather to explain our *perceptions* of the universe.

Everett revealed his attitude to this question in commenting on some views on the quantum theory expressed by Einstein [17]. In the von Neumann interpretation, observation induces a 'collapse' of the wave function that may affect the whole universe. Einstein objected to that as implausible and unsatisfactory. At another time Einstein put this feeling colourfully by saying that he could not believe that a mouse could bring about drastic changes in the universe by simply looking at it. In answer to this, Everett [18] says that from the standpoint of his theory, "it is not so much the system that is affected by an observation as it is the observer who becomes correlated to the system". Further along he states that "observation ... simply correlates the observer to the system". Furthermore he adds "Only the observer's state has changed, to become correlated ... with the system. The mouse does not affect the universe—only the mouse is affected" [19].

It is clear from the above that Everett did not contemplate the splitting of the universe. It follows then that there is a fundamental difference between the basic assumptions held by the two leading protagonists of the 'many-worlds' approach. Indeed Everett's view should not even be called the many-worlds interpretation but rather, as Albert and Loewer [20] have suggested, the many-minds interpretation, a view which we discuss later.

The key difference between DeWitt's approach and that of Everett is that DeWitt's theory is purely objective, while Everett's theory depends crucially on the inclusion of the subjective experiences of the observer (which are however explained objectively as being in correspondence with parts of the total Hilbert space vector). For DeWitt, the universe actually splits as might happen to material objects. This assumption creates a number of problems. Firstly we can ask just when does it split. Thus for example, while the spin is interacting with the measuring apparatus as described in equation(13.3), the wave packets corresponding to the two possible results begin by overlapping. In principle some overlap may persist for a long way even though this may be small. What is it that determines when two overlapping wave packets, originally in one universe, suddenly become two non-overlapping wave packets in separate universes? Indeed unless the measuring apparatus is complex enough to make possible a distinct classical-like result, the universes should never split, as the two wave packets could again come to interfere in the way we have described in chapters 6 and 7. But once the universe has split, it seems that such interferences would no longer be possible. In the usual approach to the quantum theory (as well as in our approach) it is always, in principle, possible for such interference to occur again.

One usually appeals to some feature such as the complex and chaotic phase relations of the wave function to make such interference unlikely to have significant effects (for example see Bohm [21]). From DeWitt's emphasis on 'complexity', one might suppose that he assumes that when the wave function reaches a certain degree of complication, the universe begins to split. But he does not define complexity (though some authors, notably Leggett [22], are now exploring how such a definition may be made, but in a different context from that of the many-worlds interpretation). What would be required for an adequate formulation of DeWitt's approach in this regard would be to come up with a good quantitative definition of complexity and to assume some further equations determining how the splitting of universes is related to this feature of complexity. But if he were to succeed in doing this, he would clearly have introduced new concepts as well as new equations, and could not hope to claim that the mathematics of the current quantum theory determines its own interpretation.

Related to this difficulty is the fact that DeWitt wants to do without a separate classical level and yet he does not suggest anything that could take its place. Nevertheless in describing what he means by an observing apparatus, he tacitly assumes all sorts of properties that are suggested by classical physics, such as the possibility of distinct results which are stable and do not vanish in the quantum uncertainties of the Hilbert space. Perhaps he could also deal with this problem in terms of the notion of complexity if he could show that, with a sufficient degree of the latter, something like a classical world would emerge in a certain limit as we have shown in terms of the quantum potential (see chapter 8). But in the absence of such a demonstration, we have to say that this key part of his basic intention has not been realised (as has indeed been pointed out by Gell-Mann and Hartle [23]).

13.4 Probabilities in the many-worlds interpretation

Whether we adopt the approach of Everett or that of DeWitt, we are still faced with the question of what is to be meant by probability in this general point of view. If we claim that the whole universe is completely described by a certain vector in Hilbert space, then this is unique and does not allow for a statistical theory (based for example on an ensemble of similar universes). How then are we to relate the many-worlds approach to the usual probabilities of quantum mechanics? Everett and DeWitt adopt what is basically the same idea.

They consider a long sequence of N measurements on similar systems initially in the same quantum state. Their aim is then to prove, without any extraneous assumptions about probability, that as $N \to \infty$ the relative frequency of a given result will be equal to the usual quantum mechanical probability.

For simplicity let us take the system to be a particle with spin. The relevant initial wave function is then

$$\psi_0 = \alpha\psi_+ + \beta\psi_-. \tag{13.4}$$

We now consider the whole set of systems as forming a single combined system whose initial wave function is

$$\chi_0 = \psi_0(1)\psi_0(2)\ldots\psi_0(N), \tag{13.5}$$

where the numeral indicates the particle in question. This can be written as

$$\chi_0 = (\alpha\psi_+(1) + \beta\psi_-(1))(\alpha\psi_+(2) + \beta\psi_-(2))\ldots(\alpha\psi_+(N) + \beta\psi_-(N)). \tag{13.6}$$

We now suppose that the spins of each of these particles is measured by a separate piece of apparatus whose relevant coordinate is $\boldsymbol{y}_n$. If $\phi_0(\boldsymbol{y}_n)$ represents the initial wave function of the nth apparatus, the total initial wave function is

$$\Psi_i = \phi_0(\boldsymbol{y}_1)\phi_0(\boldsymbol{y}_2)\ldots\phi_0(\boldsymbol{y}_N)\chi_0. \tag{13.7}$$

It is clear that this wave function is a sum of 2^N terms. A typical term would be

$$T_{s_i} = \phi_0(y_1)\phi_0(y_2)\ldots\phi_0(y_N)\alpha^S\beta^{N-S}\prod^{S}\psi_+\prod^{N-S}\psi_-, \tag{13.8}$$

where $\prod^S \psi_+$ is a product of a certain number, S, of particle wave functions with spin $+\frac{1}{2}$ and $\prod^{N-S}\psi_-$ represents the product for the $N-S$ remaining particles with spin $-\frac{1}{2}$.

After the measurement process is over, then, in accordance with the von Neumann theory that we have given in chapter 2, section 2.4, the term (13.8) becomes

$$T_{S_f} = \alpha^S\beta^{N-S}\prod^{S}\phi_+(y_i)\psi_+\prod^{N-S}\phi_-(y_i)\psi_-, \tag{13.9}$$

where $\phi_+(y_i)$ represents the wave function of the pointer of the ith apparatus indicating spin $+\frac{1}{2}$ and $\phi_-(y_i)$ represents the wave function of the pointer of the apparatus indicating spin $-\frac{1}{2}$. Each of the above terms may be said to represent a branch of the total wave function, which latter is

$$\Psi = \sum_S T_{S_f}. \tag{13.10}$$

It is the object of the proofs given by Everett and DeWitt to show that, except for a set of measure zero, all branches will yield a fraction $|\alpha|^2$ for spin $+\frac{1}{2}$ and $|\beta|^2$ for spin $-\frac{1}{2}$, as the number of branches becomes infinite. To do this Everett and DeWitt begin by assuming a measure associated with each vector in Hilbert space which is equal to the quantum mechanical probability that is usually associated with that branch of the wave function in standard quantum theory. For example, if the wave function is $\alpha\psi_+ + \beta\psi_-$, the measure of ψ_+ is $|\alpha|^2$ and that of ψ_- is $|\beta|^2$. But they do not begin by assuming this is a probability. Rather, they take it to be a purely abstract quantity which they have called 'measure'. They then show that in an infinite sequence of measurements the measure of branches in which the proportion of spin $+\frac{1}{2}$ differs from $|\alpha|^2$ and that of spin $-\frac{1}{2}$ from $|\beta|^2$ is zero.

As yet, however, nothing follows from this about actual relative frequencies. To make a connection they have to introduce a further assumption. This is essentially that branches of measure zero never occur. From this it follows, of course, that in an infinite sequence, we will get the right relative frequencies.

But this proof has been criticised by d'Espagnat [24] and others (see Deutsch [25]). There is no intrinsic reason why branches of measure zero should never occur, if measure is nothing more than a purely abstract mathematical quantity. Moreover we would fail to obtain an additional result of standard probability theory which is necessary for its consistency: i.e. that sequences of small measure must occur correspondingly rarely. The treatment of Everett and DeWitt would in any case apply only to an infinite sequence, but the theory must apply to finite sequences that can be carried out in actual experiments. Indeed, current probability theory implies that finite sequences of small measure *must* occur sometimes if the distribution is random. In other words, we cannot guarantee that there will *never* be large fluctuations. The proof given by Everett and DeWitt contains no way of obtaining such a result (except perhaps by adding further arbitrary assumptions). It is only if we suppose that the mathematical measure corresponds to a relative frequency in a random ensemble that we can deal with these questions adequately.

In any case a more careful analysis of this situation is made by Ballentine [26] who shows that the relative frequencies do not actually approach $|\alpha|^2$ and $|\beta|^2$ for the case of spin $\frac{1}{2}$. Rather the distribution Ballentine obtains is half and half. One can understand this by noting that the only way of bringing probability into this situation is to suppose equiprobability of the discrete alternatives, which will, of course, inevitably bring about a half-half distribution. It is clear then that the many-worlds interpretations of Everett and DeWitt have not succeeded in providing an adequate account of probability.

To deal with this Deutsch [25] assumes further modifications of the theory. Rather than supposing with DeWitt that the whole of existence is continually bifurcating into more and more universes, he postulates that there is always an infinite and essentially constant number of actual universes. For example in the case of spin that we have treated earlier in terms of equation (13.11) each of the whole set of universes initially has spin $\frac{1}{2}$ and the corresponding apparatus is in its initial state. After the measurement process, the universes are partitioned among the states indicated by (13.13).

The essential assumption of Deutsch is that there is a random distribution of these universes with probabilities corresponding to the usual quantum mechanical probabilities for the associated results. In a measurement, we begin with an initial distribution of universes in which all of them had the same initial spin state given by $\psi = \alpha\psi_+ + \beta\psi_-$ while all the apparatuses are in the same state. After the measurement is over this distribution is changed into one in which the probability that S particles in the sequence have wave function ψ_+ (while the remainder have wave function ψ_-) is just $|\alpha|^{2S}|\beta|^{2(N-S)}$. This is, of course, just the usual quantum mechanical probability for such a sequence. And from this expression it follows that as the sequence approaches an infinite number, the relative frequency of spin $+\frac{1}{2}$ will be $|\alpha|^2$ and spin $-\frac{1}{2}$ will be $|\beta|^2$.

It is clear then that Deutsch succeeds in achieving the aim that Everett and DeWitt failed to achieve. However, there are a number of serious questions that Deutsch has not yet answered. The most obvious of these is that he has chosen a certain basis for the interpretation, in this case the one in which the spin in the z-direction is diagonal. Of course it is clear that this basis corresponds to the orientation of the apparatus. But recall that one of the basic aims of the many-world interpretations is to get everything out of the formalism without further assumptions. Deutsch has not shown why it is just these pieces of apparatus that will partition the universes in the way he has assumed. If he stops here he will be in a position similar to that of von Neumann who likewise distinguished between the apparatus and the rest of the universe without deducing this distinction from the formalism. Presumably the apparatus is distinct in that it behaves apparently classically as part of the large-scale world of common experience. Deutsch does call attention to this point, but like Everett and DeWitt he does not offer any proofs that such a 'classical' basis must emerge from the formation itself. We shall discuss the question in more detail in the next section.

A yet more serious question that we shall discuss here is concerned with the significance of the random ensemble of universes that Deutsch had to assume in order to give meaning to the concept of probability. In a given branch of the wave function, after the measurement process is over, the quantum state of the whole set of particles will be well defined. But once such a quantum state is given no further differences within it are possible, at least according to the usually accepted form of the quantum theory itself which asserts that the wave function gives all possible physical information about the system. How then do the many universes in the same quantum state differ? Deutsch does not make this clear. It is difficult to see how one could avoid the further postulate that there is an additional randomly distributed parameter or set of parameters not present in the ordinary quantum theory which would describe their differences. But to do this is to eliminate the key advantage claimed for the many-worlds interpretation, i.e. that it introduces no basically new physical concepts or equations or parameters. Indeed our own interpretation has been criticised just because it does introduce particle variables as new parameters. At the very least we could claim that these variables are relatively easy to understand, whereas with Deutsch's random distribution of universes it is not clear how such parameters are to be understood.

Deutsch's approach has been applied thus far to the DeWitt interpretation in which it is supposed that a multiplicity of universes actually exists.

Whether this approach can also be applied to the Everett interpretation is not clear. Let us recall that for Everett there is only one total universe with many points of view, each of which corresponds to a particular memory or awareness that is within its total universe. What seems to be called for is a random distribution of awarenesses over various quantum states with relative frequencies equal to the usual quantum mechanical probabilities for the associated results. To put this notion more precisely, let us represent the total wave function of the universe by

$$\Psi = \sum C_i \psi_i \phi_i, \tag{13.11}$$

where ϕ_i represents the ith state of awareness and ψ_i represents the state of the rest of the universe that corresponds to it. The assumption analogous to that of Deutsch would be that $|C_i|^2$ represents the relative frequency of awarenesses of state i corresponding to the ith component of the total Hilbert space vector Ψ.

However, there appear to be difficulties with this suggestion. Consider, for example, a situation in which some fraction of the observers would be killed in a catastrophic event so that their memories would vanish. How would the statistics be formulated to take care of such a possibility?

To sum up then, it does not seem that there is as yet an adequate treatment of probability within the Everett interpretation, while even within the DeWitt interpretation, Deutsch's further assumptions seem to be able to provide this only by foregoing the claim that no new basic principles are needed beyond those of the current quantum theory.

13.5 On the preferred basis and the classical limit

One of the key remaining ambiguities in the basic formulation of the many-worlds interpretation is around the question of how the classical limit is to be treated. As we have seen, DeWitt's expressed intention was to avoid the assumption of a classical limit. Everett was less clear on this point, but the general tenor of his approach suggests a similar attitude. However, if we accept this then we are required either to show that something like a classical limit is contained implicitly in the many-worlds interpretation or else to show that such a limit is not needed. Yet nothing like this has been done so far in an adequate way. As we have pointed out, DeWitt appeals to an as yet vague notion of complexity and he seems to imply that this would take care of yielding the relatively well-defined and stable kinds of results that are usually comprehended in terms of a classical limit. On the other hand Everett explicitly denies that any such property of complexity is needed. Thus Everett [27] writes:

> We impose no restrictions on the complexity and number of degrees of freedom of measuring apparatus or observers, and if any of these processes are present (such as heat reservoirs etc.) then these systems are to be simply included as part of the apparatus or observer.

This lack of explanation of why the actual world of common experience corresponds to some kind of classical limit has indeed been noted by Deutsch [25]. As we have already pointed out in the previous section he acknowledges that there is some such favoured basis for interpretation. Nevertheless it is clear that the basis must be such as to somehow bring about the correct classical limit. But Deutsch does not provide an answer to the question of how this basis is related to the classical limit. We note that, in DeWitt's version of the many-worlds interpretation, it is also necessary to have this sort of preferred basis as has indeed been pointed out by many authors (Ballentine [2], Bell [3], Cooper and van Vechten [4], d'Espagnat [5], and Primas [6]).

To bring out what is required, we note that this basis determines the elements that correspond to the many-worlds into which a particular universe may further split. Some authors have suggested further mathematical assumptions that they hope might lead to determination of this basis. For example, Zeh [28] suggests that there is a unique analysis of the total wave function into relative states, called the Schmidt decomposition, and that this would deal with such questions. But without further concepts and assumptions one can easily see that such a treatment would still be ambiguous.

Thus consider an atom in some linear combination of p-states

$$\Psi = a_- \psi_- + a_0 \psi_0 + a_+ \psi_+,$$

where ψ_-, ψ_0 and ψ_+ represent wave functions in which the z-component of the angular momentum is respectively $-\hbar, 0$ and $\hbar$. Clearly DeWitt's splitting of universes cannot go so far as to imply that such an atom by itself would split into three copies with angular momentum well defined in the z-direction. Indeed if this were to happen, the theory would fail to agree with experiment in a domain that is well tested. We would not expect this sort of thing to happen even with medium-sized molecules. But as we approach the large-scale world of common experience the many-worlds interpretation forces us to assume that at some point the universe does begin to split in this way. Just where does this happen? As we have remarked earlier it is necessary in DeWitt's approach to develop some new parameters such as complexity and new equations that will determine where the split will occur. But as we have seen, if we do this, the basic claim of the many-worlds interpretation that it does not introduce new assumptions and principles will again fail.

A similar problem can be seen to arise in Everett's interpretation, but here it involves a certain ambiguity as to what is to be meant by awareness, memory or mind. Thus, for example, Everett's concept of relative state could be applied to a single hydrogen atom whose wave function may take the form

$$\Psi = \sum C_s \psi_s(x) \phi_s(y),$$

where x is the coordinate of the nucleus and y that of the electron. If we were to apply Everett's theory at this level, we would have to say that the electron is aware of the proton and that the proton is aware of the electron. It would then indeed follow that the overall awareness associated with the hydrogen atom as a whole would split into many correlated states of mutual awareness of the electron and the proton, which states are, however, not aware of each other. This would imply that everything was aware of everything else and that human awareness (or indeed machine awareness) were only special cases of this.

It is certainly not clear what such a theory could mean. If it were Everett's intention to propose this sort of theory, it would have been up to him to go into what he means by universal awareness of everything by everything. For such a notion is evidently far more radical than anything that arises in the commonly accepted views of the many-worlds interpretation. In the absence of any discussion of this point by Everett or by those who follow him, we must assume that such a radical version of his interpretation was not his intention.

But then this immediately raises the question of just what determines the conditions for awareness to exist. As we have seen, Everett expressly denies that the notion of complexity is basically relevant to his theory. Yet it seems evident because of his referring to computers and memories, that he has tacitly in mind some kind of complexity (as well as stability) as a basis for awareness. This means that he faces essentially the same difficulty as DeWitt. He will have to introduce some new assumptions as to what constitutes the conditions for awareness and new laws determining just when these conditions are satisfied.

Recently these questions have been re-examined by Stapp [29] and by Albert and Loewer [20] in what is called the many-minds interpretation. The key step here is to assume that awareness or consciousness is not basically a physical property, but rather a property of mind which is supposed to be entirely different from matter. Thus the theory, unlike Everett's, assumes a dualism between mind and matter along lines similar to that proposed much earlier by Descartes. The advantage of this approach is that we do not need to assume that matter has properties given to it by awareness, i.e. that physical meaning can be given in the quantum theory to all branches of a linear superposition of wave functions without contradicting appearances. It thus makes possible a more consistent way of separating one domain (matter) that satisfies the superposition principle of quantum theory and another domain (mind) that does not.

Such a dualist approach is in certain ways reminiscent of Wigner's suggestions [11], but it must be remembered that in the latter, mind and matter are supposed to interact in a two-way process. In the many-minds interpretation however there is no such interaction. Rather mind is assumed to be capable of becoming conscious of certain aspects of the state of matter without influencing the latter in any way at all.

The many-minds interpretation shares with Everett the assumption that physical reality corresponds to the total wave function of the universe. Likewise along with Everett, it assumes that on the large scale stable configurations of matter exist which make possible not only measuring instruments, but also a reliable memory of what the results of measurement are. We recall in the Everett theory that the wave function describing relevant parts of the spin measuring apparatus and the memory is given by equation (13.3).

$$\lambda = \alpha\chi_+(M)\phi_+(y)\psi_+(s) + \beta\chi_-(M)\phi_-(y)\psi_-(s).$$

The basic new assumption relative to Everett's is that each person has a total mind that can split into many sub-minds that are not aware of each other. Each mind can be aware of a particular brain state corresponding to a memory of a particular experimental result that is stable and distinct from other memories corresponding to different possible results of the experiment. Let these minds be represented by μ_+ and μ_-. If the mind were in the domain of quantum mechanics, the wave function would then be in a linear superposition

$$\Gamma = \alpha\mu_+\chi_+(M)\phi_+(y)\psi_+(s) + \beta\mu_-\chi_-(M)\phi_-(y)\psi_-(s).$$

But here is where a break is made with the basic principles of quantum mechanics by assuming that μ_+ and μ_- do not superpose. We further assume a random process in which the original mind m has split into two distinct new minds, μ_+ and μ_-. The mind μ_+ is aware of the physical situation whose wave function is $\chi_+(M)\phi_+(y)\psi_+(s)$ and mind μ_- is aware of the physical state corresponding to the wave function with the minus sign. Neither μ_+ nor μ_- are aware of each other.

In the further development of this process, there will be additional splits in the mind. For example, if some physical property corresponding to a wave function ξ_N is measured and remembered, there will then be a total of $2N$ minds aware of the N physical states, N, corresponding to measurements and memories of both the spin and the new property.

In the random process in which the mind splits, we assume that the probability of a particular mind being associated with the Nth physical state and a particular spin variable '+' is proportional to $|\alpha_N|^2$. As time goes on there will be a constant process of further splitting of the mind.

All of this is very close to the treatment given by Everett and differs only in that we have further assumed a unique last stage, i.e. mind, in which the linear superposition fails (whereas Everett says nothing about this question). By thus defining this unique last stage as a splitting of the mind and relating this split to the linear superposition of memories in the brain through random processes, the many-minds interpretation avoids difficulties of specifying the basis in which linear superposition breaks down.

On the other hand, this advantage has been gained at the expense of not only bringing in duality of mind and matter, but also of having to assume special properties of mind which are difficult to understand, i.e. the ability to be aware of matter without any interaction and the ability to split in a random way into many minds, each of which is associated to a particular physical state.

It should be pointed out here that the many-minds interpretation does not imply nonlocality. This comes about because reality is assumed to be in perfect correspondence with the total wave function of the universe and because the whole of this reality is covered completely by the perceptions of the many minds which have no effect on the wave function. Because Schrödinger's equation contains only local interactions, its development is entirely local even though it has to be expressed in configuration space. Each mind becomes aware of only one facet of this multidimensional reality.

The perceived classical world is then a result of the fact that each mind is only aware of a small part of the whole, the perceived classical world is in a sense an illusion. What is illusory is that this part is the whole. If it were seen to be merely a facet then there would be no illusion.

To conclude, the many-minds interpretation does remove many of the difficulties with the interpretations of Everett and DeWitt, but requires making a theory of mind basically to account for the phenomena of physics. At present we have no foundations for such a theory. Indeed even to consider the mere possibility of this sort of assumption, we have further to suppose that there are new kinds of laws uniquely relating real physical properties such as awareness, to parts or components of the total vector in Hilbert space. As we have pointed out in section 13.2 this is equivalent to assuming that the state of awareness is not limited by anything like the uncertainty principle—clearly a gigantic step beyond present day physics. Thus even if we succeed in developing such a theory, it is clear that we will require new principles, new concepts and new equations. This means once again, that the claim of the many-worlds interpretation to avoid the addition of speculative new assumptions will fail.

13.6 Comparison between our interpretation and that of the many-worlds

One of the first objections that tends to arise against the many-worlds interpretation is that its assumption of an infinity of universes or minds, almost all of them being unobservable, violates the principle of Occam's razor, i.e. that we should not introduce a multiplicity of unnecessary assumptions. However, the proponents of this interpretation answer such objections by saying, e.g. with d'Espagnat [5], that multiplication of worlds entitles one to economise on principles. Or as Primas [6] has said, "The Everett interpretation is superior in logical economy." As we have already indicated, they would criticise our interpretation by saying that in addition to postulating the reality of the wave function, it brings in the further concept of particles with their equations of motion. It is thus clear that they feel that the many-worlds interpretation is superior because it does not do this sort of thing.

Leaving aside this question for the moment, it should be said that our interpretation accomplishes a number of things which cannot be done in the many-worlds view. Firstly, as we have seen in chapter 8, it gives a simple and coherent account of why the large scale world of common experience should be essentially classical. This is demonstrated by showing that the quantum theory implies a relatively autonomous domain in which the quantum potential can be neglected so that an approximately classical behaviour will result.

The physical structures of observers are also manifest in this domain. As pointed out in chapter 8, it is a well-known fact that our sensory information comes from this level. From this it follows that we are *directly* aware of the particle aspect of the universe through the senses and that the more subtle wave function aspect is *inferred* by thought about our sensory experience in the domain that is manifest to the senses. We are therefore not basing our approach on the *assumption* that we are aware only of particles. Rather we are simply making use of the *fact* that at least in science as we know it thus far, all information comes ultimately through the manifest level (recalling that it comes not only through the particles, but also through the classical limit of electromagnetic field beables).

The second advantage of our interpretation is that we do not have to assume the relation $\rho = |\psi|^2$ *ab initio*, but that we are able to arrive at this as an equilibrium distribution resulting from chaotic processes.

Finally we cite what we regard as the main advantage, which is that it gives an intelligible and intuitively understandable account of how quantum processes take place.

It is clear that the many-worlds interpretation cannot do any of these things. As we have seen earlier, it does not provide an adequate account of the large scale behaviour at the classical limit. Moreover it is compelled to assume the relation $\rho = |\psi|^2$ (e.g. as was done by Deutsch). And it gives no intuitively intelligible account of how quantum processes take place.

With regard to the latter point, it defies the imagination to grasp intuitively how the universe could split, and even more as to how it could be doing so in a stupendous number of ways. It equally defies the imagination to grasp intuitively how awareness could split into many parts that are not aware of each other, and even more so how awareness could be, in effect, identified with some part of a vector in Hilbert space.

Let us then consider once again the claim that the many-worlds interpretation is the one that is most economic in concepts and principles. We have seen, however, that in order to introduce probability in a consistent way, Deutsch had to assume an infinity of universes in the same quantum state, and in effect to imply a random distribution of these universes over an undefined parameter whose nature is completely unspecified. As we have pointed out earlier, he has thus in essence replaced the new particle variables by a new parameter that would distinguish the various universes. So there is no advantage in the number of hypotheses. Moreover there is the disadvantage that his new parameters are far less clear in meaning than others.

If we adopt DeWitt's approach, then, as we have seen, it is necessary to supplement this with some further principles involving perhaps an as yet unavailable definition of complexity and further equations determining just how this would determine the splitting of universes. On the other hand, if we adopt Everett's approach or indeed that of the many-minds interpretation, we have to do something similar with regard to the splitting of awareness. In this connection a complete and consistent expression of these interpretations would, as we have seen, require new principles and assumptions of a speculative nature going beyond the current quantum theory and beyond our present knowledge of awareness and its possible relationship to Hilbert space.

In view of all these unresolved problems we have been led to ask why the many-worlds interpretation seems to be so attractive to some physicists. Of course we realise that the Hilbert space formulation of quantum mechanics has a great deal of symmetry and may therefore be felt to possess great beauty. Hence it is understandable that there may be great reluctance to give it up or to go beyond it. On this basis one can see that the greatest advantage of the many-worlds interpretation for its proponents is that it provides an ontological formulation which relates all its basic concepts only to Hilbert space. In effect its says that everything can be put in some correspondence with Hilbert space. The first priority is thus to maintain the notion that all basic structures of concepts have to be expressed in terms of Hilbert space. As long as this can be done it is felt that all sorts of further assumptions are possible and that these do not count heavily as new principles. For example the assumption that the mind can be put in correspondence with a component of a vector in Hilbert space is considered to be of minor importance compared with maintaining the universal validity of the concept of Hilbert space itself. Similarly with regard to all the other assumptions that we have outlined here.

We can however ask whether symmetry and beauty are always a sure sign that we have reached an ultimate truth that can never be altered through further enquiry. Indeed in our approach, which we shall explain in more detail in the next chapter, we argue that there is no reason to assume the ultimate truth of any particular feature of knowledge, however beautiful we may feel it to be. Moreover from our discussion of the many-worlds interpretation, which shows that it has not yet fulfilled its claims to economy of principles, it does not seem to be justified to dismiss the assumption of new concepts in our interpretation (such as particle variables) that go beyond what can be accommodated in terms of Hilbert space.

As we have already remarked earlier we feel that the many-worlds interpretation has interesting possibilities, but that its presentation would be greatly clarified it if were frankly admitted that some of its basic hypotheses are of a highly speculative nature. Indeed the main motivation behind the many-worlds interpretation is, in our view, not really the minimisation of principles and hypotheses. But rather it is to retain Hilbert space as the ground for the basic conceptual structure so that everything in the theory can be formulated in terms of Hilbert space. It could be valuable to develop such a theory but it has to be said that before this is actually done in an adequate way, many as yet unresolved difficulties have to be dealt with. We shall go into this question further in the succeeding chapters.

13.7 References

1. Y. Ben-Dov, 'Versions de la Mécanique Quantique sans réduction de la Fonction d'Onde. Le Théorie d'Everett et l'Onde Pilot', Thèse de Doctorat, Octobre 1987.
2. L. E. Ballentine, *Found. Phys.* **3**, 229 (1973).

3. J. S. Bell, *Speakable and Unspeakable in Quantum Mechanics*, Cambridge University Press, Cambridge, 1987.

4. L. N. Cooper and D. van Vechten, *Am. J. Phys.* **37**, 219–227 (1969).

5. B. d'Espagnat, *Conceptual Foundations of Quantum Mechanics*, chapter 20, Benjamin, Reading, Mass., 1976.

6. H. Primas, *Chemistry, Quantum Mechanics and Reductionism, Sec. 3.6, Lecture Notes in Chemistry*, Springer, Berlin, 1981.

7. H. Stapp, *Found. Phys.* **10**, 767 (1980).

8. H. D. Zeh, 'On the Irreversibility of Time and Observation in Quantum Theory', in *Proc. Int. School of Phys. 'Enrico Fermi', Course 49: 'Foundations of Quantum Mechanics'*, ed. B. d'Espagnat, Academic Press, New York, 1971, 263–273, and also *Found. Phys.* **3**, 109 (1973).

9. H. Everett, *Rev. Mod. Phys.* **29**, 454–462 (1957).

10. B. S. DeWitt and N. Graham, *The Many-Worlds Interpretation of Quantum Mechanics*, Princeton University Press, Princeton, New Jersey, 1973, 155–165.

11. E. Wigner, 'Remarks on the Mind-body Question', in *The Scientist Speculates*, ed. I. J. Good, Heinemann, London, 1961, 284–302.

12. H. Everett, 'The Theory of the Universal Wave Function', in *The Many-Worlds Interpretation of Quantum Mechanics*, ed. B. S. DeWitt and N. Graham, Princeton University Press, Princeton, New Jersey, 1973, 64.

13. E. J. Squires, *Eur. J. Phys.* **8**, 171 (1987).

14. B. S. DeWitt, 'Quantum Mechanics and Reality', in *The Many-Worlds Interpretation of Quantum Mechanics*, ed. B. S. DeWitt and N. Graham, Princeton University Press, Princeton, New Jersey, 1973, 160.

15. J. A. Wheeler, *Rev. Mod. Phys.* **29**, 151–153 (1957).

16. N. Graham, 'The Measurement of Relative Frequency', in *The Many-Worlds Interpretation of Quantum Mechanics*, ed. B. S. DeWitt and N. Graham, Princeton University Press, Princeton, New Jersey, 1973, 229–253.

17. A. Einstein, 'Autobiographical Notes', in *Albert Einstein. Philosopher-Scientist,* ed. P. A. Schilpp, Cambridge University Press, Cambridge, 1969, 665.

18. H. Everett, 'The Theory of the Universal Wave Function', in *The Many-Worlds Interpretation of Quantum Mechanics*, ed. B. S. DeWitt and N. Graham, Princeton University Press, Princeton, New Jersey, 1973, 116.

19. *Ibid.*, p.117.

20. D. Albert and B. Loewer, *Nous* **23**, 169 (1989).

21. D. Bohm, *Quantum Theory*, Prentice-Hall, London (1951). Also available in Dover Publications, New York, 1989.

22. A. J. Leggett, *Prog. Theor. Phys. Supp.* **69**, 80 (1980).

23. M. Gell-Mann and J. B. Hartle, 'Quantum Mechanics in the Light of Quantum Cosmology', in *Proc. 3rd Int. Symp. Found. of Quantum Mechanics*, ed. S. Kobyashi, Physical Society of Japan, Tokyo, 1989.

24. B. d'Espagnat, *Conceptual Foundations of Quantum Mechanics*, chapter 23, Benjamin, Reading, Mass., 1976.

25. D. Deutsch, *Int. J. Theor. Phys.* **24**, 1–41 (1985).

26. L. E. Ballentine, *Found. Phys.* **3**, 229 (1973).

27. H. Everett, 'The Theory of the Universal Wave Function', in *The Many-Worlds Interpretation of Quantum Mechanics*, ed. B. S. DeWitt and N. Graham, Princeton University Press, Princeton, New Jersey, 1973, 100.

28. H. D. Zeh, *Found. Phys.* **9**, 803–818 (1979).

29. H. P. Stapp, 'Quantum Measurement and the Mind-Brain Connection', in *Symposium on the Foundations of Modern Physics*, ed. P. Lahti and P. Mittelstaedt, World Scientific, Singapore, 1990, 403–424.

Chapter 14
Extension of ontological theories beyond the domain of quantum mechanics

14.1 Introduction

Thus far our discussion has centred on various interpretations of the current quantum theory. But one of the implications of an ontological approach is that within its conceptual framework it is possible to contemplate certain modifications and extensions of the mathematical laws that go beyond the domain of the current quantum theory. In this chapter we shall discuss a number of suggestions along this line made by various authors and in the last section (14.6) we shall also include some of our own proposals for doing this.

14.2 Our general world view, including our attitude to physical and mathematical aspects of basic concepts

Before going on to a detailed discussion of our proposals for the extension of quantum mechanics in new ways, we would like to give the outlines of our general world view. This will help make more clear what our motivation is in developing our theories. In addition it should help to clarify certain differences between our attitude towards basic physical and mathematical concepts and that which seems to be behind most of the other approaches.

We shall begin by discussing the latter question first. Thus we have seen in the previous chapter, for example, that the basic attitude in the many-worlds interpretation is to take the present general mathematical formulation of quantum mechanics as the essential truth and to try to derive the physical interpretation as something that is implicit in the mathematics. Many speculative new assumptions have been made within this mathematical framework, but the first priority has been to retain this framework as the ground for the general conceptual structure. Some of the proponents of the many-worlds interpretation, for example, Deutsch [1] as well as Gell-Mann and Hartle [2], whose work will be discussed further on in this chapter, are even ready, at least in principle, to consider extending their theories beyond the current quantum mechanics. In addition the same may be said of Ghirardi, Rimini and Weber [3], whose work we will also consider later in this chapter. However, even these authors are in essential agreement that what should not change is the requirement that the Hilbert space (and its natural extension to an infinity of dimensions) would be the basic structure in terms of which all other concepts must be expressed.

This attitude towards the mathematics has not always been dominant in physics. Indeed until the present century, the physical concepts were, for the most part, considered as primary, while mathematical equations were regarded as providing a more detailed and precise way of talking about these physical concepts. However, during the twentieth century with the advent of quantum theory (and to a lesser extent, relativity) the mathematics became much more highly developed and the physical interpretation much more abstract and indirect, as well as much less clear. As a result there was a constantly increasing focus on the mathematics, while the physical ideas were given less and less importance. Thus during the 1920s Sir James Jeans said that God must be a mathematician, implying by this that the universe was constructed on a mathematical plan and that its essence is best grasped in terms of the mathematics itself. Later, Heisenberg went much further along these lines and said very explicitly that the essential truth was in the mathematics. This view has become the common one among most of the modern theoretical physicists who now regard the equations as providing their most immediate contact with nature (the experiments only confirming or refuting the correctness of this contact). So without an equation there is really nothing to talk about. On the other hand, in the past we began talking about our concept of physical reality and used the equations to talk about them.

One of the reasons that our interpretation of the quantum theory may not have been so well understood is perhaps that it is not commonly realised that we have a rather different attitude to the mathematics. Our approach is not simply a return to the notion that the mathematics merely enables us to talk about the physical concepts more precisely. Rather we feel that these two kinds of concepts represent extremes and that it is necessary to be in a process of thinking that moves between these extremes in such a way that they complement each other.

We fully accept that progress can be made from the mathematical side alone, but we feel that to stick indefinitely to this approach is too limited. For example, starting with Schrödinger's equation expressed in the

Hamilton-Jacobi form, we were led in chapter 3 to a new physical concept of a particle which is affected by a quantum field in a way that depends only on the form of the latter and not on its amplitude. With the aid of the concept of active information implied by this behaviour, we can first of all obtain a coherent explanation of the quantum process which gives us new insights into the quantum theory and enables us to understand it in a different way. But we do not regard these physical concepts as mere imaginative displays of the meaning of our equations. Instead, as we shall see in this chapter, we can then move to the other side of this process of thought and take our physical concepts as a guide for the development of new equations. In this way we are engaged in at least the beginning of a kind of thinking in which the clue for a creative new approach may come from either side and may flow indefinitely back and forth between the two sides.

It should be clear from the above that we do not expect to come to the end of this process of discovery (for example, in a form that is currently called the "Theory of Everything"). Rather our view is that nature in its total reality is unlimited, not merely quantitatively, but also qualitatively in its depth and subtlety of laws and processes (see D. Bohm [4]). Our knowledge at any stage is an abstraction from this total reality and therefore cannot be expected to hold indefinitely when extended into new domains.

In any case it is evident that there is no way to prove that any particular aspect of our knowledge is absolutely correct. Indeed no matter how long it may have demonstrated its validity, it is always possible that later it may be found to have limits. For example, Newtonian physics, which held sway for several centuries, was found to be limited and gave way to quantum theory and relativity. Yet just before this revolution took place at the end of the nineteenth century, Lord Kelvin, one of the leading physicists at the time, advised young people not to go into physics because it was finished and only subject to refinement in the next decimal places. Nevertheless he did point to two small clouds on the horizon, the negative results of the Michelson-Morley experiment and the problems of black body radiation. It must be admitted that he chose his clouds correctly, though he totally underestimated their importance. Nowadays physicists are similarly talking about an ultimate unified theory including the four forces and perhaps strings, supersymmetry etc. that would explain everything. However, it can hardly be said that the clouds in this picture are only two and restricted to being small.

Our notion of the qualitative infinity of nature can usefully be brought out by considering the role of the atomic hypothesis of Democritus which we have already discussed in chapter 12, section 12.8. Not only did this hypothesis show the importance of speculative new concepts going far beyond what could be demonstrated at the time of its proposal, but it also exhibited a further key feature that we would like to emphasise here. This is that deeper explanations often imply the limited validity of what were previously accepted as basic concepts, which later are then recovered only as approximations or limiting cases.

This relationship of concepts is indeed typical of the whole of our experience. For example, as we go round a circular table, what we see immediately is an ever changing elliptical shape. But we have learned to regard this changing shape as a mere *appearance*, while the *essence*, i.e. the true being, is considered to be a rigid circular object. This latter is known first in thought, but later this thought is projected into our immediate experience so that the table even *appears* to be circular.

But further investigation shows that this object is not solid and eventually discloses an atomic structure. The solid object now reverts in our thought to the category of an appearance, while the essence is the set of atoms out of which it is constituted. However, deeper studies then showed that even the atom is an appearance and that the essence is a nucleus surrounded by moving planetary electrons. And later still even these particles were seen to be appearances, while the essence was a set of yet more fundamental particles such as quarks, gluons, preons, or else sets of excitations of strings, etc. But, in all of this development of our knowledge, it seems that whatever we have thought of as matter is turning more and more into empty space with an ever more tenuous structure of moving elements. This tendency is carried further by quantum field theory which treats particles as quantised states of a field that extends over the whole of space.

What has been constant in this overall historical development is a pattern in which at each stage, certain features are regarded as appearance, while others are regarded as of an essence which explains the appearance on a qualitatively different basis. But what is taken as essence at any stage, is seen to be appearance of a still more fundamental essence. Ultimately everything plays both the role of appearance and that of essence. If, as we are suggesting, this pattern never comes to an end, then ultimately all of our thought can be regarded as appearance, not to the senses, but to the mind.

What science is aiming for is that these appearances be correct, i.e. that the actions flowing from them, such as experiments, be coherent with what the appearances would imply. Incorrect appearances are thus either mistaken or illusory. The progress of science then implies the elimination of mistakes and illusions and the development of new sets of related appearances at various levels. Thus with the table to have the appearance of an elliptical object is correct only in a very limited sense. By bringing in the thought that the object is a rigid circle, we obtain a more nearly correct overall set of appearances. The thought of the atomic structure then makes the whole set still more nearly correct and so on. Thus what we are doing is constantly extending the appearances with the aid of thought. The ultimate reality is unlimited and unknown, but its successive appearances serve as an ever more accurate guide to coherent action in relation to this reality.

Another way of looking at the notion of appearances is to note that appearances basically are what arise in our *perception* of the world. As we have seen, the appearances in sense perception give rise to inferences about an essence that might be their origin, but this essence, which is seen in thought, turns out to be yet another appearance and therefore still part of our overall perception. No matter how far we go, we are therefore involved basically in perception. Our theories are not primarily forms of knowledge about the world but rather, they are forms of insight that arise in our attempts to obtain a perception of a deeper nature of reality as a whole. As such we do not expect their development ever to come to an end, any more than we would look forward to a final sense perception. This view of theory as perception can be found in Bohm [5].

At any stage in the development of perception as described above, there is always further room for new concepts which serve as ways of looking at reality. Such concepts may, in our approach, be either physical or mathematical or both. Thus we do not suppose that physical concepts are merely something to be put in correspondence with the basic mathematical concepts. Nor do we suppose that the mathematical concepts are necessarily subordinate to basic physical concepts. Rather both of these are appearances and their relationships may provide a yet more comprehensive appearance which guides our action towards the unlimited and unknown reality. The imaginative qualitative concept is therefore, in the long run, just as key a feature of this overall appearance as is the precise and abstract mathematical concept. The two together not only present a more comprehensive appearance than either one alone could do, but each can serve as a clue for further development in the other.

In all this, scientific theory may even go so far as to throw some light on the observer, to whom all these appearances are supposed to be taking place. But as with everything else, even such theories are only further extensions of appearances, so to speak, which give the observer a mental mirror in which he or she is reflected. There will therefore be no final theory of the observer. The ultimate nature of awareness is unlimited and unknown, like that of the universe as a whole of which it is a part. In the next chapter, we will make further suggestions as to how the universe can include observers, which are along the lines described above.

It may be readily seen that this general approach implies a certain kind of limit to determinism. For in it, each theory is evidently an abstraction from a totality that is inherently unlimited, both qualitatively and quantitatively. As such, it evidently cannot be entirely self-determined, because it always depends in various ways on what has been left out. So the necessity of its self-determination is only relative and not absolute.

Even here this self-determination may include unstable chaotic orders which approximate randomness. However, even the chaotic or stochastic character may be contingent and determined by conditions in domains not covered by the particular theory in question. For example, random variations in the trajectory of a Brownian particle may be partially or even totally determined by atomic motions at a deeper level.

So ultimately our overall world view is neither absolutely deterministic nor absolutely indeterministic. Rather it implies that these two extremes are abstractions which constitute different views or aspects of the overall set of appearances. Which view is appropriate in a given case will depend both on the unknown totality and on our particular mode of contact with it (e.g. the kinds of experiments we are able to do). So as the situation changes, different views are constantly called for. But the unknown and unlimited essence is not restricted to any of these views. Rather it may be thought of as somewhere between them and ultimately beyond them, as indeed it is beyond what can be captured in thought, which is always limited to some abstraction from the totality. (For a more detailed discussion of this approach, see Bohm [5].)

The above is a special case of how even our basic concepts and categories of thought are ultimately appearances. If this is so, what then is the point of making an ontological interpretation of the quantum theory? To answer this, we first point out once again that every theory is not only a mere appearance, but that it is more, in the sense that its basic concepts must be said in some sense to reflect reality *within its own domain*. In other ontological theories suggested thus far in physics, it is either implied or asserted that such basic concepts correspond to *independently* existent realities, for example, not dependent on context or deeper levels of being. In our approach, however, the basic concept of a given theory may reflect a reality that is inherently dependent either on context or on deeper levels or on both. For example, the elliptical appearances of a circular object may be seen as realities by considering the light coming from the object to the eye and how the eye makes an elliptical image which is in some way carried back into the brain. This we may consider to be an essential part of the reality corresponding to the elliptical appearance. Clearly this reality is not independently existent but depends on a context including the whole process that we have mentioned, as well as much more and particularly on the circular object which is taken to be the essential meaning of the elliptical appearances. But then the circular object is found to be dependent for its existence on a wide range of contextual parameters, e.g. temperature, and especially on the atomic constituents which are considered to be its essence. We find a similar behaviour as we go through a series of deeper essences and appearances.

In particular as we come into the quantum domain, we find that it is necessary to state the relationship between essence and appearance explicitly by considering a totality which includes the measuring instruments as well as what is measured. In this case, the appearances show themselves in the instrument, while the essence is the whole quantum ontology of quantum fields and particles. The reason why the totality must explicitly include the appearances is that there is an irreducible participation of object and observing apparatus in each other in the quantum domain, while in the domains treated by previous theories, the relation between essence and appearance can be neglected or else taken as only implicit.

Because essence and appearance had thus to be considered as a totality, this created a serious problem for the extension of the traditional approach into the domain of the quantum theory. When the quantum theory was first developed, people did not see how this question could be approached ontologically. Instead there was developed a basically epistemological approach of Bohr [6], Heisenberg and von Neumann in which the fundamental concepts of the quantum theory were taken as representing only knowledge. The feeling that any other approach was impossible seems to have been grounded in the assumption that any ontological approach would have to have a final essence that was not dependent on anything else at all, so that it would never be necessary to consider essence and appearance as a totality, as we have done.

To sum up, although each form of thought of the essence *is* an appearance, it also reflects a reality that is however always dependent for its existence as well as for its qualities and properties on broader contexts and deeper levels. This is one of the senses in which the theory is ontological, in spite of the fact that all thought gives appearances. In addition, our approach provides a reflection of reality that is in the *form* of a totality, including the measuring instruments along with everything else. This does not mean, of course, that it tells us everything about the universe or that it gives a final picture. Rather, as we shall bring out in more detail in the next chapter, one may here use the analogy of a hologram, in which any part of the record gives us information about the whole object rather than a part of it. This feature of our ontological approach is evidently especially important in cosmology and it is also important if it is ever our intention to understand the relation between the observer and the observed.

In addition to all of this, our particular ontological interpretation has the further advantage that it can be grasped more readily in an intuitive and imaginative way than is possible with a basically epistemological interpretation such as, for example, that of Bohr. Besides, as we shall see in section 14.6 of this chapter, it makes possible certain kinds of extensions of the basic equations which would not be possible in a purely epistemological approach. Thus it opens the mind to exploration of new domains which would not be discernible without the ontological concepts. It is not that we want to dismiss the epistemological approach totally, but rather that we feel that both approaches have to be pursued and perhaps ultimately brought into relationship.

Finally the view that our theories constitute appearances does not deny the independent reality of the universe as a whole. Rather it implies that even the appearances are part of this overall reality and make a contribution to it. What we emphasise is, however, that the content of the theory is not by itself the reality, nor can it be in perfect correspondence with the whole of this reality, which is infinite and unknown, but which contains even the processes that make theoretical knowledge possible. It is our view that this approach provides a more coherent understanding of the relationship of theory to the reality towards which it guides us, than do other approaches that are currently available.

14.3 The Ghirardi, Rimini and Weber approach

As we indicated in the introduction, a number of new ontological approaches in principle going beyond the current quantum theory have been suggested which have in common that they aim to preserve the notion that Hilbert space (with its natural extension to an infinity of dimensions) is the basis of the entire conceptual structure. One of the most interesting of these is the unified dynamics for microscopic and macroscopic systems of Ghirardi, Rimini and Weber (GRW) [3].

We give here a simplified treatment due to Bell [7]. The wave function $\psi(\boldsymbol{r}_1, \boldsymbol{r}_2, \ldots \boldsymbol{r}_n, t)$ is assumed to evolve according to Schrödinger's equation. However, there is a further process which reduces the spread of the wave function over each particle coordinate, $\boldsymbol{r}_n$, to something of the order of a new universal constant α. This is in a certain way similar to the process of collapse in the von Neumann interpretation, but the differences are that (1) this is happening in general and not just in a measurement process, and (2) this process constitutes a well-defined extension of the quantum mechanical laws of the development of the wave function.

At first sight it might seem that if α is chosen small enough to bring about the kind of localisation needed for a classical object, then this might contradict the quantum mechanical predictions of interference over fairly large distances. (For a more detailed account of this point see Shimony [8] and Weinberg [9].) However, it should be recalled that quantum mechanical interference properties over long distances have been extensively tested for the case of a single particle [10,11] and at most for a few particles [12].

We can avoid this difficulty by supposing that the rate of decay, λ, depends on the number of particles, N, involved. Thus we can write $\lambda = N/\tau$, where τ is a characteristic time which is a new constant of nature. τ can be chosen such that the rate of decay of the original wave function is negligible for a system of a few particles and quite rapid for systems with many particles, e.g. of the order of 10^{20}.

To simplify the discussion, we assume a process in which the spread of the wave function is reduced to something of the order of α around a centre $\boldsymbol{x}$ which is distributed at random. For the nth particle this process can be regarded as transforming the original wave function $\psi(\boldsymbol{r}_1, \boldsymbol{r}_2, \ldots \boldsymbol{r}_n, t)$ into a new wave function

$$\psi'(\boldsymbol{x}, \boldsymbol{r}_2, \ldots \boldsymbol{r}_n, t) = \frac{f(\boldsymbol{x} - \boldsymbol{r}_n)\psi(\boldsymbol{r}_1, \boldsymbol{r}_2, \ldots \boldsymbol{r}_n, t)}{R_n(\boldsymbol{x})}. \tag{14.1}$$

The factor $f(\boldsymbol{x})$ is normalised according to

$$\int \mathrm{d}^3\boldsymbol{x}\, |f(\boldsymbol{x})|^2 = 1. \tag{14.2}$$

GRW suggested that $f(\boldsymbol{x})$ be a Gaussian, for example

$$f(\boldsymbol{x}) = K \exp\left[-\frac{\boldsymbol{x}^2}{2\alpha^2}\right]. \tag{14.3}$$

$R_n(\boldsymbol{x})$ is a normalising factor where

$$|R_n(\boldsymbol{x})|^2 = \int \frac{\mathrm{d}^3\boldsymbol{r}_1 \ldots \mathrm{d}^3\boldsymbol{r}_n\, |f(\boldsymbol{x} - \boldsymbol{r}_n)|^2}{|\psi|^2}. \tag{14.4}$$

The centre $\boldsymbol{x}$ is distributed at random with probability $|R_n(\boldsymbol{x})|^2$.

We are assuming that each particle is undergoing this process of narrowing down the range of the wave function so that after a time τ/N, the wave function will be reduced more or less to something having a range, α, for each particle. Interference will then be negligible beyond this range, but will be as predicted by quantum mechanics within this range, provided of course that we are considering times appreciably longer than τ/N. It should be clear that the above assumptions will enable us to satisfy our requirements that large scale systems (such as the observing apparatus) will be fairly localised, while systems of a few particles need not be localised.

It is possible now to show how the GRW approach allows a consistent treatment of the measurement process as a special case of the general process in which the wave function of a system of many particles narrows down to a fairly well- defined region. As shown in chapter 2, section 2.4, the wave function, after the interaction between apparatus and the observed system, is

$$\Psi = \sum_n C_n \psi_n(\boldsymbol{x}) \phi_n(\boldsymbol{y}_1, \ldots \boldsymbol{y}_R, \boldsymbol{Y}) \tag{14.5}$$

where $\boldsymbol{x}$ represents the coordinates of the observed system (of which there are only a few), while $\boldsymbol{y}_1, \ldots \boldsymbol{y}_R$ represent all the internal co-ordinates of the apparatus. $\boldsymbol{Y}$ represents the centre of mass co-ordinate of the apparatus which is directly observed at an essentially classical level (e.g. the pointer position). After the interaction is over the ϕ_n do not overlap in the coordinate $\boldsymbol{Y}$, but as pointed out in chapter 2, section 2.4, there is still potentiality for interference as long as we have a linear superposition of this kind. However, if we bring in the GRW process of narrowing down the wave function to something with a range of α and if the separation in the centre of mass coordinate, $\boldsymbol{Y}$, of neighbouring wave packets is considerably greater than α, then this possibility of interference will be destroyed.

To see how this happens we note that the ϕ_n and ψ_n undergo independent GRW processes and that the effect of this process on ψ_n is negligible because there are only a few particles in the observed system. In general we may expect that the ϕ_n to have had a fairly narrow range of $\boldsymbol{y}_R$ even before the measurement took place and that this range will not change significantly in the measurement process (which does not in general greatly affect the internal coordinates of the apparatus).

As we have pointed out earlier, it is only the spread of the centre of mass coordinate $\boldsymbol{Y}$ that is affected. The GRW process will cause the collapse of all but one of the ϕ_n to leave a single result $\phi_m(\boldsymbol{y}_1 \ldots \boldsymbol{y}_r, \boldsymbol{Y})$ with a well-defined 'pointer' reading $\boldsymbol{Y}$. In this way we obtain a definite result. Clearly this process will not lead to paradoxes such as the Schrödinger cat paradox.

If we consider an ensemble of experiments we will find that in some cases the wave function (14.5) will collapse to one packet and in other cases to another, so that we will be left with a random distribution of well-defined results. It is easily shown that the probability of the mth result is $|C_m|^2$, in agreement with the usual quantum mechanics.

It is still necessary, however, to be sure that the GRW modification of the dynamics of microscopic constituents does not imply physically unacceptable consequences for the dynamics of the whole system. GRW go into this question in some considerable detail, and show that no such unacceptable consequences will arise. In doing this, they are led to suggest values for the universal constants

$$\alpha \simeq 10^{-5} \text{ cm.}$$
$$\tau \simeq 10^{16} \text{ sec.}$$

Objects containing many particles will then be localised to within 10^{-5} cms. A one-particle system will not be localised in less than something of the order of 10^9 years. But a system of 10^{20} particles will be localised in 10^{-4} secs.

In comparison, our own approach introduces a random distribution of particle positions, ξ, with a probability density equal to $|\psi(\xi)|^2$. Whether one regards the GRW approach as having a greater or lesser economy of assumptions depends primarily on whether or not we give first priority to retaining Hilbert space as the basic mode of expression of all concepts.

However, we can add that thus far our approach is well defined and thoroughly worked out, whereas the assumptions of GRW have evidently a great deal of arbitrariness in them. Of course one might adopt the attitude that this arbitrariness may eventually be reduced by the development of further new ideas. But here one should note that it seems difficult to include the GRW approach in a relativistic context. For it has all the nonlocality of ordinary quantum theory and possibly some additional ones. Therefore it seems likely that further new developments of this approach may lead to the need to assume a deeper non-relativistic level, rather as we have had to do in our own interpretation as explained in chapter 12, section 12.8.

Leggett [13] has yet another approach which is in considerable sympathy with that of GRW. As we have already indicated in a previous chapter, he wants to introduce a new concept of complexity in the form of what he calls disconnectivity. In a macroscopic object it is to be expected that this disconnectivity is very large and he would anticipate that in such cases Schrödinger's equation should change significantly in such a way as to remove interference between macroscopically different states. However, he has not yet proposed a more definite form for such a theory because he wants to find some experimental evidence for a breakdown of quantum mechanics in the macrodomain to guide his suggestions.

14.4 Stapp's suggestions for regarding quantum theory as describing an actual process

A rather different approach to an ontological theory of quantum mechanics has been suggested by Stapp [14]. (Stapp bases his approach on *S*-matrix theory, but there seems to be no inherent necessity for doing so.) His aim is to develop a theory in which quantum mechanics would describe an actual process taking place without any necessary reference to human observers. In doing this he is inspired by the ideas of Whitehead concerning nature as process. Whitehead [15] works in terms of the basic concept of an actual occassion, or as Stapp might call it, an actual event. In particular, that which is actual is represented, not by an infinitely thin space-like surface advancing through the space-time continuum, but rather by a sequence of actual processes, each of which refers to a bounded space-time region. In such a process a given event, n, is represented by a restriction on the set of classical fields allowed in the bounded space-time region R_n. The basic concept is therefore neither becoming in three-dimensional space nor being in the four-dimensional world, but rather it is becoming in the four-dimensional world.

We have referred to regions of space-time R_n, but ultimately Stapp wants to derive even these from the quantum theory. He suggests this may be achieved in a manner mentioned in the previous chapter (section 13.6). He has shown [16] that the 'soft' photons emitted by matter have well-defined 'classical' parts described by coherent states. As we have already said in a previous chapter (chapter 8), this shows that the quantum theory may contain a classical domain within it. Such a classical domain is what makes it possible to define the regions R_n in a physical way.

It remains to define what is to be meant by an event within the process assumed by Stapp [14]. He considers process to consist of a well-ordered sequence of actual events. Just prior to event n, there are several possibilities of what may be actualised. Let these possibilities be labelled by the set e_n. Only one of these events will become actualised. Let it be called a_n. For each a_n, there is a corresponding region R_n. Moreover for each possible event a_n, there is a projection operator $P_n(a_n)$ which restricts the classical field inside the region R_n to some subset of the set of all conceivable classical fields in R_n.

Let $S(e_n)$ be the S-matrix subject to the condition that classical fields be limited to those allowed by the product of projection operators

$$P(a_n)P(a_{n-1})P(a_{n-2})\cdots$$

Let d_{in} be the initial density matrix and let $d(e_n)$ be the density matrix transformed in the process of development described by the quantum theory. This will give $d(e_n) = S(e_n)d_{in}S^+(e_n)$. Then the probability for the event e_n is assumed to be

$$\frac{tr\ [d(e_n)]}{\Sigma\ tr\ [d(e'_n)]}.$$

It is clear that Stapp's basic idea also defines everything within the framework of Hilbert space. A basic new assumption in his approach is that certain kinds of actual events, e_n, correspond to specifiable projection operators $P(e_n)$ and that statistical averages of any particular kind of sequence of such events will be determined in the way described above by averaging these operators with d_{in}, the initial density matrix of the universe. The assumption that events correspond to projection operators, even though these are not measured, is evidently something new, not contained in the current quantum theory. In contrast, in our interpretation we assume, as we have already pointed out earlier, a particle position ξ with a probability density $|\psi(\xi)|^2$. As with the GRW approach, the decision as to which of these theories has a smaller number of hypotheses depends on whether or not we give a very high priority to taking Hilbert space as the basic concept.

One of the principal difficulties with Stapp's proposal is that the concept of an event has not been very thoroughly developed. First of all Stapp has not provided a general criterion for the kinds of projection operators that could represent events, especially of how 'fine grained' is the analysis that he will allow at this event level. (We shall discuss this question further in another context in the next section.)

Next it seems that the 'classical' parts of quantum mechanics should be described in terms of the movement of fermionic particles as well as bosons. The classical context should also include large scale systems and perhaps even large molecules as well as a theory of soft gravitons which would be needed to establish the notion of metric as contributing to events. It is conceivable, as Stapp himself hopes, that further new ideas might meet at least some of these requirements.* Indeed in the next section we shall discuss certain proposals of Gell-Mann and Hartle which do start along these lines and go some way to developing a theory so as to deal with questions of this kind.

14.5 The cosmological approach of Gell-Mann and Hartle

Gell-Mann and Hartle [2] have independently suggested a theory along lines similar to that discussed in the previous section which aims to provide an interpretation that covers the whole cosmological context.

These authors criticise traditional approaches in a way similar to that in which we do. In addition they especially emphasise the question of making an interpretation that is adequate for cosmology. In such a cosmological approach they say that there can be no *fundamental* division into observer

*Stapp [17] has made further suggestions along these lines in which the event is ultimately defined in consciousness.

and observed. Besides, there is no reason to assume a classical world as fundamental or outside the basic quantum theory.

They regard Everett [18] as the first one who suggested how to generalise the conventional interpretation so as to apply quantum mechanics to cosmology. He did this by showing how an observer can be considered as part of the system and how the observer's activities of measuring, recording and calculating probabilities could be described in quantum mechanics along lines that we have discussed in the previous chapter. However, in agreement with what we have said there, they regard the Everett analysis as incomplete. Gell-Mann and Hartle feel that it does not adequately explain the existence of the classical world or the meaning of the 'branching' that replaces measurement in assigning probabilities. It was, they say, a theory of 'many worlds', but it did not explain sufficiently how they were defined or how they arose.

In a short paper they briefly sketch their extension of the conventional formulation of quantum mechanics, which they claim meets these criticisms [2]. They regard it primarily as "an attempt at extension, clarification, and completion of the Everett interpretation". They admit the research is far from complete, but they regard their paper as a sketch of how it may become so.

We shall here give a summary of their general approach.

14.5.1 Histories

The theory of Gell-Mann and Hartle [2] has to be understood, from the very outset, as dealing with nothing less than the entire history of the universe from the beginning to the end. However, to get on with the exposition, they allow themselves in *this* paper to ignore the first few moments and to start at a point at which space, time and ordinary sorts of field variable are definable in their usual way. The initial quantum state of this universe is described by some density matrix ρ. Mathematical elements describing more specific information are represented by operators $O(t)$. They argue that it is sufficient to focus on projection operators $P(t)$ representing properties that at the time t are either present or not. In this, their position is similar to that of Stapp [14], discussed in the previous section, who also introduced projection operators, $P(e_n)$, in a similar role. Like Stapp, Gell-Mann and Hartle consider time sequences of such projection operators, averaged over the initial density matrix of the universe. And like Stapp, they suggest a means of using these sequences to define a classical domain within the structure of Hilbert space. The main difference is that they develop their views more extensively and systematically, and show in more detail how such objectives might be realised.

Stapp used projection operators corresponding to the quasi-classical coherent states of a field oscillator in order to represent possible sequences, of what he called actual events. Gell-Mann and Hartle use a different kind of projection operator for essentially the same purpose. What they do is to consider any one of the fields that are supposed to be basic in physics, e.g., as with Stapp, the electromagnetic. Their essential step is to obtain a classically adequate description of these fields represented by generalised coordinates $Q^i(t)$, not by using coherent states, but rather, by simply 'coarse-graining' the field variables defined, for example, at any point in space and time. Such coarse-graining only restricts their field values to a certain range, and says nothing about the further details within this range.

To this end they introduce sets of projection operators $[P_\alpha^k(t)^2]$ which are unity in a given range of field variables, Δ_α^k, and zero outside, where α denotes the range under consideration and k the particular set. These projection operators are expressed in terms of a 'square wave' function of $Q^i(t)$ which is zero outside the range Δ_α^k and unity inside. We denote this by $B_\alpha^k(Q^i)$. The corresponding matrix representation of the projection operator is then

$$P_\alpha^k\left(Q^i(t), Q'^i(t)\right) = B_\alpha^k(Q^i) B_\alpha^k(Q'^i).$$

The whole set of projection operators is $\left\{P_1^k(t), P_2^k(t), \ldots P_\alpha^k(t)\right\}$. If the total set of ranges, α, covers all possible values of the field variables, they call this set *exhaustive*. An exhaustive set evidently need not be a complete set of operators, indeed it may involve few of them.

They then go on to define sets of alternative histories. These consist of time sequences of exhaustive sets of alternative ranges,

$$\left\{P_{\alpha_1}^1(t_1), P_{\alpha_2}^2(t_2), \ldots P_{\alpha_n}^n(t_n)\right\}.$$

Such sequences provide coarse-grained descriptions of a set of possible histories with coarse-graining of each actual event represented by Δ_α^k and with the whole set given by

$$[\Delta_\alpha] = \left(\Delta_{\alpha_1}^1, \ldots, \Delta_{\alpha_n}^n\right).$$

A set of alternative histories is then obtained by allowing $\alpha_1 \ldots \alpha_n$ to vary over all possible values.

The extreme case of such a description is a *completely fine-grained history* which is specified by giving a complete set of operators at all times. This sort of history is used in Feynman's treatment of quantum mechanics in terms of a sum over histories, which begins by specifying the amplitude in terms of a particular basis of generalised coordinates. Gell-Mann and

Hartle suggest for this basis, the set of all fundamental field variables at all points of space. This amplitude is

$$\exp\left(iS\left[Q^i(t)\right]/\hbar\right),$$

where S is the action function which is determined by the Hamiltonian H (or alternatively by the Lagrangian L).

14.5.2 Decohering histories

It is essential that completely fine-grained histories as defined in the previous section cannot consistently be assigned probabilities. The reason is that to do this would be incompatible with fundamental quantum mechanical properties such as two slit interference. It is the basic claim of Gell-Mann and Hartle that by suitably defining coarse-grained histories they can obtain a consistent attribution of probabilities.

The condition that two projection operators can represent two distinct possibilities for actual events is not only that they commute, but that their product $P_1 \cdot P_2 = 0$. Thus we might require for alternative histories $[P_{\alpha'}] \cdot [P_\alpha] = 0$. But if we start from this requirement taken rigorously, we can never hope to achieve what Gell-Mann and Hartle set out to do because this set of operators is much too restricted. Their essential starting point is to note that all that is required is to satisfy this condition for the actual intial state of the universe. Since by definition the universe cannot be in a state other than its actual one, we are not required to satisfy this condition for any other state. A sufficient condition for the decoherence of two alternative histories α' and α may then be obtained by integrating the projection operators $[P_{\alpha'}], [P_\alpha]$ with the density matrix of the universe. This will yield what we may call the decoherence functional $D_{\alpha'\alpha}$:

$$\begin{aligned} D_{\alpha'\alpha} &= D\left([P_{\alpha'}],[P_\alpha]\right) \\ &= tr\left[P^n_{\alpha'_n}(t_n)\ldots P^1_{\alpha'_1}(t_1)\rho P^1_{\alpha_1}(t_1)\ldots P^n_{\alpha_n}(t_n)\right]. \end{aligned} \quad (14.6)$$

When the above is zero, the two sets of histories represented by α' and α will then correspond to two possible alternatives.

Gell-Man and Hartle argue however that even when $D_{\alpha'\alpha} \neq 0$, but is small, it is still possible, consistently, to regard α and α' as alternative histories. But if we do this, then the probability rules will only be approximately satisfied and the measure of the failure of the approximation will be indicated by the size of the off-diagonal elements. For example, in an arbitrary, but convenient way, we could define this measure more precisely as $Tr\left[|D - D_d|^2\right]$ where D_d is the diagonal part of $D_{\alpha'\alpha}$ and the matrices should be thought of as on the space of very fine-grained histories. A set of coarse-grained histories can then be said to decohere to a certain degree of approximation when the off-diagonal elements of $D_{\alpha'\alpha}$ are sufficiently small, i.e. when $Tr\left[|D - D_d|^2\right] \simeq 0$.

A further assumption is that under the above conditions the diagonal elements of $D_{\alpha'\alpha}$ are approximately equal to the probabilities $P([P_\alpha])$ of the corresponding histories, or

$$D([P_{\alpha'}], [P_\alpha]) \cong \delta_{\alpha_1'\alpha_1} \cdots \delta_{\alpha_n'\alpha_n} P([P_\alpha])$$

for each $[P_\alpha]$ in a set of approximate alternative histories.

To sum up then, the smallness of the off-diagonal elements is assumed to be the condition for a consistent assignment of approximate probabilities while the diagonal elements are assumed to give their approximate values.

14.5.3 On the meaning of decoherence

Gell-Mann and Hartle state that the (in general approximately) decoherent histories are what we may discuss in giving a meaning to quantum mechanics. As they further say

> decoherence thus generalises and replaces the notion of 'measurement'.... Decoherence is a more precise, more objective, more observer independent idea.... We may assign probabilities to various values of reasonable scale density fluctuations in the early universe, whether or not anything like a 'measurement' was carried out on them and certainly whether or not there was an 'observer' to do it.

Measurements will then have to be seen as special cases of the universal process in which there is a 'measurement situtation' such that certain objective and observer independent events in a given history are correlated to others in a different history, taking place in what we somewhat conventionally call a measuring apparatus. We shall return to this question later, but for the present we re-emphasise that measurements and observations play no fundamental role in the formulation of this theory.

However, if this approach is to be carried out consistently, it is necessary to deal with a number of points that require clarification. To begin with, in principle, any set of projection operators might be thought of as suitable for defining decohering sets of alternative histories. (Indeed if the subsets are exactly orthogonal, one can exhibit sets of histories that are exactly decoherent.) Which, among all these sets, are the ones that correspond to a sequence of actual events? It is important that we do not assume sets of actual events that would imply a contradiction with the quantum theory in any domain in which it is known to apply. But if we can actually find a set of decoherent operators $\{P_\alpha^k(t)\}$ that define reasonable histories then it is clear that within the approximations that are being used, no statistical results of the theory will be contradicted by the assumption that the $\{P_\alpha^k(t)\}$ represent alternative possibilities of sequences of actual events. For such a contradiction would arise only if there were significant 'interference properties' between these sequences and these would imply coherence.

This however raises the question of whether we can show that the theory actually implies sets of decoherent histories of the kind that we require. The decoherence functional depends, as has been seen, on the initial density matrix of the universe, ρ, but in addition it depends on the Hamiltonian H. For we have to follow the time-evolution of the operator $P_\alpha^k(t)$ in the Heisenberg representation back to the time of the origin of the universe, using the formula

$$P_\alpha^k(t) = e^{-iHt} P_\alpha^k(o) e^{iHt},$$

where $P_\alpha^k(t)$ is the projection operator at the time t and $P_\alpha^k(o)$ is the corresponding operator at the time t_o.

In effect the above procedure suggests a possible answer to the basic question that was raised in the many-worlds interpretation, i.e. what determines the preferred basis for the various branches that are to be realised in the process of proliferation of the many universes? The new idea here is that it is the evolution of the universe from its initial conditions which should provide this determination, and that this would remove an important ambiguity in the many-worlds approach.

It is clear that we cannot in the foreseeable future hope to carry out for the universe a calculation of all decohering sets of alternative histories. What Gell-Mann and Hartle do is to investigate specific illustrative examples of decoherence in more restrictive circumstances referring to the works of Joos and Zeh [19], Zurek [20] and Caldeira and Leggett [21] as well as Feynman and Vernon [22]. These authors investigate what is called dynamic decoherence, i.e. that which is brought about through the dynamic evolution of the quantum state. All of them deal with simplified systems such as a single harmonic oscillator with coordinates x coupled to a large set of oscillators constituting an 'environment'. The general result is that two histories corresponding to different trajectories, $x(t)$ and $x'(t)$, will decohere in very short times for typical situations such as having the 'environmental' oscillators in an initial state of thermal equilibrium.

Gell-Mann and Hartle feel that these illustrative examples made it very plausible that decoherence will be widespread in the universe for the familiar 'classical' variables. For example this would explain why we do not see the track of a projectile spreading out into quantum superpositions in its orbit.

14.5.4 The quasi-classical world

Ordinarily when we use coarse-graining, we are introducing something dependent on a context related either to our sensory perceptions or to our instruments, communications or records, all of which are characterised by some kind of ignorance resulting in some incomplete specification of the system. Yet as Gell-Mann and Hartle say, we have the impression that the universe exhibits a finer grained, presumably objective and observer-independent, set of decohering histories. These should make possible the definition of what they call a quasi-classical world governed approximately by classical laws. (This may be compared with our manifest world discussed in chapter 8.)

Gell-Mann and Hartle feel that this quasi-classical world should follow from the initial conditions of the universe and the Hamiltonian, with the aid of the projection operators that we have been discussing in the previous sections. To show this, they would like to find a way to make the question of the existence of a quasi-classical domain into a *calculable* question in quantum cosmology. For this they need criteria to measure how close a set of histories comes to constituting a quasi-classical domain. As they state in their article, they have not solved this problem to their satisfaction, but they discuss some ideas that they hope may contribute towards its solution.

To explain what Gell-Mann and Hartle have done in this regard, we note that their idea of a quasi-classical history is closely related to that of a broad wave packet in ordinary quantum mechanics. As is well known, the centre of such a packet follows the laws of classical physics, while the packet spreads out very slowly, so that it resembles somewhat a classical particle moving through space, at least for some time. However, Gell-Mann and Hartle describe this process differently by means of a set of projection operators $P^k_\alpha(t_1)P^k_\alpha(t_2)\ldots$. For the simple case of a free particle, for example, it is easy to show that the product $P^k_\alpha(t_1)P^k_\alpha(t_2)$ is large only when $P^k_\alpha(t_2)$ represents a coarse-grained region whose centre is displaced relative to that corresponding to $P^k_\alpha(t_1)$ by $p(t_2 - t_1)/m$ (where p/m is the mean velocity of the wave packet as determined by the wave function of the system and ultimately by the density matrix of the universe). It is clear then that products like $P^k_\alpha(t_1)P^k_\alpha(t_2)\ldots$ will be large only for histories in which the centres of the regions concerned are located more or less along the classical trajectories implied by the initial wave packet.

What has been gained by this procedure is that we have the possibility of obtaining a definite sequence of actual events (defined within the range Δ^k_α) in spite of the spread of the wave packets. However, what we in fact get is a whole set of possible sequences of actual events in which the probability will be large only for those sequences that are close to the classical trajectories.

It is along these lines that Gell-Mann and Hartle would like to show that one can construct a whole quasi-classical world corresponding to such histories. But there are certain problems that must be faced in order to do this. First of all, as we have already remarked, it is necessary not to make the history so fine-grained that the assumption of their existence would significantly disagree with the predictions of the current quantum mechanics in the domain that the latter has found to be valid. This can be achieved by going to extremely coarse-grained projection operators, because, as is well known, such coarse graining will in general make off-diagonal elements of matrices such as $D_{\alpha'\alpha}$ small. On the other hand, we must not have a graining so coarse that it does not define the classical domain to any significant extent. That is to say it must remain fine-grained enough to allow for a classical description that would not be lost in total ambiguity. So we have to assume that the degree of coarse graining is somehow objectively determinate at least within an order of magnitude as lying somewhere between the two extremes indicated above. This means, of course, that it is being assumed that quantum mechanics does break down when we reach this quasi-classical coarse-grained level. But we further assume that the point of breakdown is such that quantum mechanics holds for all practical purposes.

Gell-Mann and Hartle are not satisfied with assuming that there is a simple universal coarse graining that is fixed, the same for all circumstances. We can see a reason for this, for example, by considering that systems at high temperature can be given a much more fine grained description than is possible for systems at low temperatures without introducing any practically observable deviations from the quantum laws. Gell-Mann and Hartle have therefore tried to find a universal function to express their criterion in such a way that it could adapt to all changing circumstances that could arise on various branches of the possible histories of the universe. To this end they considered various information theoretic measures of what they call the 'classicity' of a given domain, which they hope will ultimately yield the precise definition of a quasi-classical world as a whole.

However, as we have already pointed out, they do not feel that, as yet, they have solved this problem to their satisfaction. But even if they could solve it, it would still require a basic new assumption in their approach going beyond current quantum theory. For no matter how subtle and variable their criterion for classicity may turn out to be, it will still have to have some kind of scale factor that determines the actual degree of objective coarse graining of the classical world under any given set of circumstances. This world corresponds in some rough way to a similar assumption that has been made by GRW. For they too have to assume that quantum mechanics breaks down at some level, but in a way that is unimportant for all practical purposes.

This sort of requirement is also characteristic of all forms of the many-worlds interpretation as Deutsch [1] has indeed suggested. For in assuming a preferred basis in which the universe actually splits or in which the awareness of the observer similarly splits, they are inevitably implying a breakdown of the quantum theory, though this may be unimportant for all practical purposes.

14.5.5 Measurement and observation

As we have pointed out earlier, Gell-Mann and Hartle regard measurement as a special case of a general process in the universe. When a correlation exists between the ranges of values of two operators of a quasi-classical world, there is what they call a 'measurement situation'. If the correlation has probability near unity, then the range of values of the one can be deduced from that of the other. All of this is, of course, objective and independent of the observer.

The above however discusses only measurements in the quasi-classical world. What about measurements of essentially quantum mechanical operators such as spin? Of course the spin states will not in general decohere by themselves. But in a measurement situation the different spin states are correlated with different classical histories. And these latter, of course, do decohere. The decoherence of the overall state will thus bring about the decoherence of the spin states that have been measured. This is very similar to what happens in our own interpretation, in which the non-overlap of the apparatus wave functions guarantees that there will be no interference between the spin states. In both cases the usual rules for probability will be recovered.

In the theory of Gell-Mann and Hartle, the observer is discussed in much the same way as was done in the Everett interpretation, except that all is now treated in a cosmological context. The observer's memory, awareness, etc. should, in principle, be included in the density matrix of the universe and it should also have some representation in terms of operators in the quasi-classical world. However, it is not enough to consider what happens at a given moment or period of history. It is necessary to consider how a set of observers might have evolved from the initial state of the universe. To discuss this question, Gell-Mann and Hartle refer to an observer as an "information gathering and utilising system (IGUS)". They claim that such a system can in principle be treated in terms of quantum mechanics as these authors have formulated it. That is to say, in some form of approximation, no matter how crude, an IGUS is said to employ this form of quantum mechanics in a self-referential way.

The possibility of this is based on the assumption that memory content, and that to which it refers, are correlated by the laws of quantum theory as applied in this cosmological context. It is further assumed by the authors that these laws provide for the evolution of IGUSs in such a way that these latter can make predictions and draw further inferences about the probabilities of coarse-grained histories. It is argued that this capacity was maintained and developed because it was favourable to their survival.

The reason why IGUSs focus on decohering variables is that, at least within this theory, these are the *only* variables for which predictions can be made. In addition they focus on the histories of a quasi-classical world, because these present enough regularity over time to permit prediction by relatively rudimentary algorithms that are easily evolved. Gell-Mann and Hartle thus effectively extend Darwinism and propose a possible basis for it in the laws of quantum mechanics.

If there are many significantly different quasi-classical worlds (as seems to be suggested in the many-universes point of view) one should better say that the IGUS evolves to exploit a particular quasi-classical world or a set of such worlds. This implies, of course, that the IGUS, including the human being, occupies no special place and plays no preferred role in the laws of physics. (This approach is in some sense closely related to certain forms of the anthropic principle [23].)

Bringing in the evolution of the IGUS plays a key part in how the theory of Gell-Mann and Hartle deals with measurement situations. For example, they consider the mutually exclusive character of simultaneous position and momentum measurements. What this means in their point of view is that the history in which an observer, as part of the universe, sets up a measurement correlation to the momentum, p, and one in which he does this for the position, q, are represented by two sets of decohering operators which are, however, mutually exclusive *alternatives*. (The decoherence is established by interacting with a thermal bath or with the $3^{\circ} K$ background radiation or with something similar.) Since these are mutually exclusive alternative possibilities for the universe as a whole, only one of them can be realised at a time in the development of the universe. Or more precisely, only one of them can arise on a particular branch of the evolution of the universe as a whole. They say however that "the choice of branch may ultimately be traceable to a quantum fluctuation that sets in motion a chain of events leading to one measurement situation or the other. (Some have even suggested that the human subjective impression of free will may be traceable to statistical or quantum fluctuations.)"

To present these ideas more vividly, we may say that, according to this theory, the universe evolved first to give rise to matter, then to life, and ultimately to human beings along with their societies, sciences, measuring instruments etc. Eventually in this evolution there appeared a physicist who set up a momentum measuring device and established a correlation between the subsequent history of this process and the momentum of a quantum system at the time t. It seems inescapable that the physicist's impression that he freely choose to set this measuring instrument up would have had to be illusory at least in terms of this theory. For in this theory nothing whatsoever can happen unless it is part of the evolution of a suitable branch of the history of the universe as a whole. (Therefore we can see no reason why the authors have attributed this kind of conclusion to 'some' and have not stated what they themselves believe in this regard.)

Similar considerations may be applied to the EPR-type of experiment. For example, at some part of the evolution of the universe, there may arise a molecule of spin zero, constituted of two atoms of spin $\hbar/2$, along with a physicist who sets up an instrument designed to disintegrate the molecule and to measure suitable components of the spins of the two atoms. The fact, for example, that he has apparently 'chosen' to measure the z-component of one of them was somehow implicit from the very early phases of the universe. Once this 'choice' began to unfold from the primeval explosion, then there could never be any question of what would happen if the universe had developed in such a way as to bring about a measurement of some other component of the spin.

Recall however that the paradoxical features of the EPR experiment arose because one assumed that before the measurement took place, it was always possible for the observer to 'choose' an arbitrary component of the spin (see chapter 2). If this choice has no meaning then all the conclusions commonly drawn from the EPR experiment will not follow. For example, the same quantum fluctuation from which emerged the physicist's choice to measure the z-component of the spin of one of the atoms, will also have given rise to the choice to measure a certain component of the spin of the second particle along with the correlation of the results of the two measurements. Or as Bell has put it, there is a complete 'rapport' established at the very beginning of the universe between the net results and the total measurement situation including the physicists who set the experiment up and carry it out. Some people have called this sort of situation the denial of the counterfactual [24]. But what the theory of Gell-Mann and Hartle adds is an explanation showing why this denial follows necessarily from their model of the universe.

14.5.6 Comparison with our approach

There are certain important similarities between our approach and that of Gell-Mann and Hartle, as well as, of course, important differences. One of the key similarities is that both approaches agree that the wave function (or the density matrix) is not a complete description of reality. Thus we bring in what we have called 'beables', which are either precisely defined particle trajectories or precisely defined motions of the field variables. Gell-Mann and Hartle bring in their coarse-grained projection operators out of which they construct sets of alternative histories. Let us recall that most of the unsatisfactory features of the conventional interpretation (which requires the complete linearity of the wave equation) arise from the assumption that the wave function (or the density matrix) provides a complete description of reality. It is this assumption that (in the context of linearity) requires us to postulate the collapse of the wave function with all its attendant paradoxical features, that have been discussed throughout the book. (Of course, if the wave equation is non-linear, as in the approach of GRW, paradox is avoided and the 'collapse' is explained in another way.)

A second key similarity in the two approaches is that, in both of them, the classical world is explained as part of the overall quantum world rather than simply assumed. Without such a possibility, we have to suppose that the universe is split into two qualitatively distinct parts even though we also tacitly assume that somehow the properties of the classical world are determined in principle from the quantum laws. To treat the classical level as flowing out of the quantum laws is clearly necessary if we are to consider the universe as a single whole. For such a theory requires that the properties of matter at the classical level shall be implicit in the basic general laws which are actually quantum mechanical. In particular such an approach is evidently needed if we are to include the process of measurement itself within the theory in a coherent way.

The main difference in the two points of view is that we bring in the concept of actual beables, e.g. particles and fields, while Gell-Mann and Hartle introduce their coarse-grained projection operators. A key point here is that although their projection operators are new concepts relative to the quantum theory in its current generally accepted form, they are nevertheless, as we have indicated earlier, still expressible as operators defined on a Hilbert space. On the other hand, our particle and field variables are not expressed as such operators in Hilbert space. Rather they are new concepts going outside this framework of Hilbert space altogether.

In our approach once we have assumed particle and field beables, no further fundamental assumptions are needed, nor as we have seen, are there serious problems involving the basic principles. Indeed even the probability assumption may be derived, as in chapter 9, either from the chaotic character of trajectories in the many-body system or from a stochastic process. The quasi-classical world (which in chapter 8 we have called the manifest world) comes out simply and naturally as that domain in which the quantum potential can be neglected. There is no sharp break between the quantum world and the classical world, but only a continual increase in the 'classicity' of the behaviour, as we approach the domains in which the quantum potential becomes less and less important.

On the other hand in the approach of Gell-Mann and Hartle, the simple assumption that sequences of projection operators correspond to actual histories is not enough. In addition, as we have seen, one has to assume an objective limit to the degree of fine grainedness of these operators. And besides, there is still the unresolved problem for finding a functional form for this limit that is suitable for all circumstances.

A further conceptual unclarity of their approach is that the reality status of the wave function or the density matrix is, at best, highly ambiguous. It would seem from what they have written that everything that is actual and real is contained in the quasi-classical world. This seems strange because all the basic physical properties do in fact come out of the wave functions (or the density matrix). Should not this therefore be regarded as a key factor of reality (as indeed it is in our interpretation)?

This ambiguity in the reality status of the wave function (or density matrix) is in a way a residue of the separation of the quantum world of the wave function and the classical world of the measuring instrument that was originally assumed by von Neumann (see chapter 2). The basic role of the observing instrument is now replaced by a similar role of the coarse-grained projection operators. Of course, the ambiguity in the cut between the measuring instrument and what is observed, has been removed because the latter has been given a universal significance (i.e. the universe contains a set of coarse-grained projection operators that in effect are constantly carrying out 'measurements' in a unique and objective way). Nevertheless there still remains the fundamental split between the two kinds of reality which many would feel to be undesirable.

A further key point which we emphasise once again here, is that our approach has the advantage of being much more capable of being grasped intuitively and imaginatively than is that of Gell-Mann and Hartle.

As we pointed out in chapter 13, section 13.2, the view shared by most of the protagonists of the many-worlds point of view is that they make no further serious assumptions. This view seems to be based on the extremely high priority that they give to retaining Hilbert space as the basic set of terms for the expression of all concepts. As a result, all the other assumptions that we have detailed in the previous chapter and in this section of the present chapter, do not seem to be very important. We acknowledge the possible value of basing a theory solely on Hilbert space concepts and although we shall indeed explore such a theory in the next chapter, we still do not feel that Hilbert space has such advantages (e.g. beautiful symmetries) as to imply the necessity of giving this concept so overriding a value. It is in our view very important to keep the options open to allow the explanation of all approaches, as long as these have certain specific advantages such as those that have been outlined here in the exposition of our interpretation. In addition it should be kept in mind that such new approaches allow new forms of idea to be explored, some of which we shall discuss in the next section, as well as in the next chapter.

Finally we would like to call attention to a more subtle difference between our approach and that of Gell-Mann and Hartle. In our interpretation, it is always, in principle, possible that there may be a modification coming from outside the total framework that we consider. For example, in the EPR experiment we supposed the observer could arbitrarily choose to reorientate the apparatus while the atoms were separating. Let us recall that it is this assumption which originally made the EPR experiment seem paradoxical and which ultimately led to the conclusion that there had to be a nonlocal connection between the atoms. In this case the modification might have been due to what has been called the 'free will' of the experimenter. But more generally, the central question here is not that of free will, but rather that of whether anything is possible that has a certain independence of the quantum mechanical framework which we are discussing. The assumption of free will would imply such a case of independence. But, of course, there could be all kinds of new levels of law, not only in physics but also, perhaps wherever life, mind, society etc. are involved. In other words, as we have brought out in section 14.2, there is no valid reason for assuming that all existence can be reduced to the laws of the quantum theory.

On the other hand, in the interpretation of Gell-Mann and Hartle it is, as we have seen, impossible for anything to happen that could not in principle be covered by some kind of extension of this theory based on the concept of Hilbert space plus the notion of the evolution of the universe that they have proposed. From this it follows that everything that could happen would, in principle, be implicit from the earliest phases of the universe. It is this feature that leads to the conclusion that nothing independent of the whole scheme can come in to alter the course of cosmic evolution (which latter leads even to the determination of the orientation of a piece of apparatus as something that was implicit in the quantum fluctuations of the early universe). As we have seen it is this feature that removes nonlocality, but the cost is a closure of our thinking, which, in our view, has no ground in fact and which may limit the development of our ideas in a counterproductive way.

14.6 Extension of our approach beyond the domain of current quantum theory

Thus far we have discussed a number of suggestions for ontological interpretations that go beyond the current quantum theory at least conceptually but nevertheless recover the latter as something that will be valid for all practical purposes. In this section we shall discuss some suggestions of our own in which we show how our proposals can also be extended beyond the current quantum theory. However, these extensions are not just purely conceptual but we show how in principle they can lead in new domains to qualitatively new theories and correspondingly new experimental implications.

The chief difficulty in carrying out this sort of extension of our approach is that there is such a wealth of possibilities and such a dearth of experimental clues that might narrow them down. For example, in our basic equations for a trajectory of a particle,

$$\frac{d\boldsymbol{p}}{dt} = -\boldsymbol{\nabla}(V+Q)$$
$$\boldsymbol{p} = \boldsymbol{\nabla}S,$$

it would be possible to bring in arbitrary forces, $\boldsymbol{F}$, to obtain

$$\frac{d\boldsymbol{p}}{dt} = -\boldsymbol{\nabla}(V+Q)+\boldsymbol{F}$$

as well as arbitrary additions, $\boldsymbol{\lambda}$, to the guidance relations to yield

$$\boldsymbol{p} = \boldsymbol{\nabla}S+\boldsymbol{\lambda}.$$

Any such changes could in principle fundamentally alter the consequences of the theory. Even though they might be chosen so that under conditions thus far investigated, the changes would be negligible for all practical purposes, the possibility would still be open that under new conditions these changes could no longer be neglected.

Other changes of this sort that might be considered would be to make the Schrödinger wave equation nonlinear and to introduce terms that would relate the Schrödinger wave function to the particle positions. (Again in such a way that significantly new results would follow only in new domains not yet investigated.)

One way to make Schrödinger's equation dependent on the particle positions (so that there would be a two-way relationship between wave and particle) can be seen by considering equation (14.1). In this equation, we can regard r_n as the actual position of the nth particle. From the same arguments as apply to the GRW approach, it would follow that the overall wave function would tend to 'collapse' towards the actual particle positions, so that, in a large scale system, the empty wave packets of our interpretation would tend to disappear.

Without some kind of experimental clue, we are at a loss in this sort of situation as to how to proceed. However, there have been a number of cases in the history of physics in which theoretical questions provided useful clues long before experimental physics was able to do so. A case in point is just the atomic theory of Democritus which we discussed earlier in chapter 2, section 2.8 as well as section 14.2 of this chapter.

The idea of Democritus was inspired by a consideration of a philosophical paradox in the very notion of movement, which was shown by Zeno and other members of the Eleatic school. They demonstrated logical inconsistencies, not only in the idea of movement itself, but also in the idea of an empty space in which movement could take place. From their demonstrations they concluded that only being *is* and that nonbeing is *not*, from which it followed that the concepts of becoming along with those of movement and empty space had no meaning.

Through contemplation of this paradox, Democritus was led to a resolution. He proposed that being was discontinuous in the form of distinct and separated atoms, each of which was an indivisible and eternal being. Democritus thus retained, in a certain restricted way, the Eleatic concept of being. But he then postulated that there was an empty space in which these atomic beings could move in such a way that the collective aspect of their motion then explained the large scale properties of apparently continuous matter with its continuous changes. On the other hand if we suppose that matter was truly continuous rather than only apparently so, then, as the Eleatics showed, there was no way in which its movement and transformation could have meaning.

One could, in principle, have seen at that time that the experimental detection of atoms depended basically on two related ways in which the laws of continuous matter could fail, firstly because atoms had a definite size and secondly because they had some non-zero free path between their successive collisions. These were, in fact, actually the ways in which the atomic theory was eventually verified experimentally.

Our proposed ontological explanation of the quantum theory has, as we have seen, also led to a certain paradox. For it implies nonlocality and this would seem to contradict relativity which is regarded as a theory that is equally as fundamental as quantum theory. But as we have seen in chapter 12, section 12.7, it is possible to propose a deeper theory of the individual quantum process which is not relativistically invariant and which nevertheless leads to Lorentz invariant consequences for all statistical results, as well as for the large scale manifest world. In this theory there is a preferred coordinate frame in which the instantaneous transmission of impulses is in principle possible, so that there is no contradiction with nonlocality for individual quantum processes. In other words, we say that underlying the level in which relativity is valid there is a subrelativistic level in which it is not valid even though relativity is recovered in a suitable statistical approximation as well as in the large scale manifest world.

In our discussion of this idea we have already suggested one way in which the theory might imply new experimental consequences. Thus, although there is no inherent limitation to the speed of transmission of impulses in this subrelativistic level, it is quite possible that quantum nonlocal connections might be propagated, not at infinite speeds, but at speeds very much greater than that of light. In this case, as explained in chapter 12, we could expect observable deviations from the predictions of the current quantum theory (e.g. by means of a kind of extension of the Aspect-type experiment).

It would seem then that already in this case a subrelativistic level would simultaneously be one that goes beyond the current quantum theory. But the notion of a general subquantum mechanical level would have to go beyond the current quantum theory in a much more thoroughgoing way. To explore such a possibility it is useful to consider once again the stochastic form of our ontological approach. We recall that in this approach a particle follows a trajectory which is random, but which is modified by a suitable osmotic velocity along with the action of the quantum potential. At the level of Brownian motion of atoms, for example, similar random trajectories are regarded as approximations to actual trajectories in which this randomness does not prevail at indefinitely short distances. Rather there exists a characteristic length, the mean free path, below which simple causal laws begin to have a dominant effect. This free path reveals itself in many experiments, e.g. those of diffusion, conductivity of heat, etc. Similarly we may propose that in our theory there is a minimum free path below which the trajectory ceases to be random. As the atomic free path is the first sign of a 'subcontinuous' domain in which the laws of continuous matter would break down, so the free path in our trajectories would be the first sign of a subquantum domain in which the laws of quantum theory would break down. It is implied of course that this would begin to occur for wavelengths of the order of the 'free path'.

The next sign of a breakdown of the quantum theory would be the discovery of some yet smaller dimension whose role might be analogous to the dimensions of an atom in the atomic explanation of continuous matter. We do not as yet know what this dimension is, but it seems reasonable to propose that it could be of the order of the Planck length, where, in any case, we can expect that our current ideas of space-time and quantum theory might well break down. The 'free path' would then be some multiple of this length.

At present further progress only seems to be possible through the use of the imagination to suggest new concepts that might permit a more precise formulation of these ideas. In this regard the situation is not very different from what it is in string theory, which is at present guided by speculative use of mathematical concepts, that also have little or no contact with experiment. As we have pointed out in section 12.8 of chapter 12, it took over 2000 years before the atomic theory of Democritus could achieve a more precise formulation. But we may reasonably hope that with the more rapid progress of science in the present era, it will take considerably less time.

14.7 References

1. D. Deutsch, *Int. J. Theor. Phys.* **24**, 1–41 (1985).

2. M. Gell-Mann and J. B. Hartle, 'Quantum Mechanics in the Light of Quantum Cosmology', in *Proc. 3rd Int. Symp. Found. of Quantum Mechanics*, ed. S. Kobyashi, Physical Society of Japan, Tokyo, 1989.

3. G. C. Ghirardi, A. Rimini and T. Weber, *Phys. Rev.* **D34**, 470–491 (1986).

4. D. Bohm, *Causality and Chance in Modern Physics*, Routledge and Kegan Paul, London, 1984.

5. D. Bohm, *The Special Theory of Relativity*, Addison-Wesley, New York, 1989.

6. N. Bohr, *Atomic Physics and Human Knowledge*, Science Editions, New York, 1961, 63–64.

7. J. S. Bell, *Speakable and Unspeakable in Quantum Mechanics*, Cambridge University Press, Cambridge, 1987.

8. A. Shimony, *Phys. Rev.* **20A**, 394–396 (1979).

9. S. Weinberg, *Phys. Rev. Lett.* **62**, 485–488 (1989).

10. C. G. Shull, D. K. Atwood, J. Arthur and M. A. Horne, *Phys. Rev. Lett.* **44**, 765–768 (1980).

11. J. J. Bollinger *et al.*, *Phys. Rev. Lett.* **63**, 1031–1034 (1989).

12. V. Paramananda and D. K. Butt, *J. Phys.* **G13**, 449–452 (1987).

13. A. J. Leggett, *Prog. Theo. Phys. Supp.* **69**, 80–100 (1980).

14. H. P. Stapp, 'Einstein Time and Process Time', in ***Physics and the Ultimate Significance of Time***, ed. D. R. Griffin, State University Press, New York, 1986, 264–270.

15. A. N. Whitehead, *QED. Process and Reality*, Princeton University Press, Princeton, New Jersey, 1985.

16. H. P. Stapp, *Phys. Rev.* **D28**, 1386 (1983).

17. H. P. Stapp, 'Quantum Measurement and the Mind-Brain Connection', in ***Symposium on the Foundations of Modern Physics***, ed. P. Lahti and P. Mittelstaedt, World Scientific, Singapore, 1990, 403–424.

18. H. Everett, *Rev. Mod. Phys.* **29**, 454–462 (1957).

19. E. Joos and H. D. Zeh, *Zeit. Phys.* **B59**, 223 (1985).

20. W. Zurek, in *Non-equilibrium Quantum Statistical Physics*, ed. G. Moore and M. Scully, Plenum Press, New York, 1984.

21. A. O. Caldeira and A. J. Leggett, *Physica* **121A**, 587 (1983).

22. R. P. Feynman and J. R. Vernon, *Ann. Phys. (N.Y.)* **24**, 118 (1963).

23. J. D. Barrow and F. J. Tippler, *The Anthropic Cosmological Principle*, Oxford University Press, Oxford, 1986.

24. B. d'Espagnat, *Phys. Reports* **110**, 201 (1984).

Chapter 15
Quantum theory and the implicate order

15.1 Introduction

Throughout this book we have thus far been concerned mainly with showing that a consistent ontological interpretation of the quantum theory as it now stands is possible. In the previous chapter, however, we raised the question of going beyond the current quantum theory and in the last section we sketched certain more definite proposals for doing this. These proposals however still involve introducing the concept of a particle, in addition to that of the wave function. As we pointed out there, this implies that we go outside the notion of expressing all basic concepts in terms of Hilbert space. In the present chapter we shall begin by exploring a new way of understanding the quantum theory and going beyond it which does not go out of Hilbert space. Nevertheless it does not set aside the work that we have done thus far, but rather it also includes the description of the particle trajectory within the framework of Hilbert space.

The basic idea is to introduce a new concept of order, which we call the implicate order or the enfolded order [1]. This is to be contrasted with our current concepts of order which are based on the ideas of Descartes who introduced coordinate systems precisely for the purpose of describing and representing order in physical process. The Cartesian grid (extended to curvilinear coordinates), which describes what is essentially a local order, has been the one constant feature of physics in all the fundamental changes that have happened over the past few centuries. In the quantum domain however this order shows its inadequacy, because physical properties cannot be attributed unambiguously to well-defined structures and processes in space-time while remaining within Hilbert space. Thus, for example, the uncertainty principle implies that it is not in general possible to give a definite space-time order to the motion of a particle in its trajectory. The fact that we nevertheless still use Cartesian coordinates, then, leads to a certain difficulty in thinking clearly about the subject. As we pointed out in the previous chapter, this is generally avoided by sticking to the mathematical equations and by saying that, in theoretical physics, these latter are basically what we are talking about, so that unclarities in physical concepts need not disturb us.

What we are proposing here is that this disparity between physical concepts (e.g. particle, wave, position, momentum) and the implications of the mathematical equations arises because the physical concepts are inseparably involved with the Cartesian notion of order, and this violates the essential content of quantum mechanics. What we need is a notion of order for all our concepts, both physical and mathematical, which coheres with this content. In this chapter we shall explore how such coherence may be realised with the aid of the notion of the implicate order which will be developed here.

In some ways this step is similar to that taken in the development of relativity, which likewise introduced new notions of order, e.g. with regard to simultaneity and also with regard to curvilinear coordinates. In relativity however strict continuity and locality are still maintained. But in the implicate order even these will have to be recovered as approximations and limiting cases.

15.2 Relativity and quantum theory as indicators of a new order for physics

It is generally agreed that neither special relativity nor general relativity have been united with quantum theory in a fully consistent way. To be sure we do have relativistic quantum field theory and with the aid of renormalisation techniques, this gives a wide range of correct predictions. Nevertheless the fact remains that strictly speaking the present theory implies various kinds of infinities and that the removal of these infinities by renormalisation algorithms is, logically considered, an arbitrary modification of the original basic principles. No new principles have been suggested up to the present that would lead to finite results. It is also very clear that general relativity introduces still further problems of principle which have not yet been resolved.

An important light can be thrown on these difficulties by noting that the basic orders implied in relativity theory and in quantum theory are qualitatively in complete contradiction. Thus relativity requires strict continuity, strict causality and strict locality in the order of the movement of particles and fields. And as we have seen throughout this book, in essence quantum mechanics implies the opposite.

This kind of contradiction in basic concepts is reminiscent of the paradoxes (e.g. Zeno's paradox) through which Democritus was led to the concept of the atom as explained in the previous chapter. Would it not be possible that the present contradiction between the basic concepts of relativity and quantum theory could similarly lead to a qualitatively new idea that would open the way to resolve all these difficulties?

As a clue to what this new idea might be, we could begin by asking, not what are the key differences between these concepts, but rather what they have in common. What they have in common is actually a quality of *unbroken wholeness.*

To see what this means, let us first consider relativity. As is well known the concept of a permanently existent particle is not consistent with this theory. But rather it is the point event in space-time that is the basic concept. In principle all structures have to be understood as forms in a generalised field which is a function of all the space-time points. In this sort of theory a particle has to be treated either as a singularity in the field, or as a stable pulse of finite extent. The field from each centre decreases with the distance, but it never goes to zero. Therefore ultimately the fields of all the particles will merge to form a single structure that is an unbroken whole.

The notion of unbroken wholeness is, however, still of limited applicability in the theory of relativity, because the basic concept is that of a point event which is distinct and separate from all other point events. In quantum theory however there is a much more thoroughgoing kind of unbroken wholeness. Thus even in the conventional interpretations, one talks of indivisible quantum processes that link different systems in an unanalysable way. In principle these links should extend to the whole universe, but for practical purposes their effects can be neglected on the large scale, so that in some suitable classical approximation we can use a simplified picture of the world as made up of separate parts in interaction. But in our interpretation there is also the fact that because the quantum potential represents active information, there is a nonlocal connection which can, in principle, make even distant objects into a single system which has an objective quality of unbroken wholeness.

In such a single system, the idea of separate point events clearly has no place. Yet as we have pointed out there is a certain residue of this notion still implicitly present in the fact that ultimately all fields are defined in terms of space-time points. These points are put in order by means of a Cartesian grid (extended where necessary to curvilinear coordinates), and it is this order that makes it possible to define key concepts such as local interaction between fields and differential operators which play an essential role in expressing the movements of these fields.

To sum up then, relativity and quantum theory share the notion of unbroken wholeness even though they achieve this in very different ways. As we have already explained in the previous section, we need a new notion of order that will encompass these different kinds of unbroken wholeness, which could open the way for new physical content that includes relativity and quantum theory but has the possibility of going beyond both.

15.3 Qualitative introduction to the implicate order

Our notions of order in physics have generally been tacit rather than explicit and have been manifested in particular forms which have developed gradually over the centuries in a somewhat fortuitous way. These latter have, in turn, come out of intuitive forms and common experience. For example, there is the order of numbers (which is in correspondence to that of points on a line), the order of successive positions in the motion of objects, various kinds of intensive order such as pressure, temperature, colour etc. Then there are more subtle orders such as the order of language, the order of logic, the order in music, the order of sensations and thought etc. Indeed the notion of order as a whole is not only vast, but it is also probably incapable of complete definition, if only because some kind of order is presupposed in everything we do, including, for example, the very act of defining order. How then can we proceed? The suggestion is that we can proceed as in fact has always been done, by beginning with our common intuitive notions and general experience of order and by letting these develop so as to extend into new domains and fields of application.

Our suggested next step is to explore how an order appropriate to wholeness can be found in our intuition and general experience. We begin by noting that the ordinary Cartesian order applying to separate points, finds one of its strongest supports in the function of a lens. What a lens does is to produce an approximate correspondence of points on an object to points on its image. The perception of this correspondence strongly brings our attention to the separate points. But as is well known, there is a new instrument used for making images called the hologram which does not do this. Rather each region of the hologram makes possible an image of the whole object. When we put all these regions together, we still obtain an image of the whole object, but one that is more sharply defined, as well as containing more points of view.

The hologram does not look like the object at all, but gives rise to an image only when it is suitably illuminated. The hologram seems, on cursory inspection, to have no significant order in it, and yet there must somehow be in it an order that determines the order of points that will appear in the image when it is illuminated. We may call this order implicit, but the basic root of the word implicit means 'enfolded'. So in some sense, the whole object is enfolded in each part of the hologram rather than being in point-to-point correspondence. We may therefore say that each part of the hologram contains an enfolded order essentially similar to that of the object and yet obviously different in form.

As we develop this idea, we shall see that this notion of enfoldment is not merely a metaphor, but that it has to be taken fairly literally. To emphasise this point, we shall therefore say that the order in the hologram is *implicate*. The order in the object, as well as in the image, will then be unfolded and we shall call it *explicate*. The process, in this case wave movement, in which this order is conveyed from the object to the hologram will be called *enfoldment* or *implication*. The process in which the order in the hologram becomes manifest to the viewer in an image will be called *unfoldment* or *explication*.

Enfoldment and unfoldment are actually encounted quite commonly in ordinary experience. For example, whenever one stands in a room, the order of the whole room is enfolded into each small region of space, and this includes the pupil of an eye which may happen to be there. This latter order is unfolded onto the retina and into the brain and nervous system, so as to give rise somehow to a conscious awareness of the order of the whole room. Similarly the order of the whole universe is enfolded into each region and may be picked up in various instruments, such as telescopes, from which they are ultimately unfolded to a conscious awareness of this order. The fact that light and other such processes are constantly enfolding and unfolding the whole is crucial for our ability to learn about any part of the universe of space and time, no matter where and when our observations may take place.

These notions of enfoldment and unfoldment can be understood in terms of the Cartesian order as being only particular cases of movements of fields not having the kind of general necessity that we would attribute, for example, to laws of nature. Rather what is significant for such laws of physics is considered to be the order of separate points. What we are proposing here is to turn this notion upside down and say that the implicate order will have the kind of general necessity that is suitable for expressing the basic laws of physics, while the explicate order will be important within this approach only as a particular case of the general order.

How can we justify such a radical change in our point of view? First of all we note that basically all the laws of movement in quantum mechanics do correspond to enfoldment and unfoldment. In particular, the relation between the wave function at one time $\psi(\boldsymbol{x}',t')$ and its form at another time, $\psi(\boldsymbol{x},t)$, is determined by the propagator or the Green's function,

$K(\boldsymbol{x}-\boldsymbol{x}', t-t')$ through the equation

$$\psi(\boldsymbol{x}, t) = \int K(\boldsymbol{x}-\boldsymbol{x}', t-t')\psi(\boldsymbol{x}', t')\, d\boldsymbol{x}'. \tag{15.1}$$

The value of the wave function at $\boldsymbol{x}, t$ is the sum of contributions from the whole of $\boldsymbol{x}'$ at the earlier time t' weighted with the factor $K(\boldsymbol{x}-\boldsymbol{x}', t-t')$. And so we may say that the region near $\boldsymbol{x}$ enfolds contributions from all over space at other times. Vice-versa, it follows that each region near $\boldsymbol{x}'$ will unfold into the whole space $\boldsymbol{x}$, with the weighting factor $K(\boldsymbol{x}-\boldsymbol{x}', t-t')$. A simple picture of the movement is that waves from the whole space enfold into each region and that waves from each region unfold back into the whole space.

Of course the Green's function has been derived ultimately from the Cartesian order by solving differential equations. But if we are questioning that this order holds fundamentally (especially at short distances), we may well adopt the point of view that the Green's function is more basic than the differential equation (which latter will only be an approximation). In this case the order of enfoldment and unfoldment will be fundamental, while the Cartesian order will have a relatively limited kind of significance.

The process of enfoldment and unfoldment was already well known implicitly in the Huygens' construction. Waves from each point unfold. But at the same time waves from many points are enfolding to give rise to a new wave front. So in the totality, the one process includes both enfoldment and unfoldment. It is only when we focus on a part that we are led to talk of these as distinct.

The Huygens' construction is actually the basis of the Feynman graphs which are widely used. To explain the connection, consider waves which start at a point P and arrive at a point Q (see figure 15.1). In the first interval of time, Δt, a possible path is from P to P', and in the second interval from P' to P'' and so on. Extending this construction, the path eventually arrives at Q. The Huygens' construction implies that the waves that arrive at Q from P are built up of contributions from every possible path. These paths are the starting point of the well-known Feynman diagrams.

Originally Feynman [2] wanted to regard these paths as representing the actual trajectories of particles. But this would not be consistent, because contributions from various paths can interfere destructively as well as constructively. Rather, each path represents a contribution to the final field amplitude as implied by the Huygens' construction. And together the whole set of paths describes a process of enfoldment of waves towards Q, or alternatively the unfoldment of waves from P.

Since all matter is now analysed in terms of quantum fields, and since the movements of all these fields are expressed in terms of propagators, it

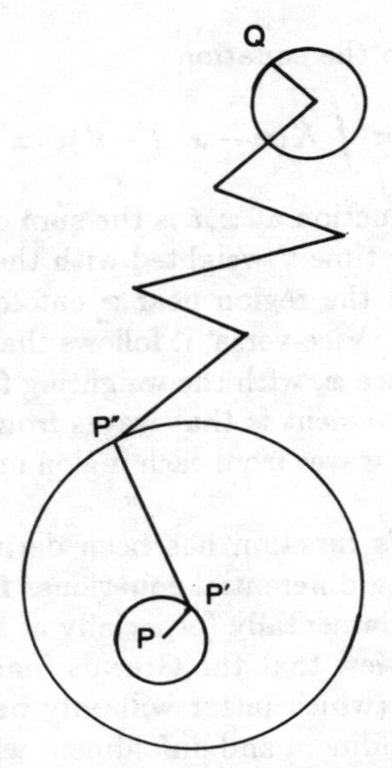

Figure 15.1: Huygens' construction and classical paths

is implied by current physics that the implicate order is universal. These fields, which are now being treated as the ground of all existence, have to be understood as being essentially in movement. This is because the field cannot be said to be an entity which would persist (as a particle for example does) in a certain form over a period of time.

Thus, for example, consider a field variable $\psi(\boldsymbol{x}, t)$ defined at the space-time point $\boldsymbol{x}, t$. This has at most a purely momentary existence and may be said to vanish at the very moment at which it would come into being. Something with persistent identity would require a unique relation between $\psi(\boldsymbol{x}, t)$ and the field at other times. If, for example, we try to relate $\psi(\boldsymbol{x}, t)$ at O in figure 15.2 to $\psi(\boldsymbol{x}, t + \delta t)$ at A and thus define a persistent form of existence, we find that this leads to something quite arbitrary. For we could equally well have considered the same field point, O, in another Lorentz frame $(\boldsymbol{x}', t')$ and then related a $\psi(\boldsymbol{x}', t')$ to $\psi(\boldsymbol{x}', t' + \delta t')$ at B. This would evidently have defined a different entity. With a particle, the path is unique and we do not have this sort of arbitrariness. What this means is that the field itself cannot be regarded as having an identity. Or to put it differently, we can never have the same field point twice. Nor is there a unique form within the field that persists. Therefore all properties that are attributed to the field have to be understood as relationships in its movement.

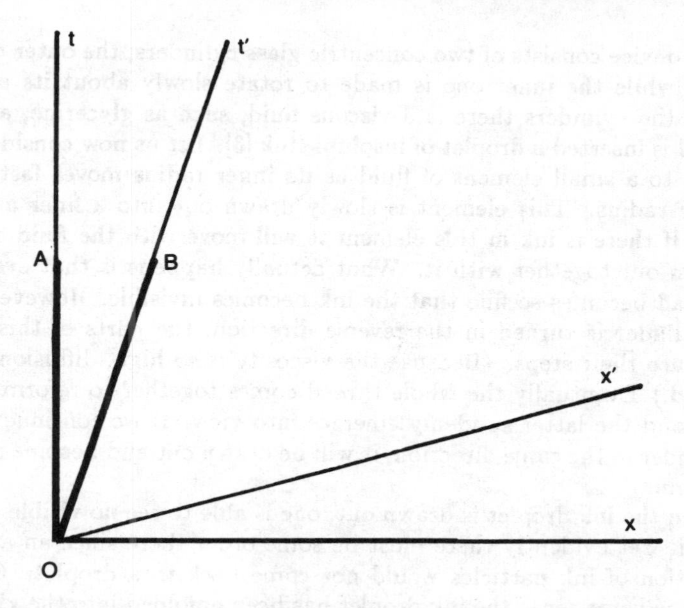

Figure 15.2: Non-uniqueness of fields in different Lorentz frames

We may suppose that the universe, which includes the whole of existence, contains not only all the fields that are now known, but also an indefinitely large set of further fields that are unknown and indeed may never be known in their totality. Recalling that the essential qualities of fields exist only in their movement we propose to call this ground the *holomovement.* It follows that ultimately everything in the explicate order of common experience arises from the holomovement. Whatever persists with a constant form is sustained as the unfoldment of a recurrent and stable pattern which is constantly being renewed by enfoldment and dissolved by unfoldment. When the renewal ceases the form vanishes.

The notion of a permanently extant entity with a given identity, whether this be a particle or anything else, is therefore at best an approximation holding only in suitable limiting cases.

15.4 Further illustrative example of the implicate order

We shall now give a further illustrative example involving a different form of the implicate order which helps us to focus our attention on general relationships that define such an order in more detail. This illustration is carried out with the aid of a device that we shall now describe.

This device consists of two concentric glass cylinders; the outer cylinder is fixed, while the inner one is made to rotate slowly about its axis. In between the cylinders there is a viscous fluid, such as glycerine, and into this fluid is inserted a droplet of insoluble ink [3]. Let us now consider what happens to a small element of fluid as its inner radius moves faster than its outer radius. This element is slowly drawn out into a finer and finer thread. If there is ink in this element it will move with the fluid and will be drawn out together with it. What actually happens is that eventually the thread becomes so fine that the ink becomes invisible. However, if the inner cylinder is turned in the reverse direction, the parts of this thread will retrace their steps. (Because the viscosity is so high, diffusion can be neglected.) Eventually the whole thread comes together to reform the ink droplet and the latter suddenly emerges into view. If we continue to turn the cylinder in the same direction, it will be drawn out and become invisible once again.

When the ink droplet is drawn out, one is able to see no visible order in the fluid. Yet evidently there must be some order there since an arbitrary distribution of ink particles would not come back to a droplet. One can say that in some sense the ink droplet has been enfolded into the glycerine, from which it unfolds when the movement of the cylinder is reversed.

Of course if one were to analyse the movements of the ink particles in full detail, one would always see them following trajectories and therefore one could say that fundamentally the movement is described in an explicate order. Nevertheless within the context under discussion in which our perception does not follow the particles, we may say this device gives us an illustrative example of the implicate order. And from this we may be able to obtain some insight into how this order could be defined and developed.

Our next step is then to consider the enfoldment of a series of ink droplets. Suppose that the first ink droplet is enfolded by n turns and that the second droplet is then inserted at the same point to be enfolded an additional n turns. In this way we enfold a series of N droplets. In this series, the droplets will be enfolded $n, 2n, 3n \dots Nn$ times respectively. Suppose then that the motion of the cylinder is reversed. These droplets will unfold and re-enfold one after another. If the cylinder is turned rapidly (the viscosity being high enough to prevent diffusion) it will appear that there is a single persistent droplet at the point in question. Nevertheless the droplet is as it were being constantly recreated and dissolved from a distribution of ink that is in some sense spread out over a whole range of fluid. We thus obtain an example of how forms that persist in the explicate order can arise from the whole background and be sustained dynamically by a movement of enfoldment and unfoldment.

If we suppose that these successive ink droplets are inserted at slightly different positions, then we will obtain what appears to be a moving ink droplet. The different rates of 'motion' correspond to different enfolded distributions of ink in the whole fluid. In a sense the 'movement' of the ink droplet is 'guided' by a general field of enfoldment belonging to the whole fluid. The attribution of this guidance to the particular place where the droplet emerges is not appropriate. Rather its behaviour is basically global in origin. This clearly can be proposed as a model of a quantum 'particle' which might be assumed to underlie the sort of particles that we have postulated in our interpretation of the quantum theory. Of course, this model is rather crude and simple, but later on we shall propose a similar one that we can take more seriously.

For the present, however, let us continue with this model. Suppose that we consider two particles initially at different positions, one represented by red ink and the other by blue ink. As we enfold these into the fluid, red ink particles and blue ink particles will eventually intermingle, so that in a given region we could not say that the ink droplets were really separate. Nevertheless when the cylinder was turned back, red and blue ink particles would begin to separate to form respectively red and blue droplets. It is clear that the ground of the particular behaviour of the visible droplets is global and that disturbances in this ground can lead to correlated changes in this behaviour.

In our model the appearance in perception of what seems to be a pair of separate ink droplets can therefore be deceiving. Actually these apparently separate ink droplets are internally related to each other. On the other hand, as conceived in the explicate order, such droplets would only be externally related. This is one of the most important new features of the implicate order. What may be suggested is that this feature may give further insight into the meaning of quantum properties such as mutual participation and nonlocality.

Another instructive illustration of the implicate order can be obtained by first considering the successive enfoldment of N droplets with the cylinder turning in a certain direction and successive droplets at a constant distance from each other. After these N droplets are enfolded the movement of the cylinder is reversed to enfold additional S droplets at the same distance from each other as in the first set. After that we turn the cylinder in the original direction and then enfold an indefinite number of further droplets in a similar way. When the cylinder is turned backwards we will then find a droplet moving as before. But as shown in figure 15.3, at the point P there will suddenly appear a pair of particles 'moving' in opposite directions and one of these will proceed to meet the original particle at the point Q. They then appear to annihilate each other and nothing is left of them thereafter. However, the other member of the original 'pair' will

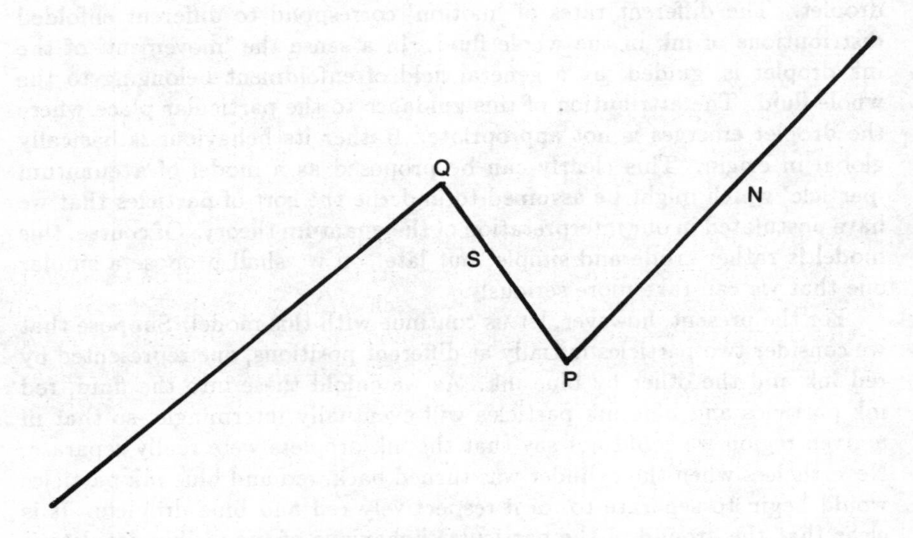

Figure 15.3: Enfoldment and pair creation

proceed onwards.

It seems that we have here a simple model of pair production and annihilation. But unlike the representation given in Feynman diagrams, we do not say that anything 'moves' backward in the time. Rather it is clear that what is significant here is the order in the degree of implication.

To bring out what this means, let us define an implication parameter, τ, of a droplet, which is proportional to the number of times the cylinder has been turned since that droplet was inserted. This implication parameter takes negative values when the cylinder is turned in the opposite direction. What happens in this example is that the implication parameter has a part that increases, another part in which it decreases and a third part in which it increases again. The entire pattern is present at each instant in the whole fluid with varying degrees of implication. All that happens with the passage of time is a change in the implication parameter which may be positive or negative.

What this suggests is that what is called the time coordinate in the Feynman's approach may actually be the degree of implication. In this interpretation, Feynman diagrams would not refer to actual processes but rather to structures in the implicate order. The meaning of time would have to be something different from τ but nevertheless related to it.

15.5 On the distinction between implicate and explicate orders

We have thus far used the distinction between implicate and explicate orders rather loosely. We shall now discuss this distinction in a more precise way.

To begin with, we might consider a form whose elements were made of ink droplets of various colours. A simple form might be a line defined by droplets at its endpoints, or a triangle defined by droplets at its vertices. Alternatively we could represent this line by a smooth thread of ink and a point by a droplet, or various other representations could be adopted.

What is characteristic of an explicate structure is that all its basic parts are separate and distinct and outside of each other, although they may be in well-defined relationships. For example, consider a two-dimensional square grid made up of points and the lines which connect them. The points are on the boundaries of the lines. Moreover each point has a definite set of four neighbours. It is clear that these requirements define the basic order of the grid which is clearly explicate in the sense defined above.

However, it is not necessary that an explicate order be equivalent to a square lattice. Indeed in section 15.10 we give a more precise mathematical definition of an explicate order through what we call a contact matrix C_{ij} which enables us to consider not only arbitrary lattices but also irregular structures such as dislocated crystals and glasses.

Let us now return to our example of the grid. We may imagine that this grid is enfolded by a certain movement of the cylinder. Nevertheless when the cylinder is reversed all the parts will unfold in such a way that they come out all together again in the same explicate structure. This follows because all the parts have the same implication parameter. What this means is that this whole set of parts retains this potential for being explicate together even when it has been enfolded. Therefore within the implicate order, there can be structures that are essentially explicate, even though they are still enfolded.

On the other hand there may be structures that are intrinsically implicate in the sense that the whole cannot be unfolded together in one single explicate order. The case of the enfoldment of successive droplets is an example of this. For only one of the droplets can be unfolded at a time.

The general case is one in which there is a succession of explicate structures enfolded to make up what is basically an intrinsically implicate order. However, this latter contains explicate suborders as aspects which are particular cases of the general notion of the implicate order. In this way we clarify our earlier statement that the implicate order is general and necessary, while explicate orders are particular and contingent cases of this.

What then is the relationship between the notion of explicate order and that of the manifest world discussed in chapter 8, section 8.6? Clearly the manifest world of common sense experience refined where necessary with the aid of the concepts and laws of classical physics is basically in an explicate order. But the motion of particles at the quantum level is evidently also in an explicate order. However, as we have suggested in chapter 8, this latter order is not always at the manifest level because it is profoundly affected by the active information represented by the quantum potential. This latter operates in a subtle way and according to what has been said in this chapter, this operation is in an implicate order. Therefore the particle movement is not understood fully as self-determined in the explicate order in which it is described. Rather, this explicate order reveals the deeper implicate order underlying its behaviour.

The movement of the beables becomes manifest only at the large scale level under conditions in which the quantum potential can be neglected. In this case the movement is not only describable in an explicate way, but it can be approximated as self-determined in the same explicate order that we use to describe it. Therefore it is essentially independent of the implicate order and can be abstracted as existing on its own. It is therefore capable of being 'grasped in the hand' without fundamental alteration (whereas at the micro-level, the movement of particles is not). Nevertheless, as we have seen in chapter 8, the implicate order is still revealed in the manifest world in some of the more subtle relationships that arise in experiments at a quantum level of accuracy (e.g. interference experiments).

15.6 More general notion of order

At this point we need to develop a more general notion of what is to be meant by order, at least within the context under discussion. To illustrate the notion that we are suggesting, we begin with the straight line divided into equal segments. What characterises the order of this line is that the differences of successive segments remain similar (and indeed equal) throughout the whole line. The difference is just that each line is displaced by its own length relative to the one before.

We may extend this suggestion to discuss the order of curves. For example, if we approximate a circle by a polygon of many sides, we see that the differences of successive sides is that not only are they displaced relative to each other, but that they have different angles. It is clear however that the successive differences are similar and that this is what makes the curve a circle. If successive elements also turned into a third dimension, we would then get a helix resulting from the similar differences. And more generally, one could complicate this construction to produce curves of arbitrary complexity (see Bohm and Peat [4]).

The next step is to consider the differences that are possible between such similarities. For example, we may arrange a set of lines in parallel. Each of these has a set of similar differences, but then each line is determined by a different set of such similar differences. If the lines are equidistant, these differences in the similarities in turn constitute a higher order of similar differences. If we then consider an additional set of such lines perpendicular to the first, the entire arrangement would have a set of similarities that were different from those in the first set.

If we go on to consider a series of such sets of lines, each having a small angle θ relative to the preceding one, we obtain a set of similar differences (of the similarities in each set), and clearly this defines an order of orientations of all these successive arrangements.

It is clear from all this that one can have an indefinitely extending hierarchy of similar differences and different similarities of these differences, constituting what may be called an order of orders. If these orders have points of contact (e.g. as there are between perpendicular sets of lines in a grid) we may say they begin to form a *structure*.

This treatment of order in terms of similar differences and different similarities appears to be broad enough not only to cover all the orders found thus far in physics, but it seems to be valid in much broader areas of experience. For example, an order of colours can be expressed as similar differences, as can an order of musical tones. The musical theme can clearly be seen in terms of similar differences and different similarities in patterns of notes, going on all the way up to the order that can be found in highly organised compositions. Similar (but different) orders can also be seen in thoughts, which often develop in patterns of similar differences and different similarities.

15.7 The algebra of the implicate order

The implicate order finds a natural mathematical expression in terms of certain kinds of algebra. We shall first illustrate this in terms of the ink droplet model and then go on to extend our treatment to the algebras of quantum mechanics.

If we start from any initial distribution of ink represented by some density function ρ_0 (e.g. such as an ink droplet) then after the cylinder has been turned a certain number of times, we will obtain a distribution

$$\rho_1 = T\rho_0, \tag{15.2}$$

where T is an operator that represents the transformation brought about by this process. If we apply a similar transformation again, we obtain $\rho_2 = T^2 \rho_0$ and if we apply it n times, we obtain

$$\rho_n = T^n \rho_0. \tag{15.3}$$

The above makes possible a simple mathematical description of the implicate order because it embodies the feature that the differences of successive elements, i.e. the transformation T, remain similar.

If we now consider the whole succession of droplets, we have to add up the contributions corresponding to the different values of n. We obtain

$$\rho = \sum_n \rho_n = \sum_n T^n \rho_0. \tag{15.4}$$

But in principle we could have put in each droplet with a different density (a_n). This would have given us

$$\rho = \sum_n a_n T^n \rho_0. \tag{15.5}$$

In this way we obtain an algebra of transformations because each term in the algebra, $T = \sum a_n T^n$, can be added to any other term and because clearly transformations can be multiplied. It is evident that by setting $n = 0$ we obtain the unit element T^0, which simply represents no transformation at all.

Let us now go on to discuss the quantum theory. We have already seen that the basic notion in the quantum theory expressed through Green's functions can be understood as a form of the implicate order. Thus we may write for the transformation between two times

$$\psi_1 = T\psi_0. \tag{15.6}$$

Here T will, in general, be complex. Successive transformations through n steps with similar differences can then be expressed as T^n, so that we obtain

$$\psi_n = T^n \psi_0. \tag{15.7}$$

Because of the linearity of the wave equation, we can, in general, add transformations with suitable coefficients to obtain new solutions of the wave equation. Thus we could have

$$\psi = \sum \alpha_n T^n \psi_0. \tag{15.8}$$

But the possibilities are much richer than this because we could bring in transformations that do not commute. For example, if T_A and T_B are two such transformations we could have

$$\psi = \sum (\alpha_n T_A^n + \beta_n T_B^n)\, \psi_0, \tag{15.9}$$

as well as more complex combinations. It is clear that at least abstractly we can form an algebra in this way.

Thus far we have given only the algebra of transformations, which in quantum mechanics is called the Heisenberg algebra (see Frescura and Hiley [5,6]). This algebra can be completed by bringing in a projection operator $P(\boldsymbol{x}, \boldsymbol{x}')$. In this way, as we shall see, we can tie up with the ideas of Stapp [7] and Gell-Mann and Hartle [8].

To simplify the discussion, let us consider an operator $P_t(\boldsymbol{x}_0, \Delta\boldsymbol{x}_0)$ which at a given time t projects out everything but a cubic region of size $\Delta\boldsymbol{x}_0$ centred at the point $\boldsymbol{x}_0$. At any other time, this operator has the form

$$P_{t'} = e^{iH(t'-t)} P_t e^{-iH(t'-t)}, \tag{15.10}$$

where $e^{iH(t'-t)}$ is the Green's function between t' and t.

Let us now give a more detailed expression for P_t. We write

$$P_t(\boldsymbol{x}_0, \boldsymbol{x}, \boldsymbol{x}') = \psi^*_{\boldsymbol{x}_0}(\boldsymbol{x}')\psi_{\boldsymbol{x}_0}(\boldsymbol{x}), \tag{15.11}$$

where $\psi_{\boldsymbol{x}_0}(\boldsymbol{x})$ is a wave function that is constant in a region of size $\Delta\boldsymbol{x}_0$, centred at $\boldsymbol{x}_0$ and zero outside. Clearly this wave function defines the projection operator that we want. We then obtain

$$P_{t'}(\boldsymbol{x}_0, \Delta\boldsymbol{x}) = e^{-iH^*(t'-t)}\psi^*_{\boldsymbol{x}_0}(\boldsymbol{x}')e^{iH(t'-t)}\psi_{\boldsymbol{x}_0}(\boldsymbol{x}). \tag{15.12}$$

It is evident that the mere fact that we have an algebra is what enables us to express the properties of quantum interference. Such interference is not described by adding projection operators. Rather to do this corresponds to what is called forming a mixed state. Quantum interference is implied by a further property of the algebra, not present in algebras of the type that describe the ink drop model. This new property involves linear superposition over both indices of the projection operators, $\boldsymbol{x}'$ and $\boldsymbol{x}$, separately. Operations carried out on each one of these correspond to what are called, in mathematical terms, *ideals* [6]. Operations from the left generate a left ideal, while those from the right generate a right ideal. Transformations of projection operators such as (15.10) involve both ideals together in a symmetric way.

It is clear that the new projection operators at the time t' can be written as

$$P_{t'}(\boldsymbol{x}_0, \Delta\boldsymbol{x}_0) = \psi^*(\boldsymbol{x}', t'-t)\psi(\boldsymbol{x}, t'-t) \tag{15.13}$$

so that what was initially a 'pure state' remains so after transformation. This means that the law of motion does not involve the superposition of projection operators, but rather, the superposition on the left and right ideals which imply contributions to the wave functions $\psi_{t'}(x')\psi_{t'}(x)$. Evidently it is this feature that makes possible the properties of quantum interference. For example, if the wave functions in (15.11) correspond to wave packets of width Δx_0, these will start to spread, at first slowly, and then more rapidly when $t' - t$ is positive. Sooner or later wave functions from packets that were originally in different regions will overlap and then we will have interference. On the other hand when $t' - t$ is negative, we can look at this situation as a packet which was originally spread out and which then converges to the region $(x_0, \Delta x_0)$. In this case all the parts of the original packet ultimately come together and interfere to produce a wave function that is constant in the region $(x_0, \Delta x_0)$ and zero outside.

If we recall the description relating the Green's function to the implicate order in section 15.3, we can see that the projection operators of Stapp and Gell-Mann and Hartle really develop according to the implicate order. An operator which corresponds to a well-defined region at the time t, corresponds to an enfoldment of this region at all other times. Their histories are therefore, in a sense, always present at any moment. But at any given time, t, only some are unfolded. It is evident that the effective assumption of these authors is that any event is actual only at the moment at which it is unfolded. Nevertheless it is always present whether unfolded or not.

In our example we have thus far been thinking of the wave function of a single particle. But suppose that there are many particles. At the time $t = 0$ we should then write the projection operator of the nth particle as

$$P_{n_0} = \psi^*_{x_0}(x'_n)\psi_{x_0}(x_n)\mathbf{1}'_n\mathbf{1}_n. \tag{15.14}$$

Here $\mathbf{1}'_n$ and $\mathbf{1}_n$ operate on all the particles other than n and where $\mathbf{1}'_n$ operates on the left ideal and $\mathbf{1}_n$ operates on the right ideal.

When the above is transformed by the Green's function, not only will the wave packet spread out as in the one-particle case, but it will start to entangle with all the other particles. So we have an implicate order in the whole configuration space of the N particles. For times in the distant past, the whole universe is thus in some way correlated in its enfoldments. As time passes, these correlations begin to disentangle and the wave functions begin to converge on the region $(x_{0_n}, \Delta x_n)$ belonging to the nth particle while it becomes a constant as a function of the coordinates of the other particles. After this the spreading and re-enfoldment process starts to take place again.

An important point here is that if there is sufficient entanglement, then sooner or later, P_{n_t} will begin to approach commutation with P_{n_0}. In the models of Stapp and Gell-Mann and Hartle, it will therefore be consistent to assume that projection operators at moderately different times will correspond to mutually compatible actual events. This will make it possible to give the history a real content as objective and independent of observers.

15.8 A Hilbert space model of the trajectories in our interpretation of quantum mechanics

We are now ready to extend the model of a particle in our interpretation so that it can be included within the framework of Hilbert space. We shall see that this model has certain similarities with the models of Stapp and Gell-Mann and Hartle, though there are important differences that we shall bring out as we go along.

We shall begin by returning to our ink droplet model of a particle as given in section 15.4. We can now relate this to the extension of our approach beyond the domain of current quantum theory as suggested in chapter 14, section 14.6. We proposed there that the stochastic trajectories in our interpretation can be thought of as arising from some random process with a certain free 'flight-time' that is related to the Planck time. So what we now suggest is to give up the idea that the particle has a continuous existence and to suppose instead that it is more like our ink droplet model, which is continually unfolding and re-enfolding. The time between these processes should be of the order of this mean free time.

Successive 'droplets' will have a random distribution. However, there will be a systematic trend in this distribution, so that the differences $\delta \boldsymbol{x}_n$ in successive positions will imply a systematic average drift velocity, as well as an osmotic velocity. This will have to be such as to correspond to the velocities assumed in our stochastic model given in chapter 12.

This model has certain features that are qualitatively right, but it does not readily fit in with quantum mechanical concepts. We therefore now propose a further new idea. This is to replace the converging and diverging droplets by converging and diverging quantum mechanical waves. In this way our concept of the particle is now included within the implicate order as a particular form in Hilbert space.

To see what the similarities are between our model and that of Stapp and Gell-Mann and Hartle, let us recall that as shown in the previous section 15.7, the projection operators of Gell-Mann and Hartle [8] are also particular cases of the implicate order. The key difference is this. Their projection operators correspond to waves that converge on extended regions and then diverge. On the other hand, ours converge to a point and then diverge. To see how the two approaches are related, we can consider a projection operator $P(\boldsymbol{x}_0, \Delta \boldsymbol{x})$ and let the range of $\Delta \boldsymbol{x}$ approach zero. It is clear that both sets of operators are contained within this framework.

Gell-Mann and Hartle propose actual histories which consist of a sequence of projection operators $P(\boldsymbol{x}_0, \Delta \boldsymbol{x})$, in which $\Delta \boldsymbol{x}$ is large enough to describe an essentially classical-type development of actual events. Our proposal is to consider a sequence of actual events forming a 'trajectory' which is essentially quantum mechanical. So in some sense we have quantum mechanical histories and they have classical histories. The advantages of our approach have been discussed earlier throughout this book, but here we recall two of them. First there is no sharp distinction of classical and quantum levels. The classical level arises naturally without further assumptions in a gradual way as the quantum potential becomes negligible. On the other hand, in the approach of Gell-Mann and Hartle, we have simply to assume such a distinction of classical and quantum levels without having, as yet, a clear criterion for what are the limits of the classical level. Secondly our approach gives a simple and fairly clear account of quantum processes as well as providing an intuitive understanding of all the transformations of the theory in terms of the implicate order. On the other hand, in their approach, the individual quantum process cannot be described and therefore there is no intuitive understanding of what it means.

The dependence of the points to which the various successive waves converge on some sort of systematic drift velocity and osmotic velocity provides an explanation of the 'inertia' of the particle. It is this dependence which also guarantees that a particle will not enter a region of zero wave function and will not leave its channel once it enters. In this way definite results can be preserved at the manifest level which includes the observable outcomes of measurements.

The notion of a particle with permanent identity has now been replaced by that of a particle with no permanent identity. Rather its basic 'elements' are constantly forming and dissolving in succession. Over times much longer than that between successive wave formations, the whole process approximates a trajectory (which may however be engaged in a stochastic process). For shorter times the notion of a trajectory dissolves away and some entirely new concepts will be needed which we shall discuss. (Further on, we shall give an example of these in terms of incoming and outgoing waves.)

15.9 An example of trajectories arising from wave structures

In order to illustrate the ideas developed in the previous section we shall now give a specific example of trajectories arising from wave structures. To simplify the discussion, we consider a particle always moving at the speed of light but with stochastic variations in direction so that we have a kind of zigzag path in space-time whose mean velocity will evidently be less than

c. When we go over to the wave model, the end points of each line of the zigzag will be the points towards which our waves converge and from which they then diverge. It seems natural to assume that these waves also move at the speed of light so that to each point P of convergence and divergence there is an incoming and outgoing light cone. The trajectories in question will then correspond to a suitable sequence of such light cones.

This sort of structure of trajectories can be clarified with the aid of the theory of the higher spherical geometry of Lie and Klein (see Klein [9]). To relate the light cones to this geometry, we first consider any convenient hyperplane, say $t = 0$. Each light cone intersects this hyperplane in a sphere. Those light cones whose vertices have $t < 0$ will be defined to produce spheres of positive radius $r = t$. The light cones whose vertices have $t > 0$ will also produce spheres but these will be assigned negative radii. So there is a $1 : 1$ correspondence between light cones and their intersections with any hyperplane $t =$ const.

What Lie and Klein proposed was an interesting way of parameterising these spheres. To obtain their parameterisation, let us first write the equation of a sphere

$$(x-a)^2 + (y-b)^2 + (z-c)^2 = r^2, \tag{15.15}$$

where the coordinates of the centre are (a, b, c) and the radius is r. Equation (15.15) reduces to

$$x^2 + y^2 + z^2 - 2(ax + by + cz) = r^2 - a^2 - b^2 - c^2. \tag{15.16}$$

We now regard this sphere as parameterised by the five quantities $a = x_1$, $b = x_2$, $c = x_3$, $r = -x_4$ and ξ. But these parameters are related by

$$\xi = x_4^2 - x_1^2 - x_2^2 - x_3^2. \tag{15.17}$$

We introduce homogeneous coordinates by writing

$$\xi = \frac{X_6 - X_5}{X_6 + X_5}$$

and

$$x_i = \frac{X_i}{X_6 + X_5}.$$

The condition (15.17) reduces to

$$X_6^2 - X_5^2 - X_4^2 + X_1^2 + X_2^2 + X_3^2 = 0. \tag{15.18}$$

The six parameters we have introduced in addition to equation (15.18) now parameterise the sphere.

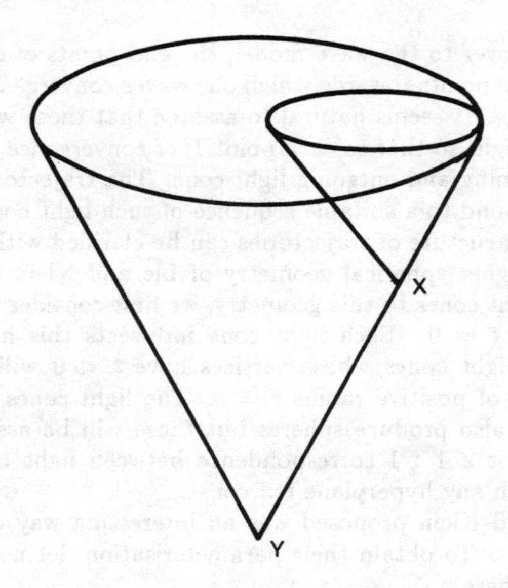

Figure 15.4: Light cones on one light ray **yx**

The advantage of this parameterisation is that the conformal invariance of the description is now evident. For equation (15.18) implies the invariance to six dimensional pseudo-orthogonal transformations which are easily seen to be isomorphic to the conformal group. By distinguishing a light cone at infinity, we can reduce this to the Poincaré group and this may be for example done by defining this light cone through the relationship

$$X_6 + X_5 = 0. \tag{15.19}$$

One can readily show that the condition for two spheres to come into contact, i.e. to touch each other at single point and have a common tangent plane, is

$$X_6Y_6 - X_5Y_5 - X_4Y_4 + X_1Y_1 + X_2Y_2 + X_3Y_3 = 0, \tag{15.20}$$

where X and Y refer respectively to the spheres in question. One can also see that if two spheres are in contact, then the vertices of their corresponding light cones are connected by a null ray (see figure 15.4). We now proceed to show how the trajectories are described in terms of these spheres. First of all consider a sphere of radius r corresponding to a light cone whose vertex is in the future at $t = r$. Now consider a sphere of radius $r' < r$ corresponding to a light cone with vertex at $t = r'$. Suppose the

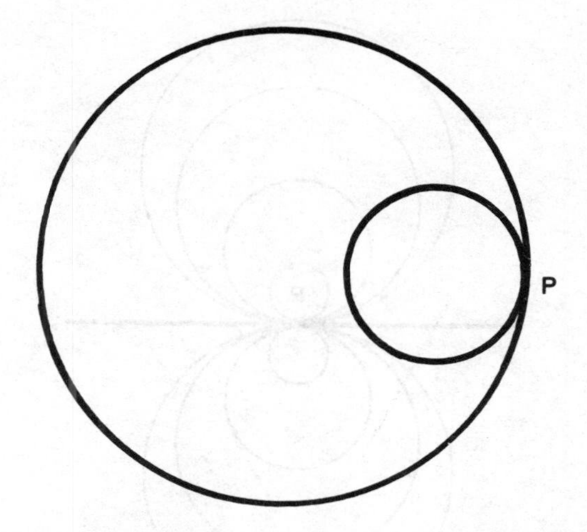

Figure 15.5: Light ray as point of contact, P, of spheres

two spheres are to be in contact as shown in figure 15.5. Between them lie other spheres with the same point of contact P and with intermediate radii. All of these correspond to light cones with a null ray in common. But further spheres of larger radii lie beyond the outer sphere going to a sphere of infinite radius which represents a plane wave. Beyond this lie spheres of negative radii corresponding to the light cones whose vertices have negative values of t (see figure 15.6). As the radius approaches zero from either positive or negative direction, we approach the point P. So a point is a sphere of zero radius.

Let us now consider our trajectory starting from a sphere of zero radius representing a light cone with vertex at $t = 0$. The first step in our trajectory (see figure 15.7) is represented by the point of contact P and a sphere of radius r. The corresponding null ray is in the direction of the radius of the sphere at the point P. The next null ray will be represented by a larger sphere contacting the first sphere at the point Q. The next null ray will correspond to a still larger sphere contacting the second sphere at yet another point R. This procedure is to be continued indefinitely so that we obtain a complete description of the zigzag trajectory.

The key point here is that we have described our trajectory entirely in terms of wave forms at any given time. It is easy to show that this description is conformally invariant. For every transformation that must turn spheres into spheres (and if we restrict ourselves to satisfying (15.19), we will have only Poincaré invariance). But now we can see the meaning

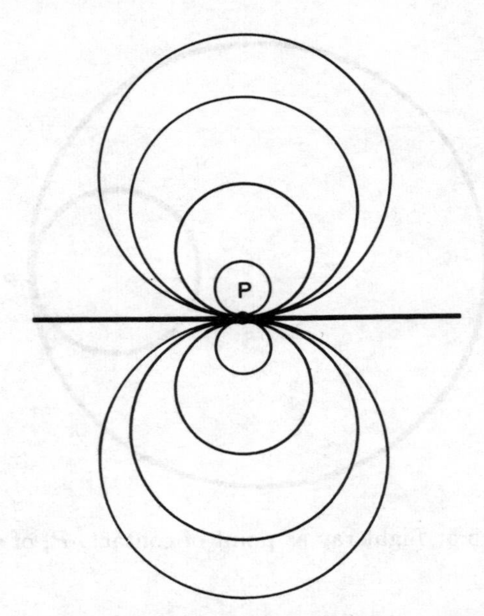

Figure 15.6: Illustration of spheres with positive and negative radii

of these transformations without explicit reference to time. For example in a time displacement, Δt, the radius of each sphere changes by Δt. In a Lorentz transformation, however, the radii and centres change together so that initially concentric spheres cease to be concentric.

From what we have been saying, it follows that our trajectory is now described in an implicate order. As this order unfolds in time, the radii of the spheres corresponding to incoming waves decrease while those corresponding to outgoing waves increase. The entire trajectory is now shown as a kind of enfolded geometric structure whose meaning can be seen all at once as a 'chain' of successively contacting spheres. The law of development is implicit in this structure.

This way of looking at the subject is close to what is implied by Feynman's approach as well as to the Huygens principle. Feynman [2] describes his trajectories as a sequence of points in time. We describe them as sequences of spheres all enfolded at a given time. The Lagrangian is thus, in our approach, a property of the implicate order which holds at any given moment. And as seen in section 15.8, backward tracks in time are replaced by tracks in which the implication parameter is decreasing. In our model, the implication parameter is just the radius of the sphere. Thus in figure 15.6, we could have the radii of spheres decrease for a while and then

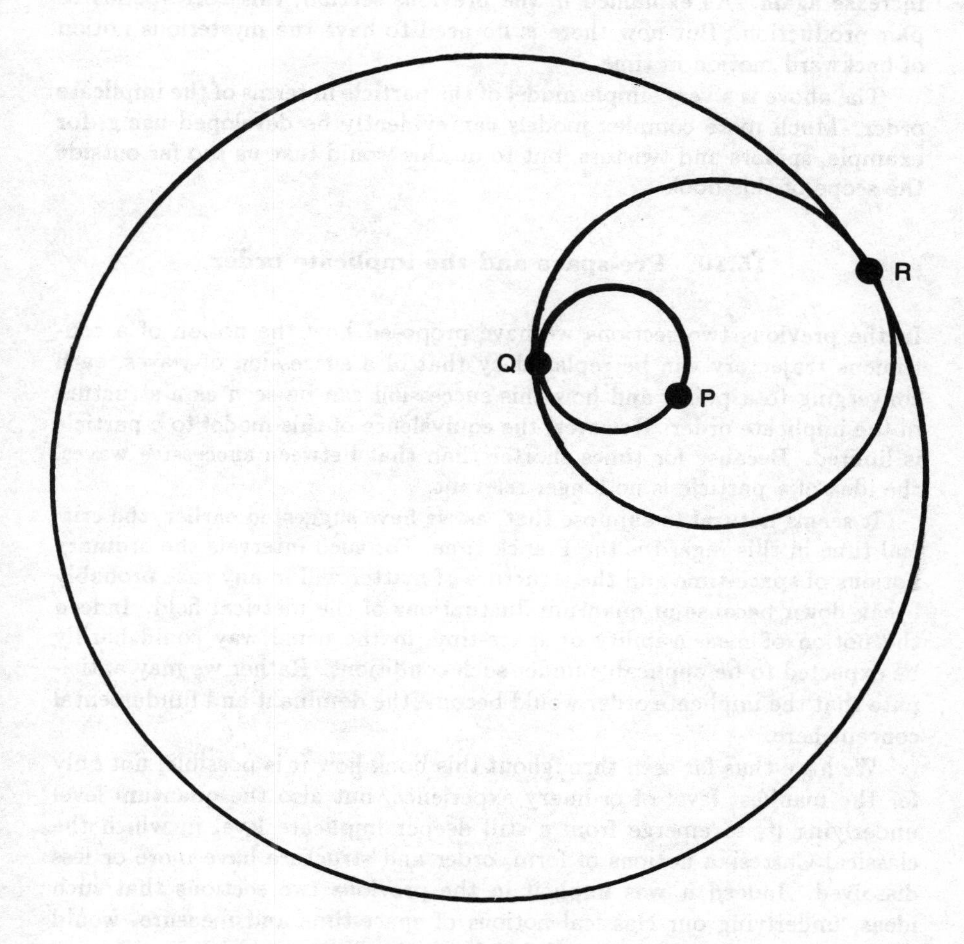

Figure 15.7: Trajectory in terms of spheres

increase again. As explained in the previous section, this corresponds to pair production. But now there is no need to have the mysterious notion of backward motion in time.

The above is a very simple model of the particle in terms of the implicate order. Much more complex models can evidently be developed using, for example, spinors and twistors, but to do this would take us too far outside the scope of this book.

15.10 Pre-space and the implicate order

In the previous two sections we have proposed how the notion of a continuous trajectory can be replaced by that of a succession of waves, each converging to a point, and how this succession can be seen as a structure in the implicate order. However, the equivalence of this model to a particle is limited. Because for times shorter than that between successive waves, the idea of a particle is no longer relevant.

It seems natural to suppose that, as we have suggested earlier, the critical time in this regard is the Planck time. For such intervals the ordinary notions of space-time and the structure of matter will in any case probably break down because of quantum fluctuations of the metrical field. Indeed the notion of measurability of space-time in the usual way could hardly be expected to be applicable under such conditions. Rather we may anticipate that the implicate order would become the dominant and fundamental concept here.

We have thus far seen throughout this book how it is possible, not only for the manifest level of ordinary experience, but also the quantum level underlying it, to emerge from a still deeper implicate level in which the classical Cartesian notions of form, order and structure have more or less dissolved. Indeed it was implicit in the previous two sections that such ideas, underlying our classical notions of space-time and measure, would have to arise as limiting cases of this deeper implicate order. In many ways this suggestion is close to one that has been under consideration recently by physicists, i.e. that of 'pre-space' [10]. The difference is that we want to go much further than has generally been contemplated in the discussion of pre-space and to propose more definite lines along which we could actually enquire into developing an overall mathematical formulation of these ideas.

We have space in this book only for a brief sketch of our proposals. A number of preliminary papers on this subject have already been published [11,12] and further work is proceeding.

The principal current difficulty in the attempt to make theoretical physics coherent is the notion of a mathematical point in space-time without extension or duration. This has been very useful as an abstraction, but we feel that there is a great deal of evidence (some of which we have discussed in this book) that this notion has reached its limits of validity and usefulness. The idea of beginning with an extended region has indeed been long in the air and has been explored by many people [13]. Indeed both Stapp [7] and Gell-Mann and Hartle [8] have at least implicitly begun with this notion by considering projection operators that cover regions.

As a point of departure let us therefore begin by considering integrals of fields over regions. These may be the ordinary fields, like the electromagnetic field, or they may be quantum fields represented by the wave function. To simplify the discussion we restrict ourselves to a set of scalar fields ϕ. Let ϕ^p denote any particular total configuration of the field. Different values of the index p then denote different total configurations. We further consider a set of regions R_i and we define the integral of ϕ^p over such a region as

$$e_i^p = \int_{R_i} \phi^p. \tag{15.21}$$

The above types of integral have been extensively studied in algebraic topology [14] and they are called p-sets. These p-sets may include integrals of vectors, tensors etc. over regions of various dimensionalities. One of the main concepts of the theory is that these p-sets can be put together to describe the composition of regions into larger regions and that this can be comprehended within an appropriate algebra. In this algebra, boundary relationships can be expressed as a discrete structure of matrix elements and the entire content of typical physical equations (such as Laplace's equation, Maxwell's equations, etc.) can be replaced by purely algebraic relationships among the p-sets. All of this has been done classically, but we are now exploring how this can be extended to a quantum context.

The crucial point that is relevant for our purposes is that these algebraic relationships are invariant under two closely related transformations,

$$e_i^p = \alpha_i^j e_j'^p \tag{15.22}$$

and

$$e_i^p = e_i'^q \beta_q^p, \tag{15.23}$$

or the two together give

$$e_i^p = \alpha_i^j e_j'^q \beta_q^p. \tag{15.24}$$

The similarity of the above to the transformations of quantum theory is evident. Thus (15.22) describes a transformation in which the *field* in one region is 'exploded' into the whole space (or else 'imploded' from the whole space into any one region). It therefore corresponds to a wave, which now comes in, not to a point, but to a region. It is therefore simply an extension of the implicate order to a theory in which we do not need to assume dimensionless points.

The equations (15.23) however describe a transformation in which the *regions* rather than the fields are 'exploded' or 'imploded'. There is a complete symmetry between the implicate order as applied to fields and the implicate order applied to regions. This symmetry is similar to that between points and planes in projective geometry or to that of any element and its dual in ordinary geometry. It is different in that it is far more general and comprehensive in its possibilities.

This symmetry gives another intuitive image of the quantum mechanical symmetry in Hilbert space and of the symmetry of the bra and the ket of Dirac. As the invariant is a scalar product of a bra and a ket, so here the invariant p-set is the scalar product of a field and a region. Indeed from this symmetry it becomes clear that there is ultimately no distinction between the fields and space-time itself. Each is a dual description of the other. Physical quantities will then be invariants which can be interpreted as relations between fields or between regions.

In general these relationships will be nonlocal. Locality must be a special case. How then can we define locality? To do this, we must have an explicate order, i.e. one in which the elements are outside of each other and yet cover the whole space. A simple illustrative example of this would be a set of square elements in contact in such a way as to make up a square grid. To express their local contact, we can define a contact matrix C_{ij} which is zero when the elements ij are not in contact and unity when they are. In this example, corresponding to each specific region R_i, C_{ij} would have a certain four elements that were not zero, and these elements would define the neighbours of the region R_i. Each value of j would also define a direction from which could be built a line ordered according to the index j (with similar differences between successive elements).

But the whole theory must be invariant to an explosive transformation, i.e. an enfoldment into the implicate order. When this is done, the elements C_{ij} will no longer have the simple structure of zeros and ones described above, but will be very complex. Yet they are in some sense implicitly describing the same order. Moreover if we simultaneously transform the field ϕ^p in a dual way then the quantity e_i^p remains invariant, i.e.

$$e_i^p = \int_{R_i} \phi^p = \int_{R_{i'}} \phi^{p'}. \tag{15.25}$$

The above is a kind of generalisation of the invariance that arises in general relativity.

Given a set of contact matrices, C_{ij}, we can ask whether or not it could be transformed completely into an explicate order. Of course this need not be a simple square lattice. Nevertheless more generally after the transformation each index i must be non-zero for a relatively small number of indices j. It is not even necessary however that the pattern be the same for all indices i. Changes in the pattern may for example correspond to dislocations in the lattice.

Structures of the above general kind will be called 'locally connected'. If there exist a transformation that reduces a given matrix, C_{ij}, to a locally connected matrix, C'_{ij}, we shall say C_{ij} is in essence locally connected even though it appears to be non-local in the particular representation that we have chosen.

An interesting question arises if there is no matrix C'_{ij} which is locally connected. A similar question arises in general relativity when the metric tensor cannot be made diagonal over the whole space. In this case we have a new property called curvature. If C_{ij} cannot be transformed into a locally connected matrix, this means that we have a 'space' which may be 'locally explicate' but which gradually enfolds as we move along the locally ordered lines. To picture this we can say that 'wherever you are' you may represent this as an 'ordinary local space' but that there will be a horizon in which this space 'dissolves' into an implicate order. Such a 'universe' would not have a definite boundary, but would simply fade into something that does not manifest to us beyond its horizon. Nor would it have a beginning or an end. But rather it would similarly fade in the distant past and in the distant future. Yet, to anyone that was beyond our horizon, everything would be as definite (or indefinite) as it is to us. That is to say, there is a kind of relativity of the implicate order. Some new mathematical concept corresponding to curvature would have to be developed to express more precisely the distinction between an irreducibly implicate order and one that could be transformed as a whole into an explicate order.

Another interesting example of an irreducibly implicate order is one that could not be put in explicate form locally, but that could be approximately explicate on the larger scale. That is to say, the C_{ij} matrices could not be reduced to some canonical form implying only local contact. This would mean, in general, that the non-zero elements of the contact matrix would spread out over whatever parameters with which we try to label the regions constituting our space. This would of course imply a general nonlocal contact between elements and would mean that the ordinary ideas of space as constituted of localisable elements would no longer hold. So processes that were associated with what we now call the Planck length would have to be described in what is an essentially quantum mechanical kind of space in which Cartesian notions of order would not even arise. Nevertheless the

C_{ij} matrices might have such properties that large sets of elements could be grouped together and that a transformation could be found in which these groups would have a simple contact matrix of a local nature that would give rise to an explicate order. In such a theory an ordinary geometrisable space could emerge in the 'large', while it had no meaning in the 'small'. Indeed these very words 'large' and 'small' would be somewhat misleading, since measure could prevail only in relation to the elements that are grouped together to give rise to a local contact matrix implying an explicate order.

We have been discussing how explicate orders, or approximations to such orders, could emerge from a ground that was irreducibly implicate as a whole. This is an example of how an explicate order arises as a particular case of the implicate order. We have here distinguished this case by requiring certain properties of the contact matrix. A similar procedure has already been explored in ordinary geometry through the Klein Erlanger programme [15]. Thus if we begin with projective geometry as our most general concept we can obtain affine geometry as a particular case by distinguishing a plane at infinity. To obtain ordinary metric geometry, we merely have to distinguish a certain quadric on this plane and call this the circle at infinity. But from our starting point, even projective geometry is very special. Our first step is to obtain a simple explicate order without which the idea of geometry itself would have very little significance. We do this by distinguishing a particular implicate order as the explicate order. But this is possible only if the contact matrix has a suitable form. From here on we could try to go further in defining the equivalent to points, lines, planes etc. in terms of p-sets, and thus we would distinguish a particular case of the explicate order, i.e. one that is geometrisable as a whole. From here we could go on to projective, affine and metric geometries by further distinctions. If this were done then the whole of geometry would be invariant to enfoldment transformations in the implicate order. We could then say that the order of space (and ultimately of time) flowed out of a deeper implicate order of pre-space. Moreover this pre-space would have a structure that was inherently conformable to the laws of the quantum theory. We would thus be free of the present incoherence of trying to force quantum laws into a framework of a Cartesian order that is really only suitable for classical mechanics.

15.11 The super implicate order

Thus far we have been considering the implicate order mainly in relation to particle theories, but further interesting insights can be obtained by extending the notion of implicate order to quantum field theories.

The classical field, which generally obeys a linear equation, has already been understood in terms of the implicate order, e.g. as in the example of the hologram. But when this field is quantised, a further kind of implicate order is introduced. We shall call this the super implicate order. The super implicate order is related to the implicate order as the implicate order, in particle theories, is related to the particles. Consider, for example, a particle which classically executes simple harmonic motion. Its equation of motion is linear

$$m\ddot{x} = -\alpha^2 x.$$

When this theory is quantised by bringing in a wave function $\psi(x) = R(x)\exp[iS(x)]$ the equation of motion becomes

$$m\ddot{x} = -\alpha^2 x - \nabla Q$$

where

$$Q = -\frac{1}{2m}\frac{\nabla^2 R}{R}.$$

As we have seen in earlier chapters, this not only makes the equation nonlinear, but also introduces a force that depends on the whole quantum field and that reflects the activity of information (so that if we have several harmonic oscillators, we also obtain nonlocality).

Similarly consider a classical linear field equation

$$\frac{\partial^2\phi}{\partial t^2} = \nabla^2\phi.$$

As pointed out in chapter 11, when the theory is quantised, we bring in a wave functional

$$\psi = \psi(\ldots\phi(x^\mu)\ldots) = R\exp[iS]$$

and the field equation becomes

$$\frac{\partial^2\phi}{\partial t^2} = \nabla^2\phi - \frac{\delta Q}{\delta\phi},$$

where Q is given by (11.32). The wave equation now becomes nonlinear as well as nonlocal, as happened in the particle theory.

The connection between wave and particle theories can be made yet more evident by considering the normal modes of the field theory. If q_k is such a normal mode corresponding to wave number k, then classically it satisfies the harmonic oscillator equation

$$\ddot{q}_k + k^2 q_k = 0.$$

All the harmonic oscillators are independent. On quantisation, the wave functional can be written as

$$\psi = \psi(\ldots q_k \ldots ; q_k^* \ldots t) = R \exp[iS] .$$

If we develop the field equations along the lines given in chapter 11, we obtain

$$\ddot{q}_k + k^2 q_k = -\frac{\partial}{\partial q_k} Q(\ldots q_k \ldots ; q_k^* \ldots t)$$

where

$$Q = -\frac{1}{2} \sum_k{}' \frac{1}{R} \frac{\partial^2 R}{\partial q_k^*} .$$

Each of the oscillators now obeys a nonlinear equation and it is in general coupled to all the others.

As in the particle theory, the implicate order manifests in the activity of the particle through the quantum potential. It is clear then that in field theory there is a super implicate order that manifests in the field beables.

The effect of the implicate order on the particle is to make its behaviour more wave-like than it would be in classical physics. However, as we have seen in chapter 11, section 11.7, the effect of the super implicate order on the field beables is to oppose the classical tendency for waves to spread and to cause the wave field to be 'swept in' towards atoms, for example. The beables may now be regarded as the first implicate order. The super implicate order thus tends to organise the first implicate order into something more particle-like than it would be in classical physics.

At this point a little reflection shows that the whole idea of implicate order could be extended in a natural way. For if there are two levels of implicate order, why should there not be more? Thus if we regard the super implicate order as the second level, then we might consider a third level which was related to the second as the second is to the first. That is to say, the third implicate order would organise the second which would thereby become nonlinear. (For example there might be a tendency for the whole quantum state to collapse into something more definite.)

Evidently we could go on indefinitely to higher levels of implicate order. Since each is related to the one below, as the one below is related to the one still further below, we have a sequence of similar differences. In other words we have an order of implicate orders.

Thus far we have considered only a one-way connection of implicate orders in which the higher affects the lower but not the other way round. However, we could quite readily extend the idea to allow lower orders to affect the higher ones. At present we have very little to guide us as to how to do this, but we have indeed already considered a possibility of this kind in chapter 14, section 14.6. We proposed there to change Schrödinger's wave equation so as to depend on the particle positions in a way that is similar to what is done in the approach of Ghirardi, Rimini and Weber [16] with regard to the position towards which the wave function is made to collapse. As a result of this change the wave function in our theory will tend to remain loosely associated with the particles and the empty channels would disappear so there would be a two-way relationship between wave function and particle. A similar tendency could be assumed such that all levels of implicate order would tend to be loosely associated with those below. Thus the order of implicate orders would be based on sequences of two-way relationships.

To develop such an order of implicate orders would require that we first obtain some notion of the third implicate order. At present there is very little to guide any speculations about this. It is however an interesting possibility that may be kept in mind for further exploration. The main suggestion that we have, at present, is that a third implicate order would require a superwave functional. This would imply that the wave functional is spread over a range of possibilities. But the notion of a density matrix can be thought of as implying the same sort of thing, i.e. a statistical spread over wave functions or over wave functionals. If the density matrix were thus connected to the superwave functional, this would mean, for example, that a state of thermal equilibrium spreading over a range of energies is actually a reflection of some new property that flows out of the superwave functional.

Another possible area in which the super implicate order may be relevant is in consciousness as will be brought out in the next section of this chapter.

15.12 The implicate order and consciousness

Several physicists have already suggested that quantum mechanics and consciousness are closely related and that the understanding of the quantum formalism requires that ultimately we bring in consciousness in some role or other (e.g. Wigner [17], Everett [18] and Squires [19]). Throughout this book it has been our position that the quantum theory itself can be understood without bringing in consciousness and that as far as research in physics is concerned, at least in the present general period, this is probably the best approach. However, the intuition that consciousness and quantum theory are in some sense related seems to be a good one, and for this reason we feel that it is appropriate to include in this book a discussion of what this relationship might be.

Our proposal in this regard is that the basic relationship of quantum theory and consciousness is that they have the implicate order in common. (For an extensive discussion of this approach see Bohm [20].) The essential features of the implicate order are, as we have seen, that the whole universe is in some way enfolded in everything and that each thing is enfolded in the whole. However, under typical conditions of ordinary experience, there is a great deal of *relative* independence of things, so that they may be abstracted as separately existent, outside of each other, and only externally related. However, more fundamentally the enfoldment relationship is active and essential to what each thing is, so that it is internally related to the whole and therefore to everything else. Nevertheless, the explicate order, which dominates ordinary 'common sense' experience as well as classical physics, appears to stand by itself. But actually this is only an approximation and it cannot be properly understood apart from its ground in the primary reality of the implicate order, i.e. the holomovement. All things found in the explicate order emerge from the holomovement and ultimately fall back into it. They endure only for some time, and while they last, their existence is sustained in a constant process of unfoldment and re-enfoldment, which gives rise to their relatively stable and independent forms in the explicate order.

It takes only a little reflection to see that a similar sort of description will apply even more directly and obviously to consciousness, with its constant flow of evanescent thoughts, feelings, desires, urges and impulses. All of these flow into and out of each other and, in a certain sense, enfold each other (as, for example, we may say that one thought is implicit in another, noting that this word literally means enfolded).

A very clear illustrative example of this enfoldment can be seen by considering what takes place when one is listening to music. At a given moment, a certain note is being played, but a number of the previous notes are still 'reverberating' in consciousness. Close attention will show that it is the simultaneous presence and activity of all these related reverberations that is responsible for the direct and immediately felt sense of movement, flow and continuity, as well as for the apprehension of the general meaning of the music. To hear a set of notes so far apart in time that there is no consciousness of such reverberation will destroy altogether the sense of a whole unbroken living movement that gives meaning and force to what is being heard.

The sense of order in the above experience is very similar to what is implied in our model of a particle as a sequence of successive incoming and outgoing waves. Thus in section 15.9, we saw how a particle trajectory could be expressed at a given moment in terms of an order of spherical waves that are present at that moment. Here the outgoing waves would correspond to 'reverberations', while the incoming waves would correspond to anticipations (which are also evidently present as we listen to the music). The essential point is that the whole movement is contained in this way at any given moment.

Of course not all of it is conscious at any given moment and a great deal may be unconscious. But even that which is unconscious and vaguely anticipatory may contain a whole movement. The most striking example of this is Mozart saying that the whole composition came to him almost in a flash and that from there on he was simply able to play it or to write it. Thus the whole was unfolded rather as the sequence of spheres is unfolded in our model.

Further similar examples can be obtained from visual experiences (e.g. as a series of similar frames in a movie sensed simultaneously can give rise to a feeling of movement). What all this suggests is that our most primary experience in consciousness actually *is* of an implicate order. And our perception of the explicate order is constituted mostly by a series of abstractions from this. To go into all this here would take us too far afield, but an account of it can be found elsewhere (Bohm [21]). To sum up then, in consciousness, as in quantum theory, the explicate order emerges from the implicate order as a relatively stable and self-determined domain and ultimately flows back into the implicate order.

The implicate order is not only the ground of perception, but also of the actual process of thought. For thought is based on information contained in the memory. We do not know a great deal of how memory works at present, but it does not seem that it operates through a one to one correspondence between, for example, an object and an image stored somewhere in the brain. Rather what is relevant for thought is the storage of many abstracted items of information, for example, about shape, colour, the uses of the object, the various things with which it is associated, etc. Indeed in our customary use of language we say that a certain content is re-collected, thus suggesting that this content is collected together from the whole of memory. The fact that it does seem to come from the whole can be seen with the aid of an elementary example. Thus English nouns ending in '-ation' generally come from verbs ending in '-ate'. Thus 'rotation' comes from 'rotate'. But this rule does not always hold. For example 'inclination' does not come from 'inclinate'. The crucial point here is that we can tell immediately that 'inclinate' is not an accepted word, whereas 'rotate' is. This strongly suggests that somehow our 're-collection' is directly from a whole rather than the result of an algorithm for searching the memory in detail.

All of this is clearly compatible with the notion that the basic order of the mind is implicate and that the explicate arises as a particular case of this implicate order in much the way that we have suggested in this chapter for the quantum theory and for pre-space. So the various items of information that have been stored up may be regarded as somehow enfolded in the brain and indeed probably in the rest of the nervous system as well as in the muscles and organs.

It is clear moreover that this information is basically active rather than passive. We may easily verify this in our subjective experience. Suppose, for example, that on a dark night one encounters some shadows. If the memory contains information that there may be assailants in the neighbourhood this can give rise immediately to a sense of danger, with a whole range of possibilities (fight, flight, etc.). This is not what could be called purely a mental process, but includes an involuntary and essentially unconscious process of hormones, heartbeat and neuro-chemicals of various kinds, as well as physical tensions and movements. However, if we look again and we see that it is only a shadow that confronts us, this thought has a calming effect and all this activity ceases. Such a response to information is extremely common (e.g. information that 'X' is a friend or enemy, good or bad etc.). More generally in consciousness, information is seen to be active in all these ways physically, chemically, electrically etc.

This sort of activity is evidently similar to that which was described in connection with quantum processes involving particles such as electrons, for which the information in the wave function was also active in the movement of the particle. At first sight, however, there may still seem to be a significant difference between these two cases. Thus, in our subjective experience, action can, in some cases at least, be *mediated* by reflection in conscious thought. On the other hand, in the case of the electron, the action of information is *immediate*. But actually, even if this happens, the difference is not as great as it might appear to be. For such reflection follows on the suspension of physical action. This gives rise to a train of thought. However, both a suspension of physical action and the resulting train of thought follow immediately from a further kind of active information implying the need to do this.

It seems clear from all this that at least in the context of the processes of thought, there is a kind of active information that is operating which is in key ways rather similar to that which operates in the action of the quantum potential. To clarify what this means in the present context, it is convenient to introduce the notion that consciousness shows or manifests on two sides which may be called the physical and the mental. Active information can serve as a kind of link or 'bridge' between these two sides. These latter are however inseparable, in the sense, for example, that information contained in thought, which we feel to be on the mental side, is at the same time a related neurophysiological, chemical and physical activity (which is clearly what is meant by the 'material' side of thought).

We have however up to this point considered only a small part of the significance of thought. Thus, our thoughts may contain a whole range of information content of different kinds. This may in turn be surveyed by a higher level of mental activity, as if it were a material object at which one were 'looking'. Out of this may emerge a yet more subtle level of information, whose meaning is an activity that is able to organise the original set of information into a greater whole. But even more subtle information of this kind can, in turn, be surveyed by a yet more subtle level of mental activity, and at least in principle this can go on indefinitely. Each of these levels may then be seen from both sides. From the mental side, it is a potentially active information content. But from the material side, it is an actual activity that operates to organise the less subtle levels, and the latter serve as the 'material' on which such operation takes place.

The proposal is then that a similar relationship holds at indefinitely great levels of subtlety. It is being suggested that this possibility of going beyond any specifiable level of subtlety is the essential feature on which the possibility of intelligence is based.

It is interesting in this context to consider again the meaning of the word 'subtle'. As stated in chapter 8, section 8.6, this is "rarefied, highly refined, delicate, elusive, indefinable"; its Latin root is ***sub-texere***, which means "finely woven". This suggest a metaphor for thought as a series of more and more closely woven nets. Each can 'catch' a certain content of a corresponding 'fineness'. The finer nets cannot only show up the details of form and structure of what is 'caught' in the coarser nets; they can also hold within them a further content that is implied in the latter. We have thus been led to an extension of the notion of implicate order, in which we have a series of inter-related levels. In these, the more subtle–i.e. 'the more finely woven' levels including thought, feeling and physical reactions–both unfold and enfold those that are less subtle (i.e. 'more coarsely woven'). In this series, the mental side corresponds, of course, to what is more subtle and the physical side to what is less subtle. And each mental side in turn becomes a physical side as we move in the direction of greater subtlety.

Let us now return to a consideration of the quantum theory. We have already seen that there is a basic similarity between the quantum behaviour of a system of electrons for example and the behaviour of mind. But if we wish to relate mental processes with quantum theory, this similarity will have to be extended. The simplest way of doing this is to use the idea of a series of implicate orders for matter in a way that has already been discussed in the previous section. So we are supposing that, like mental processes, physical processes are also capable of extension to indefinitely great levels of subtlety in a series of implicate orders with two-way relationships.

One may then ask what is the relationship between the physical and the mental processes? The answer that we propose here is that there are not two processes. Rather, it is being suggested that both are essentially the same. This means that that which we experience as mind, in its movement through various levels of subtlety, will, in a natural way ultimately move the body by reaching the level of the quantum potential and of the 'dance' of the particles. There is no unbridgeable gap or barrier between any of these levels. Rather, at each stage some kind of information is the bridge. This implies that the quantum potential acting on atomic particles, for example, represents only one stage in the process.

The content of our own consciousness is then some part of this overall process. It is thus implied that in some sense a rudimentary mind-like quality is present even at the level of particle physics, and that as we go to subtler levels, this mind-like quality becomes stronger and more developed. Each kind and level of mind may have a relative autonomy and stability. One may then describe the essential mode of relationship of all of these as *participation*, recalling that this word has two basic meanings, to 'partake of' and to 'take part in'. Through enfoldment, each relatively autonomous level of mind partakes of the whole to one degree or another. Through this it partakes of all the others in its 'gathering' of information. And through the activity of this information, it similarly takes part in the whole and in every part. It is in this sort of activity that the content of the more subtle and implicate levels is unfolded (e.g. as the movement of the particle unfolds the meaning of the information that is implicit in the quantum field and as the movement of the body unfolds what is implicit in subtler levels of thought, feeling, etc.).

For the human being, all of this implies a thoroughgoing wholeness, in which mental and physical sides participate very closely in each other. Likewise, intellect, emotion, and the whole state of the body are in a similar flux of fundamental participation. Thus, there is no real division between mind and matter, psyche and soma. The common term psychosomatic is in this way seen to be misleading, as it suggests the Cartesian notion of two distinct substances in some kind of interaction.

Extending this view, we see that each human being similarly participates in an inseparable way in society and in the planet as a whole. What may be suggested further is that such participation goes on to a greater collective mind, and perhaps ultimately to some yet more comprehensive mind in principle capable of going indefinitely beyond even the human species as a whole.

Finally, we may ask how we can understand this theory if the subtle levels are carried to infinity. Does the goal of comprehension constantly recede as we try to do this? We suggest that the appearance of such a recession is in essence just a feature of our language, which tends to give

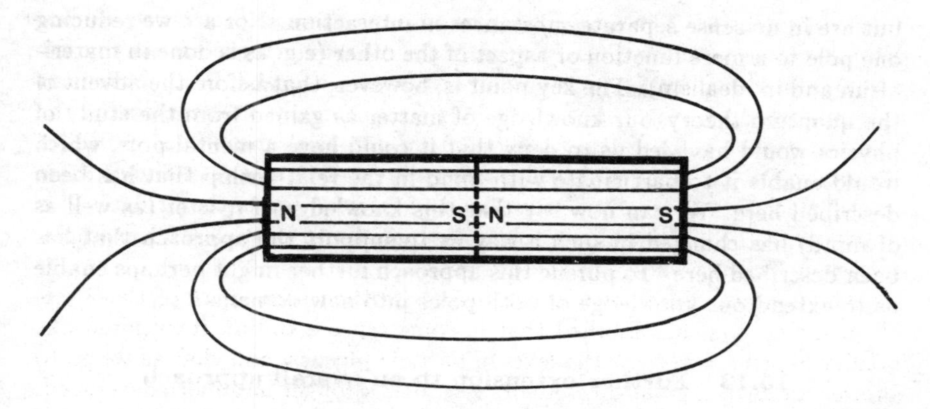

Figure 15.8: Magnetic poles as abstractions from an overall magnetic field

too much emphasis to the analytic side of our thought processes.

To explain what is meant here, one may consider the analogy of the poles of a magnet, which are likewise a feature of linguistic and intellectual analysis, and have no independent existence outside such analysis. As shown in figure 15.8, at every part of a magnet there is a potential pair of north and south poles that overlap each other. But these magnetic poles are actually abstractions, introduced for convenience of thinking about what is going on, while the whole process is a deeper reality–an unbroken magnetic field that is present over all space.

Similarly, we may for the sake of thinking about the subject, abstract any given level of subtlety out of the unbroken whole of reality and focus our attention on it. At each such level, there will be a 'mental pole' and a 'physical pole'. Thus as we have already implied, even an electron has at least a rudimentary mental pole, represented mathematically by the quantum potential. Vice versa, as we have seen, even subtle mental processes have a physical pole. But the deeper reality is something beyond either mind or matter, both of which are only aspects that serve as terms for analysis.*These can contribute to our understanding of what is happening

***See Marshall [22] for an account of an idea having important similarities with what has been proposed here. He, too, uses the notion of a general quantum reality as a basis for the bodily and mental realms, considered as inseparable sides or aspects. But he proposes to explain this from the quantum theory as it now stands in its usual interpretation. However, we have here used the ontological interpretation of the quantum theory with its additional concepts of particle trajectories and active information, and have assumed that ultimately the relationship of mental and material sides can be understood only by**

but are in no sense separate substances in interaction. Nor are we reducing one pole to a mere function or aspect of the other (e.g. as is done in materialsim and in idealism). The key point is, however, that before the advent of the quantum theory, our knowledge of matter as gained from the study of physics would have led us to deny that it could have a mental pole, which would enable it to participate with mind in the relationship that has been described here. We can now say that this knowledge of matter (as well as of mind) has changed in such a way as to support the approach that has been described here. To pursue this approach further might perhaps enable us to extend our knowledge of both poles into new domains.

15.13 Further extension to an overall approach

What we have done here opens up the possibility of an overall approach that encompasses all aspects of objective nature and of our subjective experience (e.g. as an observer and a thinker). To develop such an approach, we first note that not only inanimate nature, but also living nature, is to be understood in terms of the implicate order. Consider a tree, for example, which grows from a seed. Actually the seed makes a negligible contribution to the material substance of the plant and to the energy needed to make it grow. The substance comes from the air, water and soil, while the energy comes from the sun. In the absence of a seed, all these move in an implicate order of a relatively low level of subtlety (based, for example, on the order of the constituent particles). This brings about the constant re-creation of inanimate matter in various forms. However, through the information contained in the DNA molecule, the overall process is subtly altered so that it produces a living tree instead. All the streams of matter and energy which had hitherto developed in an unorganised way, now begin to bring substance and energy to the plant, which grows and eventually decays to fall back into inanimate matter.

At no stage is there a break in this process. For example one would not wish to say that an 'inanimate' molecule of carbon dioxide enters the tree and suddenly becomes 'alive', or that the oxygen molecule suddenly 'dies' when it leaves the tree. Rather, life is eternally enfolded in matter and more deeply in the underlying ground of a generalised holomovement as is mind and consciousness. Under suitable conditions, all of these unfold and

---

extending the scheme beyond the domain in which the current quantum theory is valid.

For other recent attempts to consider the mind-matter relation in the light of the quantum theory, see Penrose [23] and Lockwood [24]. For a discussion of the notions of active information and implicate order by a number of authors, see Pylkkänen [25].

evolve to become manifest in ever higher stages of organisation.

The notion of a separate organism is clearly an abstraction, as is also its boundary. Underlying all this is unbroken wholeness even though our civilisation has developed in such a way as to strongly emphasise the separation into parts.

In this whole process, it is clear that the observer along with his thoughts is contained in an inseparable way. For example, the general theories of physics (including the very one we are proposing here) are an extension of this process. What such a theory produces is a kind of reflection of reality as a whole. As explained in chapter 14, section 14.1, this provides another aspect or appearance of the whole. We may usefully regard this appearance as a kind of mirror through which reality is perceived. Classical physics provided a mirror that reflected only the objective structure of the human being who was the observer. There is no room in this scheme for his mental process which is thus regarded as separate or as a mere 'epiphenomenon' of the objective processes. As we have seen throughout this book, however, quantum physics interpreted ontologically and extended in a natural way makes possible a reflection even of the subtle mental aspects of the observer. Thus, through this mirror the observer sees 'himself' both physically and mentally in the larger setting of the universe as a whole. There is no need, therefore, to regard the observer as basically separate from what he sees, nor to reduce him to an epiphenomenon of the objective process. Indeed the notion of separateness is an abstraction and an approximation valid for only certain limited purposes. More broadly one could say that through the human being, the universe is making a mirror to observe itself. Or vice versa the universe could be regarded as continuous with the body of the human being. After all, this latter, like the plant, gets all its substance and energy from the universe and eventually falls back into it. Evidently the human being could not exist without this context (which has very misleadingly been called an environment).

Although the implicate order is a theory of the whole, it is in no sense a 'theory of everything'. Or to put it differently we do not expect any 'mirror' to provide a complete and perfect reflection of reality as a whole. Every mirror abstracts only some features of the world and reflects them only to a limited degree of accuracy as well as from a limited perspective. We emphasise again that physical theories share in this sort of limit as much as all other forms of knowledge. For our thought is always a limited abstraction from the indefinitely great subtlety of the implicate order as a whole. We can always extend it, but then it will always still be limited. And beyond these limits, there is always indefinitely more that has not been grasped in thought.

At the very least, the above is what would be implied by our proposal of an order of ever increasing orders of subtlety in the implicate order. But probably the whole scheme of the implicate order will be found to be limited too, and will eventually have to be contained in some yet more comprehensive idea. More generally, we are not asserting finality for any of the ideas that we propose here. All proposals are points of departure for exploration. Eventually some of them may be developed so far that we may take them as working hypotheses. But in principle we will be ready to modify them or even to drop them whenever we find evidence for doing this. This applies even to the very idea that we are discussing at this moment. We suggest that the approach outlined above is more coherent than one in which we either tacitly or explicitly take specified features of our ideas as finally fixed and absolutely true.

In particular we would say here that certain features of theories that have attempted to cover both mind and matter, for example, those proposed by Everett [18] and by Gell-Mann and Hartle [8], do imply that mind is a rather limited sort of system (that could be completely covered for example by the notion of an 'information gathering and utilising system' (IGUS)). No doubt mind is capable of doing this sort of thing. But a little observation shows that it is much richer in its capabilities. In particular, if a human being is nothing more than such a system he is effectively reduced to a kind of machine. How then would we understand creativity? Perhaps we could say it is an illusion brought about by our not having complete knowledge of how the machine works. But we can ask how can it be known that this is all that we are? Thus, in the time of Newton, people compared everything including living beings to a clockwork and at later times the brain and nervous system were compared to a telephone exchange. It is now being compared to a super computer. In each age people use the latest technological developments as models for the mind. Very probably in another century or so the whole notion of computers will be seen as rather limited and perhaps even as outmoded as nineteenth-century machines are now thought to be. Doubtless we will then have a still more intricate machine with which one could compare the mind if one chose to do so.

This whole approach seems to be affected with irreducible contingency. There is really no necessity for it at all. We are therefore free, if we so choose, to explore the broader approach to mind and matter, and therefore to the whole of reality, that we have been discussing here and throughout the rest of the book.

15.14 References

1. D. Bohm, *Wholeness and the Implicate Order*, Routledge and Kegan

Paul, London, 1980.

2. R. F. Feynman, *QED. The Strange Theory of Light and Matter*, Princeton University Press, Princeton, New Jersey, 1985.

3. J. P. Heller, *Am. J. Phys.* **28**, 348–353 (1960).

4. D. Bohm and F. D. Peat, *Science, Order and Creativity*, chapter 3, Bantam Books, Toronto, 1987.

5. F. A. M. Frescura and B. J. Hiley, *Found. Phys.* **10**, 705–722 (1980).

6. F. A. M. Frescura and B. J. Hiley, *Rev. Brasileira de Fisica* **Vol. Especial**, 49–86 (1984).

7. H. P. Stapp, 'Einstein Time and Process Time', in *Physics and the Ultimate Significance of Time*, ed. D. R. Griffin, State University Press, New York, 1986, 264–270.

8. M. Gell-Mann and J. B. Hartle, 'Quantum Mechanics in the Light of Quantum Cosmology', in *Proc. 3rd Int. Symp. Found. of Quantum Mechanics*, ed. S. Kobyashi, Physical Society of Japan, Tokyo, 1989.

9. F. Klein, *Elementary Mathematics from an Advanced Standpoint: Geometry*, Dover, New York, 1939.

10. C. W. Misner, K. S. Thorne and J. A. Wheeler, *Gravitation*, W. H. Freeman, San Francisco, 1973.

11. D. Bohm, B. J. Hiley and A. E. G. Stuart, *Int. J. Theor. Phys.* **3**, 171–183 (1970).

12. D. Bohm and B. J. Hiley, *Rev. Brasileira de Fisica* **Vol. Especial**, 1–26 (1984).

13. K. Menger, *The Rice Institute Pamphlet* **27**, 80–107 (1940).

14. W. V. D. Hodge, *The Theory and Applications of Harmonic Integrals*, Cambridge University Press, Cambridge, 1952.

15. F. Klein, *Bull. N.Y. Math. Soc.* **11**, 215–249 (1893).

16. G. C. Ghirardi, A. Rimini and T. Weber, *Phys. Rev.* **D34**, 470–491 (1986).

17. E. P. Wigner, 'Remarks on the Mind-body Question', in *Symmetries and Reflections*, M.I.T. Press, Massachusetts, 1967.

18. H. Everett, *Rev. Mod. Phys.* **29**, 454–462 (1957).

19. E. Squires, *The Conscious Mind in the Physical World*, Adam Hilger, Bristol, 1990.

20. D. Bohm, *Phil. Psych.* **3**, 271 (1990).

21. D. Bohm, *The Special Theory of Relativity*, Addison-Wesley, New York, 1989.

22. I. N. Marshall, 'Consciousness and Bose-Einstein Condensates', *New Ideas in Psychology* **7**, 73–83 (1989).

23. R. Penrose, *The Emperor's New Mind: Concerning Computers, Minds and the Laws of Physics*, Oxford University Press, Oxford, 1989.

24. M. Lockwood, *Mind, Brain and the Quantum*, Basil Blackwell, Oxford, 1989.

25. P. Pylkkänen, *The Search of Meaning*, Thorsons, Wellingborough, 1989.

Author index

Albert, D. 303, 311, 347
Aspect, A. 144, 145, 294

Ballentine, L.E. 269, 307, 310
Belinfante, F.J. 116
Bell, J.S. 1, 7, 26, 41, 114, 117, 118, 122, 134, 140, 145, 146, 205, 220, 296, 310, 326
Ben-Dov, Y. 296
Bloch, F. 278
Bogolubov, N. 66
Bohm, D. 4, 5, 18, 22, 26, 39, 142, 161, 163, 170, 194, 205, 238, 290, 304, 321, 323, 324, 363, 382
Bohr, N. 2, 13, 15, 19, 24, 38, 41, 49, 104, 114, 122, 123, 127, 130, 136, 137, 325
Brown, H.R. 118

Caldeira, A.O. 336
Chew, G.F. 166, 287, 288
Chusing, J.T. 4
Clauser, J. 143, 144
Cooper, L.N. 296, 310
Cufaro-Petroni, N. 233

de Broglie, L. 3, 4, 38, 39, 198, 216, 233
d'Espagnat, B. 296, 306, 310, 314
Deutsch, D. 306, 307, 310, 315, 320, 339
Dewdney, C. 288
DeWitt, B.S. 8, 296, 300, 302, 315
Dirac, P.A.M. 7, 17, 41
Droz-Vincent, P. 286

Eberhard, P.H. 139
Eccles, J.C. 179
Einstein, A. 19, 26, 57, 136, 137, 138 163, 164, 174, 196, 292, 303
Everett, H. 8, 296, 297, 302, 309, 332, 381, 390

Fenech, C. 273
Feynman, R.P. 66, 67, 68, 190, 336, 355, 372
Folse, H.J. 16
Freedman, S. 144
Frescura, F.A.M. 365

Gell-Mann, M. 1, 9, 10, 161, 320, 331, 365, 367, 375, 390
Ghiradi, G.C. 9, 161, 320, 326, 381
Glauber, J. 287
Gleason, A.M. 118
Graham, N. 302
Greenberger, D.M. 118

Hartle, J.B. 9, 10, 161, 320, 331, 365, 367, 378, 390
Hawking, S.W. 188
Heisenberg, W. 2, 13, 15, 18, 20, 320, 325
Hiley, B.J. 38, 142, 194, 365
Horn, M. 118

Jauch, J.M. 118
Jeans, J. 320
Joos, E. 336

Klein, F. 369
Kochen, S. 118, 122

Landau, L.D, 66, 67
Leggett, A.J. 163, 304, 329, 336
Lehr, W.J. 273
Lie, S. 369
Lockwood, M. 388
Loewer, B. 303, 311

Mandel, L. 232
Marshall, I.N. 387
Mermin, N.D. 118, 120, 121, 122, 146, 147

Nambu, Y. 235

Nelson, E. 194, 197, 202
Newton, I. 157

Park, J.L. 273
Paul, H. 260
Pauli, W. 7, 23, 39
Peat, F.D. 363
Penrose, R. 163, 388
Peres, A. 118
Pfleegor, R.L. 232
Pinch, T.J. 4
Pines, D. 66
Piron, C. 118
Podolsky, B. 19, 134
Popper, K.R. 288
Prigogine, I. 188
Primas, H. 296, 310, 314
Prosser, R.D. 234
Pylkkänen, P. 388

Renninger, M. 231, 259
Rimini, A. 9, 161, 320, 326, 381
Rosen, N. 19, 134, 163

Schrödinger, E. 7, 125
Schweber, S.S. 277
Shiller, R. 205
Shimony, E.P. 118, 327
Specker, E.P. 18, 122
Squires, E.J. 299, 303, 361
Stapp, H.P. 9, 10, 161, 287, 288, 296, 311, 329, 330, 331, 332, 365, 367, 375
Svetlichny, G. 118

Tiomno, J. 205
Tolman, R.C. 193
Tonomura, A. 164

van Vechten, D. 296, 310
Vernon, J.R. 336
Vigier, J.P. 194, 238, 273, 286
von Neumann, J. 3, 17, 19, 23, 98, 102, 104, 116, 117, 118, 125, 137, 139, 325

Weber, T. 9, 161, 320, 326, 381
Weinberg, S. 327
Weisskopf, V. 88, 90
Wentzel, G. 278
Wesley, J.P. 234
Wheeler, J.A. 38, 114, 127, 128, 130
Whitehead, A.N. 330
Wigner, E. 3, 88, 90, 297, 312, 381

Zeh, H.D. 296, 310, 336
Zeilinger, A. 118
Zurek, W. 336

Subject index

action-at-a-distance 57, 157; 'spooky' 136
actual event 320, 337
Aharonov-Bohm effect 50
angular momentum measurement 148, 190
anti-symmetric wave function 153
Auger-like effect 83, 131
awareness 301, 303, 311, 315, 339

barrier penetration 74, 79
beables, in Dirac theory 276; in field theory 230, 239, 259, 287, 342, 380; in particle theories 41, 120, 342
Bell inequalities 134, 140, 142, 147, 150, 293
bifurcation 75, 89
boson fields 8, 230
bullet-like photon *see* photon, bullet-like properties
brain 36, 179, 381

cat paradox, Schrödinger 7, 125, 128, 328
Cartesian order 10, 350, 353, 374, 377, 386
Casimer effect 38
causal interpretation 2, 28, 30, 46, 202
causality 29
channel 6, 78, 81, 102, 107, 121, 124, 155, 202, 257
characteristic length 347
chemical bond 63
classical limit 4, 7, 13, 20, 31, 47, 62, 151, 160, 174, 242, 308; of fields 287
classicity 338
coarse graining 337, 342
coherent state 252
collapse 6, 20, 22, 73, 79, 82, 95, 104, 115, 123, 125, 138, 149, 297, 380
collapse of wave function of universe 303
complementarity 16, 107, 136
complexity 304, 309
conformal group 370
consciousness 10, 126, 312, 381, 388
contact matrix 361, 376
context dependence 108, 114, 118, 120, 122, 147
correspondence principle 4
cosmology 165, 338
cut, von Neumann's 20, 178

decay of the wave function 327
decohering histories *see* histories, decohering
delayed choice 127
density matrix 188, 190, 343
determinism 24, 324
DNA 36, 388
Dirac equation 7, 40, 199, 214, 272
Dirac vacuum 276
dispersion-free 117
double solution 39

Eleatics 346
energy swept out of field 264
energy-momentum conservation 39, 47, 70
enfoldment 354, 359, 376, 386
entropy 108
epistomology 2, 13, 19, 26, 97, 154
EPR 1, 7, 19, 91, 134, 141, 151, 190, 202, 225, 282, 341
EPR for fields 264
explicate order 350, 354, 361, 382
'exploding' transformation 375

Fabry-Perot interferometer 163
faithful measurement 110
faster-than-light signals 57, 139, 282
Feynman diagrams 360
field: ground state 242; first excited state 247

fine-grained histories *see* histories, fine-grained
fission-fusion 92
form 31, 39, 325

gauge transformation 51
Gaussian wave packet 46, 75, 252
general relativity 351
Green's function 354, 364
GRW theory 326, 346, 381
guidance condition 39, 81, 149, 198, 218, 221, 234, 282

Hamilton-Jacobi theory 29, 56, 62, 199, 241, 315
Heisenberg algebra 365
Heisenberg microscope 14, 22
hidden variables 2, 19, 117, 122, 142, 145
Hilbert space 9, 297, 313, 320, 329, 342, 350, 367
histories 332, 365; decohering 334; fine-grained 333
hologram 353
holomovement 357, 382
Huygens construction 355, 372
hyperspherical geometry 369

ideal 365
implicate order 10, 350, 354, 361, 378, 382
implication parameter 360
indistinguishable particles 154
indivisibility of the quantum of action 14, 16
information 38, 42, 48, 61, 71, 94, 105, 200, 388; active 5, 36, 41, 49, 62, 68, 71, 78, 94, 379, 384; inactive 79, 99; passive 37
information gathering and utilising system (IGUS) 339, 390
ink-droplet model 358, 363
interference: two-slit experiment 32, 42, 49, 61, 199; two-slit using lasers 261
intrinsic property 114
irreversibility 82

Klein Erlanger programme 378
Klein-Gordon equation 233; and proper time 235

limit cycle 91, 94

localisation of atoms 166
locality in explicate order 377
Lorentz ether 290
Lorentz invariance 271, 282, 288, 346; in Dirac theory 276, 279; and nonlocality 286

Mach's principle 109
magnetic moment of electron 218
manifest world 176, 191, 288, 314, 362, 368
many-body system 8, 40, 56, 182, 201, 208, 222
many-body Dirac 274
many-minds interpretation 313
many-worlds interpretation 8, 296, 336
meaning 36
measurement problem 18, 20, 47, 97, 109, 114, 120, 131, 138, 297, 328, 334, 342
measurement at space-like separation 144
memory 300, 303, 311, 339
metric tensor 289, 292
mixed state 181, 185
momentum measurement 110, 266
multi-time formalism 278

negative measurement 123
non-linearity 39, 78, 91, 95, 251
nonlocality 1, 6, 49, 57, 60, 70, 106, 134, 145, 151, 157, 179, 201, 292, 346, 379; in field theory 250, 255; and special relativity 136, 158; in time 238
non-stationary states 45, 47

observables 41
ontological interpretation 31, 40, 47, 128, 140, 166, 325, 346; of Dirac electron 279; of fields 259
ontology, 2, 13, 16, 18, 23, 137, 154, 324; of boson field 232
organic wholeness 177
osmotic velocity 197, 199, 201, 347

participation 26, 107, 114, 222, 284, 386
Pauli equation 7, 40, 204, 214
Pauli exclusion principle 153, 156
perception 323
Pfleegor-Mandel experiment 262
photon: concept of 255; bullet-like properties 231, 259
pilot wave 38

Planck length 9, 289, 377
Planck time 367, 374
Poincaé group 370
position measurement 109, 115
potentiality 18, 36, 79, 97, 107, 115, 218
preferred basis in many-worlds
interpretation 309
preferred frame 292
pre-space 374, 378
probability 40; in ontological
interpretation 181, 184
projection operators 330, 365
P-sets 375
P-state of atom 44, 58, 91
pure state 41, 181, 365

quantum algebra 363
quantum algorithm 17, 23
quantum cosmology 4, 338
quantum jump 25
quantum potential 29, 37, 40, 48, 52, 57,
65, 75, 81, 92, 95, 98, 102, 108, 112,
149, 161, 199, 221, 379
quasi-classical world 338

rapport 341
relation between the physical and the
mental 385
relative state 299
rotons 67

Solvay Congress 3, 39
spin measurement 119, 135, 138, 141,
299
splitting of universe 310
spreading wave function 73
stationary states 42, 49, 97
statistics 181, 194
statistical mechanics 7, 191
Stern-Gerlach experiment 101, 107, 117,
121, 221
S-state of atom 42, 58, 92, 200
stochastic theory 7, 30, 43, 145, 194,
196, 202; of Dirac electron 293; and
Lorentz invariance 273
sub-quantum level 9
subtle world 177
superfluidity and superconductivity 65,
134
super-implicate order 378, 380
super-quantum potential 240, 246

"Theory of Everything" 321, 389
trajectory 32, 75, 77, 105, 175, 209, 227,
347, 368, 373; arising from waves 369
transitions, atomic 95

unanalysable whole 23, 94, 137, 176
unbroken wholeness 352
uncertainty principle 15, 43, 46, 114,
116, 194, 251, 313, 350
universal time 288

vacuum fluctuations 49
varifyability 288
vector potential 50, 54

watchdog effect 131
wholeness 6, 15, 38, 42, 58, 65, 95, 122
WKB approximation 28, 105, 161, 164

Zeno paradox 131, 352